Progress in Mathematics
Volume 109

Series Editors
J. Oesterlé
A. Weinstein

Computational
Algebraic Geometry

Frédéric Eyssette
André Galligo
Editors

Birkhäuser
Boston · Basel · Berlin

Frédéric Eyssette
Université de Nice-Sophia Antipolis
U.F.R. Faculté des Sciences
Mathématiques
06108 Nice Cedex 2
France

André Galligo
Université de Nice-Sophia Antipolis
U.F.R. Faculté des Sciences
Mathématiques
06108 Nice Cedex 2
France

Library of Congress Cataloging-in-Publication Data

Computational algebraic geometry / edited by F. Eyssette, A. Galligo.
    p.   cm. -- (Progress in mathematics ; v. 109)
    Papers from a symposium held in Nice, France, Apr. 21-25, 1992.
    Includes bibliographical references.
                    (alk. paper). --   (alk.
paper)
    1. Geometry, Algebraic--Data processing--Congresses.
I. Eyssette, F. (Frédéric), 1952- . II. Galligo, A. III. Series:
Progress in mathematics (Boston, Mass) ; vol. 109.
QA564.C6564 1993   92-44350
516.3'5'0285--dc20   CIP

Printed on acid-free paper.

ISBN-13: 978-1-4612-7652-4   e-ISBN-13: 978-1-4612-2752-6

DOI: 10.1007/978-1-4612-2752-6

Camera-ready copy prepared by the Editors.

9 8 7 6 5 4 3 2

Table of Contents

Preface

The theory and practice of computation in algebraic geometry and related domains, from a mathematical point of view, has generated an increasing interest both for its rich theoretical possibilities and its usefulness in applications in science and engineering. In fact, it is one of the master keys for future significant improvement of the computer algebra systems (e.g., Reduce, Macsyma, Maple, Mathematica, Axiom, Macaulay, etc.) that have become such useful tools for many scientists in a variety of disciplines.

The major themes covered in this volume, arising from papers presented at the conference MEGA-92 were:

— Effective methods and complexity issues in commutative algebra, projective geometry, real geometry, and algebraic number theory

— Algebro-geometric methods in algebraic computing and applications.

MEGA-92 was the second of a new series of European conferences on the general theme of Effective Methods in Algebraic Geometry. It was held in Nice, France, on April 21–25, 1992 and built on the themes presented at MEGA-90 (Livorno, Italy, April 17–21, 1990). The next conference — MEGA-94 — will be held in Santander, Spain in the spring of 1994. The Organizing committee that initiatiod and supervises this bienniel conference consists of A. Conte (Torino), J. H. Davenport (Bath), A. Galligo (Nice), D. Yu. Grigoriev (Petersburg), J. Heintz (Buenos Aires), W. Lassner (Leipzig), D. Lazard (Paris), H. M. Möller (Hagen), T. Mora (Genova), M. Pohst (Düsseldorf), T. Recio (Santander), J. J. Risler (Paris), M. F. Roy (Rennes), R. Schoof (Utrecht), and C. Traverso (Pisa).

During the conference, an informal session was organized, in which some participants were given the opportunity to give short talks on current research, of too informal a nature for inclusion in the volume. In addition, the decision since was made to publish only in English, some excellent papers do not appear here. The following papers given at the conference were omitted from the volume:

Assi (Grenoble): Homogenization and standard bases with minimal ecart

Bjorck (Stockholm): There are $\infty + 1152$ cyclic 8-roots

Burgisser, Lickteig (Bonn): On verification complexity of linear prime ideals

De Concini (Roma): Deformations of quantum groups and representations

Gaeta (Madrid): Applications of associated forms (generalized chow forms) to elimination theory

Galligo (Nice), Vorobjov (Petersburg): Complexification of real algebraic varieties

Giusti (Paris), Heintz (Buenos-Aires): La détermination des points isolés et de la dimension d'une variété algébrique peut se faire en temps polynomial

Greuel (Kaiserslautern), Pfister (Berlin), Schoenemann (Kaiserslautern): The computer algebra system singular

Iarrobino (Boston), Yameogo (Nice): Partitions and ideals in $k[x, y]$

Jouanolou (Strasbourg): Formes d'inertie et applications

Labhalla (Marrakech), Lombardi (Besançon), Marlin (Nice): Computation of hermite normal form

Le Prevost (Paris): Famille de courbes hyperelliptiques de genre g munies d'une classe de diviseurs rationnels d'ordre $2g^2 + 4g + 1$

Mazurowskii (Ivanovo): Kauffman polynomial of non singular configurations of lines in RP^3

Mora (Genova): The complexity of the tangent cone algorithm

Morain (Paris): Primality proving using elliptic curves

Traverso et al. (Pisa): Natural representation of algebraic numbers

Van Effelterre (Leuven): Aspect graphs of solids of revolution

Yakoubsohn (Toulouse): The sturm theorem in the complex case

We wish to express our appreciation to these people whose presentations, though not included in the volume, made valuable contributions to the spirit and content of the conference.

To the organizers of MEGA-92 and the 114 participants, the conference was a gratifying success. The sessions were held in the renovated building of the Old Seminary at the seaside. The weather was excellent and the working atmosphere warm and stimulating. We wish to thank the members of the program committee and the referees whose work made the conference and the volume possible. A special votes of thanks goes to Odile Goëpp whose efficient cooperation made everything run more easily.

As a word of epilogue, an indirect consequence of the two symposia, MEGA-90 and MEGA-92, was the submission by a group of ten Euro-

pean teams led by members of our program committe of an ESPIRIT
— Basic Research Action proposal called POSSO (Polynomial Systems
Solving) to the European Community. The project is designed to provide realization in practice (including computer programmed aspects) of
some mathematical ideas developed in MEGA and similar conferences.
In the face of severe competition, the funding was offered and an of-
ficial organization meeting held after MEGA-92 to plan MEGA-94 and
beyond.

Fréderic Eyssette
André Galligo

Computation of Real Radicals of Polynomial Ideals

E. Becker   R. Neuhaus

Abstract. We describe an algorithm for the computation of the τ-radicals of ideals in polynomial rings over rational function fields $k(T_1, \ldots, T_m)$ where (k, τ) is a preordered field satisfying certain computational conditions.

1 Introduction

Let k denote a field and $I \lhd k[X_1, \ldots, X_n]$ an ideal. If $L|k$ is any extension field we set

$$\mathcal{V}_L(I) := \{x \in L^n \mid \forall f \in I : f(x) = 0\}$$

for the set of zeros of I over L. Classical Algebraic Geometry is basically concerned with the zeros of I over the algebraic closure $\bar{k}$ of k, i.e. $\mathcal{V}_{\bar{k}}(I)$. If $X \subseteq \mathcal{V}_L(I)$ is given we set

$$\mathcal{I}_k(X) := \{f \in k[X_1, \ldots, X_n] \mid f = 0 \text{ on } X\}$$

for the associated vanishing ideal. The fundamental Hilbert's Nullstellensatz states

$$\mathcal{I}_k(\mathcal{V}_{\bar{k}}(I)) = \sqrt{I} = \{f \in k[X_1, \ldots, X_n] \mid f^r \in I \text{ for some } r\}.$$

Moreover, the (geometric Zariski-) dimension of $\mathcal{V}_{\bar{k}}(I)$ equals the Krull-dimension of I.

We will be concerned with a formally real base field k and its set $X(k)$ of orderings α, β, ... In this situation new phenomena occur. Each $\alpha \in X(k)$ gives rise to a real closure $R_\alpha \subset \bar{k}$ which is unique up to k-conjugacy. It is one main objective of Real Algebraic Geometry to study the so called real algebraic sets $\mathcal{V}_{R_\alpha}(I) \subseteq R_\alpha^n$ as well. Since two real closures R_α and R'_α are k-conjugate we get a bijection $\mathcal{V}_{R_\alpha}(I) \xrightarrow{\sim} \mathcal{V}_{R'_\alpha}(I)$ induced by a k-automorphism of $\bar{k}$. Therefore, setting

$$\mathcal{V}_\alpha(I) := \bigcup_{R \text{ real closure of } \alpha} \mathcal{V}_R(I)$$

we see that $\mathcal{I}_k(\mathcal{V}_{R_\alpha}(I)) = \mathcal{I}_k(\mathcal{V}_\alpha(I))$. The elements of

$$\mathcal{V}_{real}(I) := \bigcup_{\alpha \in X(k)} \mathcal{V}_\alpha(I)$$

are called the *real points* of I.

The Real Nullstellensatz (cf. [BCR, p. 76 ff]), independently proved by Krivine, Dubois, Risler (cf. [K], [D], [R]) has the same fundamental importance for Real Algebraic Geometry as Hilbert's Nullstellensatz does for classical Algebraic Geometry. It describes the vanishing ideals of $\mathcal{V}_\alpha(I)$ and $\mathcal{V}_{real}(I)$ in terms of I. To cover both cases we introduce the following concept: let $\tau \subset k$ be any preordering, i.e. τ a subset of k satisfying $\tau + \tau \subseteq \tau$, $\tau \cdot \tau \subseteq \tau$, $k^2 \subseteq \tau$, $-1 \notin \tau$. Then we introduce the τ-radical of I:

$$\sqrt[\tau]{I} := \left\{ f \in k[X_1, \ldots, X_n] \ \mid \ f^{2r} + \sum_{i=1}^N a_i g_i^2 \in I \text{ for some } r, N \in \mathbb{N}, \right.$$
$$\left. a_i \in \tau,\ g_i \in k[X_1, \ldots, X_n] \right\}.$$

The ideals $\sqrt[\tau]{I}$, where τ is any preordering in k, are referred to as *real radicals* of I. If we are dealing with $\tau_0 = \sum k^2$, the smallest preordering of k, we accentuate this by setting

$$\sqrt[re]{I} := \sqrt[\tau_0]{I}$$

and calling $\sqrt[re]{I}$ *the* real radical of I. In our situation the Real Nullstellensatz states

$$\mathcal{I}_k(\mathcal{V}_{R_\alpha}(I)) = \mathcal{I}_k(\mathcal{V}_\alpha(I)) = \sqrt[\alpha]{I}, \qquad \mathcal{I}_k(\mathcal{V}_{real}(I)) = \sqrt[re]{I}.$$

From the definition one deduces

$$\sqrt[re]{I} = \mathcal{I}_k(\mathcal{V}_{real}(I)) = \bigcap_{\alpha \in X(k)} \mathcal{I}_k(\mathcal{V}_\alpha(I)) = \bigcap_{\alpha \in X(k)} \sqrt[\alpha]{I}.$$

As in the case of $\mathcal{V}_{\bar{k}}(I)$ the dimensions of (the Zariski–closures of) $\mathcal{V}_{R_\alpha}(I)$, $\mathcal{V}_\alpha(I)$ and $\mathcal{V}_{real}(I)$ equal, in all cases, the Krull–dimensions of their vanishing ideals, i.e. $\sqrt[\alpha]{I}$ and $\sqrt[re]{I}$. See section 2 for details.

It is the main object of this paper to present algorithms for the computations of the τ-radicals $\sqrt[\tau]{I}$, in particular of $\sqrt[\alpha]{I}$ and $\sqrt[re]{I}$. These algorithms work if k satisfies certain computational requirements which are specified in section 3. Our first step is to transfer the method of localizing, introduced by Gianni, Trager and Zacharias, [GTZ], to our situation. This method reduces the general situation to the univariate case $k(T_1, \ldots, T_m)[X]$ which can be handled by factorization and quantifier elimination (see section 3). However, after localizing and contracting ideals we cannot restore our real radicals: the zero–dimensional components get lost (see section 5). Handling these zero–dimensional ideals is the content of the sections 6 and 7, where two distinct methods are presented, and of section 4 where the case of an arbitrary zero–dimensional ideal is dealt with.

We express our thanks to the referees of this paper whose comments helped to improve the exposition, clarify some concepts and encourage us to extend our previous results.

2 Real radicals

The following more or less known results on real radicals are listed for the convenience of the reader since we couldn't find a place in the literature where our special needs are met exactly. Let A denote an arbitrary commutative ring and $\sigma \subset A$ a preordering, i.e. a subset of A satisfying

$$\sigma + \sigma \subseteq \sigma, \quad \sigma \cdot \sigma \subseteq \sigma, \quad A^2 \subseteq \sigma, \quad -1 \notin \sigma.$$

If $I \lhd A$ is an ideal we define the σ–radical by

$$\sqrt[\sigma]{I} := \left\{ f \in A \mid f^{2r} + s \in I \text{ for some } r \in N, \ s \in \sigma \right\}.$$

If $P \in Spec(A)$ then $k(P) := Quot(A/P)$ is the residue field of P. We write $\bar{a} := a + P \in k(P)$ and denote by $\bar{\sigma}$ the set $\left\{ \sum_{\text{finite}} x_i^2 \bar{s} \mid x_i \in k(P), \ s \in \sigma \right\}$. We readily check $\bar{\sigma} = \left\{ \frac{\bar{s}}{\bar{a}^2} \mid s \in \sigma, \ a \in A \setminus P \right\}$. We directly see that $\bar{\sigma}$ is a preordering of $k(P)$ iff $-1 \notin \bar{\sigma}$ iff $\sqrt[\sigma]{P} = P$.

An ideal I is called σ–real if $\sqrt[\sigma]{I} = I$. If $\sigma = \sum A^2$ then we just say *real* instead of $\sum A^2$–real.

Proposition 1   *i) $\sqrt[\sigma]{I}$ is an ideal and every minimal prime ideal P of $\sqrt[\sigma]{I}$ is σ–real.*

ii) $\sqrt[\sigma]{I} = \bigcap P$ where P ranges over all σ–real prime ideals over I.

iii) A prime ideal P of A is σ–real iff $\bar{\sigma}$ is a preordering in $k(P)$.

Proof: i) That $\sqrt[\sigma]{I}$ is closed under addition is the only non–trivial step in showing $\sqrt[\sigma]{I}$ to be an ideal. If $f^{2r} + s \in I$, $g^{2r'} + t \in I$ where $s, t \in \sigma$ then, by raising to appropriate powers, we may assume that $r = r'$. Then $(f + g)^{4r} + (f - g)^{4r} + u \in I$ for some $u \in \sigma$. Hence $f + g \in \sqrt[\sigma]{I}$. Let P be a minimal prime ideal of $\sqrt[\sigma]{I}$ and $a \in \sqrt[\sigma]{P}$, i.e. $a^{2r} + s \in P$ for some $r \in N$, $s \in \sigma$. From the minimality of P we deduce the existence of $l \in N$ and $x \notin P$ such that $(a^{2r} + s)^l x \in \sqrt[\sigma]{I}$. This means $(a^{2r} + s)^{2ll'} x^{2l'} + t \in I$ for some further $l' \in N$, $t \in \sigma$. Multiplying by powers of x^2 if necessary the factor $x^{2l'}$ can be replaced by some y^{2r} and t by $t' \in \sigma$ accordingly. Now, after expanding the power of the sum we end up having a statement of the type $(ay)^{2r} + s' \in I$ where $s' \in \sigma$. This shows $ay \in \sqrt[\sigma]{I}$, hence $ay \in P$. But $y \notin P$ thus $a \in P$, and $\sqrt[\sigma]{P} = P$ is shown.

ii) $\sqrt[\sigma]{I}$ is clearly a radical ideal. iii) has already been noticed. $\square$

We obviously have $\sqrt[\sigma]{I \cap J} = \sqrt[\sigma]{I} \cap \sqrt[\sigma]{J} = \sqrt[\sigma]{I \cdot J}$.

The general results above will be specialized in two steps. We first assume k to be a field equipped with a preordering τ, A a k-algebra and σ to be generated by τ, i.e. $\sigma = \{\sum_{\text{finite}} x_i^2 a_i \mid x_i \in A, a_i \in \tau\}$. By abuse of notation and terminology we write $\sqrt[\tau]{I}$ instead of $\sqrt[\sigma]{I}$ and call an ideal I of A τ-real instead of σ-real.

Proposition 2 *Assuming the hypothesis above the following statements hold:*

i) $\sqrt[\tau]{I} = \bigcap_{\alpha \in X(k),\ \alpha \supseteq \tau} \sqrt[\alpha]{I}$

ii) *A prime ideal P of A is τ-real iff some $\alpha \in X(k)$, $\alpha \supseteq \tau$ is extendible to $k(P)$ iff P is α-real for some $\alpha \in X(k)$, $\alpha \supseteq \tau$.*

Recall that $X(k)$ is the set of orders of k.

Proof: ii) By proposition 1, iii) $\sqrt[\tau]{P} = P$ iff τ generates a preordering $\bar{\tau}$ in $k(P)$. By the Artin–Schreier theory, cf. [P, Cor. 1.6] this holds iff some order $\alpha \supseteq \tau$ extends to $k(P)$. Applying this argument to the order α, being extendible to $k(P)$, we conclude $\sqrt[\alpha]{P} = P$. From $P \subseteq \sqrt[\tau]{P} \subseteq \sqrt[\beta]{P}$ for any order $\beta \supseteq \tau$ the remaining statement is deduced.

i) follows from the proof above and proposition 1, ii). $\square$

Corollary 3 *Retaining the assumption above assume in addition that A is noetherian. Then given an ideal I and a preordering $\tau \subset k$ there are finitely many orders $\alpha_1, \ldots, \alpha_N \supseteq \tau$ such that*

$$\sqrt[\tau]{I} = \sqrt[\alpha_1]{I} \cap \ldots \cap \sqrt[\alpha_N]{I}.$$

Proof: $\sqrt[\tau]{I}$ has only finitely many minimal prime ideals. They all satisfy $\sqrt[\tau]{P} = P$. Now apply proposition 2. $\square$

Remark 1 *It seems a very delicate computational task to determine the needed orders $\alpha_1, \ldots, \alpha_N$ given the input of I and τ. In the case of affine k-algebras the open-mapping theorem of Elman–Lam–Wadsworth [ELW] as well as a method suggested by Recio and Dubois [DR] will play important roles. We hope to come back to this topic in the near future.*

As it is well known in Real Algebraic Geometry the general results above together with the Artin–Lang homomorphism theorem, cf. [BCR, 4.1], yield the Real Nullstellensatz. We will present the arguments in a ring theoretic fashion. To this end we restrict ourselves to the case of affine k-algebras.

Proposition 4 *Let k, τ be as above and let A be any affine k–algebra such that τ generates a preordering in A. Assume that $P \in Spec(A)$ satisfies $\sqrt[r]{P} = P$. Then $P = \bigcap M$ where M ranges over all maximal ideals over P satisfying $\sqrt[r]{M} = M$.*

Proof: We have to show that, given any $f \notin P$, we can find a maximal ideal $M \supseteq P$ such that $f \notin M$, $\sqrt[r]{M} = M$. In fact, consider the affine k–algebra $B := A/P\left[\frac{1}{f}\right] \subseteq k(P)$. We know that $k(P)$ admits an order $\tilde{\alpha}$ extending some order $\alpha \supseteq \tau$ of k. By the Artin–Lang theorem, e.g. in the version of [B], there exists a homomorphism $\varphi : B \longrightarrow R_\alpha$ which in turn induces the required maximal ideal of A.   □

As a consequence we now derive the Real Nullstellensatz. Its proof is an immediate consequence of proposition 4 once we have observed the following fact: if M is any maximal ideal of $k[X_1, \ldots, X_n]$ then $\sqrt[r]{M} = M$ iff the field $k[X_1, \ldots, X_n]/M$ can be embedded in some real closure R_α of k where $\alpha \supseteq \tau$. Also note that every preordering $\tau \subset k$ generates a preordering in $k[X_1, \ldots, X_n]$.

Proposition 5 (Real Nullstellensatz) *Let k be a formally real field and τ a preordering of k. If I is any ideal of $k[X_1, \ldots, X_n]$ then setting*

$$V_\tau(I) := \bigcup_{\alpha \in X(k),\; \alpha \supseteq \tau} V_\alpha(I)$$

we have

$$\mathcal{I}_k(V_\tau(I)) = \sqrt[r]{I}.$$

Corollary 6   i) $\mathcal{I}_k(V_{R_\alpha}(I)) = \sqrt[\varrho]{I}$

ii) $\sqrt[\varrho]{I \cdot R_\alpha} \cap k[X_1, \ldots, X_n] = \sqrt[\varrho]{I}$

Proof: i) Obviously $\mathcal{I}_k(V_{R_\alpha}(I)) = \mathcal{I}_k(V_\alpha(I))$.
ii) This follows from the relation $\mathcal{I}_k(V_{R_\alpha}(I)) = \mathcal{I}_{R_\alpha}(V_{R_\alpha}(I)) \cap k[X_1, \ldots, X_n]$ and $\mathcal{I}_{R_\alpha}(V_{R_\alpha}(I)) = \mathcal{I}_{R_\alpha}(V_{R_\alpha}(I \cdot R_\alpha)) = \sqrt[\varrho]{I \cdot R_\alpha}$, the latter equation by i).   □

It remains to determine the dimensions of $V_{R_\alpha}(I) \subseteq R_\alpha^n$ and $V_\tau(I) \subseteq V_{\bar{k}}(I)$. The R_α–dimension of $V_{R_\alpha}(I)$ as a R_α–algebraic set, defined in [BCR, 2.8], equals the Krull–dimension of $\mathcal{I}_{R_\alpha}(V_{R_\alpha}(I)) = \sqrt[\varrho]{I \cdot R_\alpha}$ which coincides with the dimension of the Zariski–closure of $V_{R_\alpha}(I)$ in $V_{\bar{k}}(I)$ (since $\mathcal{I}_{\bar{k}}(X) = \mathcal{I}_{R_\alpha}(X) \cdot R_\alpha(\sqrt{-1})$ if $X \subseteq R_\alpha^n$). Thus there is no ambiguity in studying the concept of dimension for $V_{R_\alpha}(I)$.

Proposition 7 *Under the hypothesis above*

i) $\dim V_\tau(I) = \dim \sqrt[r]{I}$

ii) $\dim V_{R_\alpha}(I) = \dim \sqrt[\varrho]{I}$.

Proof: We have $\sqrt[r]{I} = \mathcal{I}_{\bar{k}}(\mathcal{V}_r(I)) \cap k[X_1, \ldots, X_n]$ and $\sqrt[r]{I} = \sqrt[r]{I \cdot R_\alpha} \cap k[X_1, \ldots, X_n]$. Moreover, the geometric dimensions are given by the Krull–dimensions of $\mathcal{I}_{\bar{k}}(\mathcal{V}_r(I)) \lhd \bar{k}[X_1, \ldots, X_n]$ and $\sqrt[r]{I \cdot R_\alpha} \lhd R_\alpha[X_1, \ldots, X_n]$. Therefore, we are facing the following situation: $L|k$ an algebraic field extension, $J \lhd L[X_1, \ldots, X_n]$ a radical ideal and $J^c = J \cap k[X_1, \ldots, X_n]$. We claim: $\dim J^c = \dim J$. To prove this let $J = P_1 \cap \ldots \cap P_r$ and $J^c = Q_1 \cap \ldots \cap Q_s$ be the primary decomposition of the radical ideals J^c and J. Set $A := k[X_1, \ldots, X_n]$. From $Q_1 \cap \ldots \cap Q_s = (P_1 \cap A) \cap \ldots \cap (P_r \cap A)$ one first deduces that for every Q_i there exists j with $Q_i = P_j \cap A$, hence $\dim J^c \leq \dim J$ since $L[X_1, \ldots, X_n]$ is an integral extension of A. Secondly, every $P_i \cap A$ contains some Q_j which, again by the theory of integral extensions, implies $\dim J \leq \dim J^c$. Thus the claim follows. $\square$

Proposition 8 *A prime ideal $P \lhd k[X_1, \ldots, X_n]$ is τ-real if and only if there exists an order $\alpha \supseteq \tau$ such that P extends to a real prime ideal Q of $R_\alpha[X_1, \ldots, X_n]$.*

Proof: If Q is as stated and $P = Q \cap k[X_1, \ldots, X_n]$ then the chain of inclusions $P \subseteq \sqrt[r]{P} \subseteq \sqrt[r]{P} \subseteq \sqrt[r]{Q} \cap k[X_1, \ldots, X_n]$ yields $P = \sqrt[r]{P}$. Conversely assume $\sqrt[r]{P} = P$. By proposition 2, ii), there is an order $\alpha \supseteq \tau$ such that $P = \sqrt[\alpha]{P}$. Statement ii) of corollary 6 yields $P = \sqrt[\alpha]{P \cdot R_\alpha} \cap k[X_1, \ldots, X_n]$. Now the arguments in the proof of the previous proposition show that P extends to a minimal prime ideal of $\sqrt[\alpha]{P \cdot R_\alpha}$ which has to be real by proposition 1, i). $\square$

A final remark seems in order. The real algebraic sets $\mathcal{V}_\alpha(I)$ may vary very strongly if the order α changes. Consider the circle over k defined by the equation $x^2 + y^2 = a$ where $a \in k^\times$. If $a \in \alpha$ then $\mathcal{V}_{R_\alpha}$ is a circle over k. On the other hand, if $a \in -\alpha$ then $\mathcal{V}_{R_\alpha} = \emptyset$. The variation of $\mathcal{V}_\alpha(I)$ or $\mathcal{V}_{R_\alpha}(I)$ as a function of α will be the subject of subsequent work.

3   The univariate case

In this and the following sections we keep the convention of the previous one. This means we start with a preordered field (k, τ) and if I is any ideal of a k-algebra A then $\sqrt[r]{I}$ and the notation "τ-real" replace $\sqrt[\sigma]{I}$ and "σ-real", where σ is the preordering of A generated by τ. If F is any extension field and $f \in F[X]$ then f is called τ-real over F if $\sqrt[r]{fF[X]} = fF[X]$. In the case $\tau = \sum k^2$ then $\sqrt[r]{fF[X]} = \sqrt[\infty]{fF[X]}$ and we call f *real over* F. Note that a polynomial f, real over a field F, need not be real over a proper extension field. E.g. $\sqrt[\infty]{(X^3 - 2)Q[X]} = (X^3 - 2)$, but, setting $L = Q(\sqrt[3]{2})$, we see $\sqrt[\infty]{(X^3 - 2)L[X]} = (X - \sqrt[3]{2})$.

In this section we are concerned with the computation of the *τ-real part* of $f \in F[X]$, i.e. a polynomial $\tilde{f}$ subject to $\sqrt[r]{fF[X]} = \tilde{f}F[X]$. The τ-real part is unique up to a nonzero constant factor in F.

The computation of the ordinary radical $\sqrt{fF[X]}$ amounts to the determination of the squarefree part of f which is easily achieved by taking derivatives and gcd-calculations (in fields of characteristic zero). In contrast to this, the computation of τ–real parts is much more difficult. Moreover, the simple example

$$\sqrt[re]{(X^3 - 2)R[X]} = \left(X - \sqrt[3]{2}\right) R[X]$$

shows that in general one has to pass to proper algebraic extensions of the field of coefficients of the input–polynomial. Note that the computation of the squarefree part is performed without leaving this field of coefficients.

Later on, in the course of the algorithm we will have to pass to a rational function field $F = k(T_1, \ldots, T_m)$. For these fields we are going to derive a criterion by means of which one can identify τ–real polynomials.

Proposition 9 *Let* $p \in k[T_1, \ldots, T_m, X]$ *be irreducible in* $F[X]$ *where* $F = k(T_1, \ldots, T_m)$. *Then the following statements are equivalent:*

i, $\sqrt[re]{pF[X]} = (p)$, *i.e.* p *is* τ–*real over* $F = k(T_1, \ldots, T_m)$.

ii, *There exists an order* $\alpha \supseteq \tau$ *such that the polynomial* $-lc_X(p) \cdot p$ *takes on a strictly positive value in the real closure* R_α *of* k, *where* $lc_X(p)$ *denotes the leading coefficient of* p *w.r.t.* X.

Proof: i) $\Longrightarrow$ ii): From section 2, proposition 2, ii) we get that some $\alpha' \in X(F)$, $\alpha' \supseteq \tau$ extends to an ordering $\bar{\alpha}$ of the field $l := F[X]/(p)$. Consider the p-adic valuation on $F(X)$ and use $\pi := -lc_X(p) \cdot p$ as a uniformizing element. Every element $f \in F(X)^\times$ has a unique representation $f = \pi^r \cdot g(X)$, $g \in F[X]$, $\pi \nmid g(X)$. Set

$$\tilde{\alpha} := \{0\} \cup \{\pi^r \cdot g(X) \mid r \in Z, g \in F[X], \pi \nmid g(X), g(u) \in \bar{\alpha}\}$$

where $u = X + (p(X)) \in l$. We readily check that $\tilde{\alpha}$ is an order on $F(X) = k(T_1, \ldots, T_m, X)$ with $\pi = -lc_X(p) \cdot p \in \tilde{\alpha}$. By the Artin–Lang theorem, e.g. in the version of [B], π takes on a strictly positive value in R_α where $\alpha = \tilde{\alpha} \cap k$.

ii) $\Longrightarrow$ i): By assumption $\pi(u_1, \ldots, u_m, v) > 0$ in some appropriate real closure R_α, $\alpha \supseteq \tau$. The kernel of the homomorphism

$$\varphi : k[T_1, \ldots, T_m, X] \longrightarrow R_\alpha, \ (T_1, \ldots, T_m, X) \longmapsto (u_1, \ldots, u_m, v)$$

is a regular ideal since $k[T_1, \ldots, T_m, X]$ is a regular ring. Now the Artin–Lang theorem shows the existence of an order $\tilde{\alpha}$ on $k(T_1, \ldots, T_m, X)$ which extends α such that $\pi = -lc_X(p) \cdot p \in \tilde{\alpha}$. Set $\tilde{\alpha}_0 := \tilde{\alpha} \cap k(T_1, \ldots, T_m)$. The Artin–Lang theorem, applied to $k(T_1, \ldots, T_m)$ and $\tilde{\alpha}_0$, yields the existence of a real closure $R_{\tilde{\alpha}_0}$ of $k(T_1, \ldots, T_m)$ such that $\pi(T_1, \ldots, T_m, \tilde{v}) > 0$ for some $\tilde{v} \in R_{\tilde{\alpha}_0}$. Clearly, π takes on a negative value in $R_{\tilde{\alpha}_0}$. By the intermediate value theorem we find a zero of π in $R_{\tilde{\alpha}_0}$, hence of $p(T_1, \ldots, T_m, X)$.

This means that $F[X]/(p)$ admits an order $\bar{\alpha}_0$ extending $\tilde{\alpha}_0$. Therefore, by section 2, proposition 2, ii), $\sqrt[\alpha]{pF[X]} = pF[X]$. From $\tau \subseteq \bar{\alpha}_0$ we derive $\sqrt[\tau]{pF[X]} = pF[X]$ as desired. □

To cope with an arbitrary polynomial $f \in k(T_1, \ldots, T_m)[X]$ we now introduce the following assumptions. The preordered field (k, τ) should allow:

- factorization of univariate polynomials (over k)

- an algorithm to decide whether a multivariate polynomial over k takes on a strictly positive value in some real closure R_α of k where α is an order containing τ.

Proposition 10 *Under these assumptions on the preordered field (k, τ) there is an algorithm to compute the τ-real part of every polynomial $f \in k(T_1, \ldots, T_m)[X]$ over $k(T_1, \ldots, T_m)$.*

Proof: First of all, the factorization of f in $k(T_1, \ldots, T_m)[X]$ can be carried out using the first assumption (cf. [Se, p. 289]). Next, let p be an irreducible factor of f in $k(T_1, \ldots, T_m)[X]$. W.l.o.g. we may assume $p \in k[T_1, \ldots, T_m, X]$. In view of proposition 9, the second assumption allows us to decide whether the τ-real part of p equals p or 1, i.e. whether p is τ-real or not. If $f = \prod p_i$ is the prime factorization in $k(T_1, \ldots, T_m)[X]$ then

$$\sqrt[\tau]{fF[X]} = \left(\prod_{p_i\ \tau-\text{real}} p_i \right). \quad □$$

In section 8 a list of preordered fields (k, τ) satisfying the assumptions is presented.

We stress the fact that the reduction method described in the following sections which reduces the multivariate to the univariate case does not need the above additional requirements on k. Thus the strong additional computational conditions imposed on k are exclusively demanded for the handling of the univariate case. However we do not know of any other method for the computation of τ-radicals in $k(T_1, \ldots, T_m)[X]$ working under weaker computational assumptions on k.

4 The zero–dimensional case

For a preordered field (k, τ) and a zero–dimensional ideal $I \lhd k[X_1, \ldots, X_n]$ the computation of $\sqrt[\tau]{I}$ can be reduced to the univariate case by the Shape-Lemma. Because of $\sqrt{I} \subseteq \sqrt[\tau]{I}$ we may assume that I is already radical (for the computation of $\sqrt{I}$ see [EH], [GM], [KMH], [L], [Se, Lemma 92]). Then for "almost all" vectors $c = (c_2, \ldots, c_n) \in k^{n-1}$, i.e. excluding the points of a finite union of proper affine linear subspaces of k^{n-1}, the automorphism

$$\varphi := \varphi_c : \left\{ \begin{array}{ccc} k[X_1, \ldots, X_n] & \longrightarrow & k[X_1, \ldots, X_n] \\ f(X_1, \ldots, X_n) & \longmapsto & f(X_1 + \sum_{i=2}^{n} c_i X_i, X_2, \ldots, X_n) \end{array} \right. \quad (1)$$

puts $\varphi(I)$ into general position w.r.t. X_1 (cf. [GM], [GTZ]). Alternatively, it is well known that by introducing a new variable a suitable vector c can be computed deterministically. However, from a practical point of view the strategy of "guessing and testing" seems to be superior.

In accordance with the Shape-Lemma the reduced Gröbner basis G of $\varphi(I)$ w.r.t. the pure lexicographical ordering with $X_1 < \ldots < X_n$ has the form

$$G = \{X_j - g_j(X_1) \, (j = 2 \ldots n), \, g_1(X_1)\}$$

with a squarefree polynomial $g_1(X_1)$ (cf. [GM]). If $g_1 = \prod p_i$ is the prime factorization of g_1 in $k[X_1]$, then the

$$M_i := (X_j - g_j(X_1) \, (j = 2 \ldots n), \, p_i(X_1))$$

are the maximal ideals in the primary decomposition $\varphi(I) = \bigcap_i M_i$. We have $k[X_1, \ldots, X_n]/M_i \simeq k[X]/(p_i(X))$ via

$$f(X_1, \ldots, X_n) + M_i \mapsto f(X, g_2(X), \ldots, g_n(X)) + (p_i(X)).$$

Hence M_i is a τ–real ideal of $k[X_1, \ldots, X_n]$, i.e. $\sqrt[\tau]{M_i} = M_i$, iff $(p_i(X))$ is a τ–real ideal of $k[X]$. Therefore we obtain

$$\sqrt[\tau]{\varphi(I)} = \bigcap_{M_i \, \tau\text{–real}} M_i = (X_j - g_j(X_1) \, (j = 2 \ldots n), \, \tilde{g}_1(X_1)),$$

where $\tilde{g}_1(X_1)$ is the τ–real part of $g_1(X_1)$ in $k[X_1]$ (see section 3). Finally we have

$$\sqrt[\tau]{I} = \varphi^{-1}\left(\sqrt[\tau]{\varphi(I)}\right).$$

5 Localization

The real radical operator $\sqrt[\tau]{}$ is well behaved under localization: Let S be a multiplicatively closed subset of a k–algebra A with $1 \in S$, $0 \notin S$ and $J \lhd A$. Then

$$\sqrt[\tau]{J_S} \cap A = \sqrt[\tau]{J}_{(S)}, \quad (2)$$

where J_S denotes the extension ideal of J in the quotient ring A_S and $\sqrt[\tau]{J}_{(S)} = \{a \in A \mid \exists s \in S : sa \in \sqrt[\tau]{J}\}$ the S-component of $\sqrt[\tau]{J}$.

Proposition 11 *Let (k, τ) be a preordered field, $I \lhd k[X_1, \ldots, X_n]$ and $\sqrt[\tau]{I} = \bigcap_i P_i \cap \bigcap_j M_j$ the reduced primary decomposition of $\sqrt[\tau]{I}$, the P_i,*

M_j τ–real prime ideals with $\dim P_i > 0$ and $\dim M_j = 0$. Defining $S_l :=$ $k[X_l] \setminus 0$ for $l = 1 \ldots n$ we have

$$\bigcap_{l=1}^{n} \left(\sqrt[\tau]{I_{S_l}} \cap k[X_1, \ldots, X_n] \right) = \bigcap_{i} P_i. \tag{3}$$

Proof: Since $\dim P_i > 0$ there exists l such that $(P_i)_{(S_l)} = P_i$. Since $\dim M_j = 0$ we have $(M_j)_{(S_l)} = k[X_1, \ldots, X_n]$ for all l, and therefore, using equation (2),

$$\bigcap_{l=1}^{n} \left(\sqrt[\tau]{I_{S_l}} \cap k[X_1, \ldots, X_n] \right) = \bigcap_{l=1}^{n} \sqrt[\tau]{I}_{(S_l)}$$

$$= \bigcap_{i} \bigcap_{l=1}^{n} (P_i)_{(S_l)} \cap \bigcap_{j} \bigcap_{l=1}^{n} (M_j)_{(S_l)}$$

$$= \bigcap_{i} P_i. \quad \square$$

For the computation of the contraction ideals in (3) see [GTZ, Corollary 3.8].

Our reduction process, which reduces the computation of τ–real radicals from the multivariate to the univariate case, is based on proposition 11: Using $\sqrt[\tau]{\cdot}$–computations in $k(X_1)[X_2, \ldots, X_n]$ we are able to compute the intersection of the τ–real primes of positive dimension of any ideal $I \lhd k[X_1, \ldots, X_n]$. Now it becomes clear, too, why we had to consider function fields of type $k(T_1, \ldots, T_m)$ as coefficient domains in section 3 even if we are only interested in the computation of τ–real radicals in polynomial rings over k.

Equation (3) shows that after localization we will have lost the intersection $\bigcap_j M_j$ of the zero–dimensional components of $\sqrt[\tau]{I}$. In the following we will present two different methods for catching this missing intersection. In section 6 we proceed by determining topologically isolated real points using projections. The method described in section 7 is based on the equi–dimensional decomposition and the computation of singular loci via Jacobians.

6   Real isolated points

In the sequel, every real closed field R will be understood as a topological field under its order topology, and the affine space R^n will be endowed with the product topology.

Definition 1 *Let k be a formally real field with algebraic closure $\bar{k}$ and $I \lhd k[X_1, \ldots, X_n]$. A point $x \in V_{\bar{k}}(I)$ will be called a* real (topologically) isolated point *(for short: isolated) if there exists a real closure R such that*

x is isolated in the topological space $\mathcal{V}_R(I) \subseteq R^n$. $\mathcal{V}_{iso}(I)$ denotes the set of all isolated points.

The set $\mathcal{V}_{iso}(I)$ is closed under the natural Galois–action of the Galois group $Gal(\bar{k}|k)$, hence $\mathcal{I}_{\bar{k}}(\mathcal{V}_{iso}(I)) = \mathcal{I}_k(\mathcal{V}_{iso}(I)) \otimes_k \bar{k}$. We set $I^{iso} := \mathcal{I}_k(\mathcal{V}_{iso}(I)) \lhd k[X_1, \ldots, X_n]$.

Proposition 12 *Let k be a formally real field and $I \lhd k[X_1, \ldots, X_n]$. Then:*

i) $I^{iso} = k[X_1, \ldots, X_n]$ *or I^{iso} is a zero–dimensional real ideal.*

ii) *If M is any zero–dimensional component of $\sqrt[r]{I}$ then $I^{iso} \subseteq M$ and hence M is also a component of I^{iso}.*

Proof: i) It is readily checked that I^{iso} is a real ideal. If $\dim I^{iso} \geq 1$ there would be an isolated point x in some $\mathcal{V}_R(I)$ which is at the same time a regular point of $\mathcal{V}_{\bar{k}}(I^{iso})$ of local dimension at least 1. The implicit function theorem over R (cf. [BCR, Corollary 2.9.6]) shows that x is not isolated in $\mathcal{V}_R(I^{iso})$, hence not isolated in the bigger variety $\mathcal{V}_R(I)$.

ii) Let M be as above. Then $\sqrt[r]{M} = M$ by section 2, proposition 2. Using section 2, proposition 8 we can extend M to a (necessarily) maximal real ideal $\tilde{M} \lhd R_\alpha[X_1, \ldots, X_n]$ for some $\alpha \supseteq \tau$. The reality of $\tilde{M}$ shows that it has a zero $x \in R_\alpha^n$. We next claim that $\tilde{M}$ is a zero–dimensional component of $\sqrt[r]{I \cdot R_\alpha}$. If not we would find a further real prime ideal $\tilde{Q}$ between $I \cdot R_\alpha$ and $\tilde{M}$: $I \cdot R_\alpha \subseteq \tilde{Q} \subset \tilde{M}$. But then $I \subseteq \tilde{Q} \cap k[X_1, \ldots, X_n] =: Q \subset M$. Since $\tilde{Q}$ is real and $\tau \subseteq \alpha$ we get $\sqrt[r]{Q} = Q$. Hence $\sqrt[r]{I} \subseteq Q \subset M$: a contradiction. Thus, $\tilde{M}$ occurs in the primary decomposition of $\sqrt[r]{I \cdot R_\alpha} = \bigcap_i P_i \cap \tilde{M}$. Now choose $f \in (\bigcap_i P_i) \setminus \tilde{M}$. Then $\mathcal{V}_{R_\alpha}(I) \cap \{f \neq 0\} = \{x\}$, i.e. x is isolated yielding $I^{iso} \subseteq M$. $\square$

The algorithmic importance of the previous proposition is obvious: if we were able to determine a zero–dimensional ideal $J \subseteq I^{iso}$ then, keeping the notation of proposition 11, $\sqrt[r]{I} = \bigcap_i P_i \cap \sqrt[r]{I + J}$ where $\bigcap_i P_i$ is obtained by localization and $\sqrt[r]{I + J}$ by using the method above for the zero–dimensional case. In the following a construction of such an ideal will be given. But before beginning this let us remark that in general there are more isolated points than those given by the components M_j of $\sqrt[r]{I}$. To see this consider

$$I = \left((Y^2 - X^2(X - 1))(X + 1), (Y^2 - X^2(X - 1))Y\right) \lhd R[X, Y].$$

Then $\mathcal{V}_{iso}(I) = \{(-1, 0), (0, 0)\}$ but only $(-1, 0)$ corresponds to a component of $\sqrt[r]{I} = I$.

Consequently, by constructing a zero–dimensional ideal $J \subseteq I^{iso}$ we are dealing with more points than really necessary. No way of finding exactly the M_j's is known to us.

Given an ideal $I \lhd k[X_1, \ldots, X_n]$ we will now construct via projections a zero–dimensional $J \lhd k[X_1, \ldots, X_n]$ such that $J \subseteq I^{iso}$ (see proposition 15 below). The projection method relies on the following fact which seems to be known:

Proposition 13 *Let k be an arbitrary field, $I \lhd k[X_1, \ldots, X_n]$ and $\{X_1, \ldots, X_m\}$ a maximal subset of $\{X_1, \ldots, X_n\}$ independent modulo I (cf. [KrW]) with $m \leq n - 2$. If the extension ideal $I^e := Ik(X_1, \ldots, X_m)[X_{m+1}, \ldots, X_n]$ is in general position w.r.t. X_{m+1} then we can construct $s \in k[X_1, \ldots, X_m] \backslash 0$, $r_{m+2}, \ldots, r_n \in k[X_1, \ldots, X_{m+1}]$ and $J \lhd k[X_1, \ldots, X_{m+1}]$ such that for all $x \in \bar{k}^n$ with $s(x_1, \ldots, x_m) \neq 0$ the following equivalence holds:*

$$x \in V_{\bar{k}}(I) \Leftrightarrow (x_1, \ldots, x_{m+1}) \in V_{\bar{k}}(J), \ x_i = \frac{r_i(x_1, \ldots, x_{m+1})}{s(x_1, \ldots, x_m)} \ (i = m+1 \ldots n)$$

Proof: First of all we may assume that I is a radical ideal (for the computation of $\sqrt{I}$ see e.g. [AMR], [EH], [GTZ]), [KL], [V]). Let G be a Gröbner basis of I w.r.t. the pure lexicographical ordering with $X_1 < \ldots < X_n$ and $S := k[X_1, \ldots, X_m] \setminus 0$. Then G is also a Gröbner basis of the radical zero–dimensional ideal I^e w.r.t. the pure lexicographical ordering with $X_{m+1} < \ldots < X_n$. According to the Shape-Lemma there are $g_{m+2}, \ldots, g_n \in G$ of type $g_i = h_i \cdot X_i + r_i$ for some $h_i \in S$ and $r_i \in k[X_1, \ldots, X_{i-1}]$ $(i = m + 2 \ldots n)$. We define $s := \prod_{i=m+2}^{n} h_i \in S$. Pseudoreduction of G by multiplication with $h_{m+2}, \ldots, h_n$ yields a finite subset $G' = \{g'_{m+2}, \ldots, g'_n, g'_{n+1}, \ldots\}$ of $k[X_1, \ldots, X_n]$ satisfying

- $g'_i = h_{m+2}^{n_{i,m+2}} \cdots h_{i-1}^{n_{i,i-1}} h_i \cdot X_i + r'_i$, where $n_{i,m+2}, \ldots, n_{i,i-1} \in N_0$ and $r'_i \in k[X_1, \ldots, X_{m+1}]$ for $i = m + 2 \ldots n$

- $g'_i \in k[X_1, \ldots, X_{m+1}]$ for $i > n$

- $V_{\bar{k}}(G') \cap \{s \neq 0\} = V_{\bar{k}}(I) \cap \{s \neq 0\}$.

Defining $J := \big(g'_i \ (i > n)\big) k[X_1, \ldots, X_{m+1}]$ we obtain for all $x \in \bar{k}^n$ with $s(x_1, \ldots, x_m) \neq 0$ that $x \in V_{\bar{k}}(I)$ iff $(x_1, \ldots, x_{m+1}) \in V_{\bar{k}}(J)$ and

$$\big(h_{m+2}^{n_{i,m+2}} \cdots h_{i-1}^{n_{i,i-1}} h_i\big)(x_1, \ldots, x_m) \cdot x_i + r'_i(x_1, \ldots, x_{m+1}) = 0$$

for $i = m + 2 \ldots n$. $\square$

Remark 2 *Let k be an infinite field (e.g. k formally real), $I \lhd k[X_1, \ldots, X_n]$, $m := \dim I \leq n - 2$ and w.l.o.g. $\{X_1, \ldots, X_m\}$ a maximal subset of $\{X_1, \ldots, X_n\}$ independent modulo I. For $c = (c_{m+2}, \ldots, c_n) \in k^{n-m-1}$ we define the automorphism*

$$\varphi_c : \begin{cases} k[X_1, \ldots, X_n] & \to & k[X_1, \ldots, X_n] \\ f(X_1, \ldots, X_n) & \mapsto & f\left(X_1, \ldots, X_m, X_{m+1} + \sum_{i=m+2}^{n} c_i X_i, X_{m+2}, \ldots\right) \end{cases}.$$

*Then for "almost all" $c = (c_{m+2}, \ldots, c_n) \in k^{n-m-1}$ (i.e. excluding the
points of a finite union of proper affine linear subspaces of k^{n-m-1}) $\varphi_c(I)$
fulfills the assumptions of proposition 13, i.e. $\{X_1, \ldots, X_m\}$ is a maximal
subset of $\{X_1, \ldots, X_n\}$ independent modulo $\varphi_c(I)$ and the extension ideal of
$\varphi_c(I)$ in $k(X_1, \ldots, X_m)[X_{m+1}, \ldots, X_n]$ is in general position w.r.t. X_{m+1}.
As mentioned already in section 4 a suitable vector c can be computed deterministically.*

Proposition 13 can be applied only to ideals $I \lhd k[X_1, \ldots, X_n]$ with
$\dim I \leq n - 2$. However, in connection with the computation of isolated
points we can restrict ourselves to ideals complying with this condition by
using the theorem of implicit functions (cf. [BCR, Corollary 2.9.6]):

Lemma 14 *Let k be a formally real field, $I = (f_1, \ldots, f_r) \lhd k[X_1, \ldots, X_n]$
with $\dim I = n - 1$ and f the squarefree part of $\sum_{i=1}^r f_i^2$ in $k[X_1, \ldots, X_n]$.
If R is a real closed field with $k \subseteq R \subset \bar{k}$ then every isolated point of $V_R(I)$
is a zero of the ideal*

$$J := \left(I, \frac{\partial f}{\partial X_i} \ (i = 1 \ldots n) \right) k[X_1, \ldots, X_n],$$

where $\dim J \leq n - 2$.

Now we are able to prove:

Proposition 15 *Let k be a formally real field and $I \lhd k[X_1, \ldots, X_n]$.
Then an ideal $J \lhd k[X_1, \ldots, X_n]$ can be constructed satisfying $\dim J \leq 0$
and $J \subseteq I^{iso}$.*

Proof: We will proceed by induction on the number n of variables and
then, for fixed n, by noetherian induction in $k[X_1, \ldots, X_n]$. If I is a trivial
ideal or zero–dimensional the assertion is evident. Hence we may assume
$m := \dim I \in \{1, \ldots, n-2\}$ (see lemma 14) and $n \geq 3$ with $\{X_1, \ldots, X_m\}$
being a maximal subset of $\{X_1, \ldots, X_n\}$ independent modulo I. By induction assumption we consider the proposition to be proved for ideals in
$k[X_1, \ldots, X_{n'}]$ with $n' < n$ and also for ideals in $k[X_1, \ldots, X_n]$ properly
containing I.
Choose $c = (c_{m+2}, \ldots, c_n) \in k^{n-m-1}$ such that proposition 13 can be applied to $\varphi(I)$ ($\varphi := \varphi_c$ as in remark 2) yielding some polynomials
$s \in k[X_1, \ldots, X_m] \setminus 0$, $r_{m+2}, \ldots, r_n \in k[X_1, \ldots, X_{m+1}]$ and an ideal $J \lhd$
$k[X_1, \ldots, X_{m+1}]$. Since $m + 1 < n$ we use the induction hypothesis to con-
struct an ideal $J_1 \lhd k[X_1, \ldots, X_{m+1}]$ satisfying $\dim J_1 \leq 0$ and $J_1 \subseteq J^{iso}$.
Let be $J_2 \lhd k[X_1, \ldots, X_{m+1}]$ with $V_{\bar{k}}(J_2) = V_{\bar{k}}(J_1) \cap \{s \neq 0\}$. Then the
ideal
$$I_1 := (J_2, \ s \cdot X_i - r_i \ (i = m+2 \ldots n)) k[X_1, \ldots, X_n]$$
has dimension 0 or -1 (i.e. is the unit ideal). Since I is a proper subideal of
(I, s) we may construct by noetherian induction an ideal $I_2 \lhd k[X_1, \ldots, X_n]$

with dim $I_2 \leq 0$ and $I_2 \subseteq (\varphi I, s)^{iso}$. Then $V_{iso}(I) \subseteq V_{\bar{k}}(\varphi^{-1}(I_1 \cdot I_2))$, and we have dim $\varphi^{-1}(I_1 \cdot I_2) \leq 0$.   □

Remark 3 *To obtain an explicit bound on the recursion depth of the method described above the noetherian induction argument can be replaced by induction on the dimension $m := \dim I$ and then, for fixed m, by induction on the number of subsets $S \subseteq \{X_1, \ldots, X_n\}$ independent modulo I with $|S| = m$.*

After these preparations we finally obtain :

Proposition 16 *Let (k, τ) be a preordered field satisfying the computational conditions stated in section 3 and $T_1, \ldots, T_m, X_1, \ldots, X_n$ ($m \in N_0$, $n \in N$) algebraically independent over k. Then the $\sqrt[\tau]{\cdot}$-operator is computable in the polynomial ring $k(T_1, \ldots, T_m)[X_1, \ldots, X_n]$.*

Proof: For arbitrary m we proceed by induction on the number n of variables. The case $n = 1$ was discussed in section 3. Now let be $n \geq 2$, $F := k(T_1, \ldots, T_m)$, $I \vartriangleleft F[X_1, \ldots, X_n]$ and $\sqrt[\tau]{I} = \bigcap_i P_i \cap \bigcap_j M_j$ the reduced primary decomposition of $\sqrt[\tau]{I}$, where the P_i, M_j are τ-real primes with dim $P_i > 0$ and dim $M_j = 0$. Following propositions 12 and 15 we are able to construct an ideal $J \vartriangleleft F[X_1, \ldots, X_n]$ satisfying dim $J \leq 0$ and $J \subseteq \bigcap_j M_j$. Then

$$\sqrt[\tau]{I} \subseteq \bigcap_i P_i \cap \sqrt[\tau]{I+J} \subseteq \bigcap_i P_i \cap \bigcap_j M_j = \sqrt[\tau]{I},$$

thus $\sqrt[\tau]{I} = \bigcap_i P_i \cap \sqrt[\tau]{I+J}$. By induction assumption $\bigcap_i P_i$ can be constructed (cf. section 5) and since $\dim(I + J) \leq 0$ we are able to compute $\sqrt[\tau]{I+J}$ (cf. section 4). But $\bigcap_i P_i$ and $\sqrt[\tau]{I+J}$, as shown, yield $\sqrt[\tau]{I}$ as their intersection.   □

7   Equi–dimensional decomposition and singular points

In this section we will present a second method to deal with the intersection $\bigcap_j M_j$ where we keep the notation of section 5, proposition 11. Instead of dealing with $V_{\bar{k}}(I)$ it is more appropriate to use $Spec\left(k[X_1, \ldots, X_n]/\sqrt[\tau]{I}\right)$ instead.

Let $\sqrt[\tau]{I} = \bigcap_i Q_i \cap \bigcap_j N_j$ be the primary decomposition with dim $Q_i \geq 1$, dim $N_j = 0$. Then

$$\sqrt[\tau]{I} = \bigcap_i \sqrt[\tau]{Q_i} \cap \bigcap_j \sqrt[\tau]{N_j}.$$

Either $\sqrt[r]{N_j} = k[X_1,\ldots,X_n]$ or $\sqrt[r]{N_j} = N_j$. Thus some of the M_j's occur as zero–dimensional components of $\sqrt[r]{I}$ and the remaining ones arise as follows. Let $\sqrt[r]{Q_i} = \bigcap_j Q_{ij} \cap \bigcap_k M_{ik}$ where $\dim Q_{ij} \geq 1$, $\dim M_{ik} = 0$. Then we find the other M_j's among these M_{ik}'s. We claim that an M_j occuring in this way is a singular prime ideal of $k[X_1,\ldots,X_n]/\sqrt[r]{I}$. This is a special case of the following lemma:

Lemma 17 *Let $Q \lhd k[X_1,\ldots,X_n]$ be a prime ideal and let P be a minimal prime ideal of $\sqrt[r]{Q}$ (hence, P is τ–real). If $\dim P < \dim Q$ then P/Q is a singular prime ideal of $k[X_1,\ldots,X_n]/Q$.*

Proof. Set $A := k[X_1,\ldots,X_n]/Q$, $K := Quot(A)$ and consider the local ring $A_{P/Q}$. The natural map $A \xrightarrow{\pi} k[X_1,\ldots,X_n]/P$ gives rise to the epimorphism $A_{P/Q} \xrightarrow{\pi} Quot(k[X_1,\ldots,X_n]/P) =: L$. If $A_{P/Q}$ were regular then, e.g. by [B, Lemma 1.4], π would extend to a place $K \xrightarrow{\lambda} L \cup \infty$. So assume P/Q to be regular. As P is τ–real there is some order $\alpha \supseteq \tau$ extendible to an order $\bar{\alpha}$ on L, by section 2, proposition 2. Using the place $K \xrightarrow{\lambda} L \cup \infty$ we can pull back the order $\bar{\alpha}$ to an order $\tilde{\alpha}$ on K (for details of this process see e.g. [P, §7]). Clearly, $\tilde{\alpha}$ extends α as well since λ is trivial on k. Therefore $\sqrt[r]{Q} = Q$ by section 2, proposition 2. Thus $P = Q$, a contradiction. $\square$

As a consequence of the arguments above we see that the zero–dimensional components M_j of $\sqrt[r]{I}$ arise from two sources: they are either

i) zero–dimensional components of $Spec\left(k[X_1,\ldots,X_n]/\sqrt[r]{I}\right)$ or

ii) singular points of one of the at least one–dimensional components of $Spec\left(k[X_1,\ldots,X_n]/\sqrt[r]{I}\right)$.

Hence, if we were able to find an ideal $J \supseteq I$ such that $Spec\left(k[X_1,\ldots,X_n]/J\right)$ comprises all the prime ideals in i) and ii) then, again using the notation of proposition 11, we would have

$$\sqrt[r]{I} = \bigcap_i P_i \cap \sqrt[r]{J}.$$

The construction of J runs as follows:

1. Compute $\sqrt{I}$ (note that $\sqrt{}$-computations are intensively used in section 6 as well)

2. (Equi–dimensional components) Compute $I^{(e)} = \bigcap Q$ where Q ranges over all the minimal prime ideals of I of dimension e, for $e = 0,\ldots,\dim I$. This can be achieved by using certain minors of the Jacobian matrix $\left(\frac{\partial f_i}{\partial X_j}\right)$, where $\sqrt{I} = (f_1,\ldots,f_r)$, as shown below after proposition 18.

3. For all $e \geq 1$ compute the singular locus $Spec\left(k[X_1,\ldots,X_n]/I^{(\tilde{e})}\right)$ of $Spec\left(k[X_1,\ldots,X_n]/I^{(e)}\right)$, $I^{(\tilde{e})} \supseteq I^{(e)}$ by using the Jacobian criterion (cf. [M, p. 213 ff]). Note that $\dim I^{(\tilde{e})} < \dim I^{(e)} = e$.

Summarizing we get:

Proposition 18 *One can construct an ideal J such that*

i) $\dim J < \dim I$, *if* $\dim I \geq 1$

ii) $\sqrt{I} = \bigcap_i P_i \cap \sqrt{J}$.

Proof: $J := I^{(0)} \cap I^{(\tilde{1})} \cap \ldots I^{(\tilde{d})}$, where $d := \dim I$.   □

Remark 4 *It is not necessary to determine first the dimension of I since the following algorithm will recognize it.*

The algorithm itself is based on the Jacobian criterion for regular prime ideals in affine rings $A = k[X_1,\ldots,X_n]/I$, $I = (f_1,\ldots,f_r)$ (cf. [M, p. 213 ff]). So, let $Q \supseteq I$ be any prime ideal, $J = \left(\frac{\partial f_i}{\partial X_j}\right)_{i=1\ldots r, j=1\ldots n}$ the Jacobian matrix of $(f_1,\ldots,f_r)$ and let $J(Q)$ be the image of J over $k(Q) = Quot\left(A\Big/Q/I\right)$. Then: Q/I regular in A iff

$$rg_{k(Q)}J(Q) = ht(Ik[X_1,\ldots,X_n]_Q) = ht(Q) - \dim A_{Q/I}.$$

In the sequel, I will always denote a radical ideal of $k[X_1,\ldots,X_n]$. Now, if Q is a minimal prime ideal of I then Q/I is a minimal prime ideal of the reduced ring A, hence $A_{Q/I}$ a field, in particular a regular ring. Thus, under these assumptions:

$$rg_{k(Q)}J(Q) = n - \dim Q. \tag{4}$$

Furthermore, we set $I^{(e)} := \bigcap Q$, where Q ranges over all e–dimensional minimal prime ideals of I. If there are none we set $I^{(e)} := k[X_1,\ldots,X_n]$. Clearly, $I = \bigcap_{0 \leq e \leq d} I^{(e)}$ where $d := \dim I$. We assume $I = (f_1,\ldots,f_r)$. Then $rgJ(Q) \leq \min\{n,r\}$ and, by (4), $I^{(e)} = k[X_1,\ldots,X_n]$ if $e < n - r$. For any e satisfying $0 < n - e \leq \min\{n,r\}$ we define

$$J_e := \text{ ideal of } k[X_1,\ldots,X_n] \text{ generated by all}$$
$$(n-e) \times (n-e)\text{–minors of } J(f_1,\ldots,f_r),$$

if $\min\{n,r\} < n - e \leq n$ then $J_e := (0)$. We then have

Proposition 19 *Let $I = (f_1,\ldots,f_r)$ be a radical ideal of $k[X_1,\ldots,X_n]$ and $0 \leq e < n$. Then*

$$(I : J_e) = I^{(e)} \cap I^{(e-1)} \cap \ldots \cap I^{(0)}$$
$$(I : (I : J_e)) = I^{(d)} \cap I^{(d-1)} \cap \ldots \cap I^{(e+1)}.$$

Proof: If $n - e > \min\{n, r\}$ then, by definition, $J_e = (0)$, hence $(I : J_e) = k[X_1, \ldots, X_n]$, $(I : (I : J_e)) = I$. This complies with (4) which states $\dim Q \geq n - \min\{n, r\}$ for all minimal prime ideals of I. Next, let $n - e \leq \min\{n, r\}$. Then, (4) yields for any minimal prime ideal Q of I

$$J_e \not\subseteq Q \Longleftrightarrow \dim Q \leq e.$$

In particular, $J_e \subseteq I^{(d)} \cap \ldots \cap I^{(e+1)}$ and $J_e \cdot \left(\bigcap_{f \leq e} I^{(f)}\right) \subseteq \bigcap_{f \leq d} I^{(f)} = I$ which means $\bigcap_{f \leq e} I^{(f)} \subseteq (I : J_e)$.

Conversely, if $f \in (I : J_e)$, i.e. $f J_e \subseteq I$, then, in particular, $f J_e \subseteq Q$ for all minimal prime ideals $Q \supseteq I$ of $\dim Q \leq e$. From the arguments above one concludes $f \in \bigcap_{f \leq e} I^{(f)}$ showing $(I : J_e) = \bigcap_{f \leq e} I^{(f)}$.

The remaining statement follows from two facts: that I is radical and $(I : J_e) \not\subseteq Q$ for every minimal prime ideal over I of dimension at least $e + 1$. □

For the computation of ideal quotients see [GTZ, Corollary 3.2].

Proposition 19 allows a recursive determination of the ideals $I^{(f)}$ by the computation of $I^{(0)}, \ldots, I^{(e-1)}$ and $I^{(\hat{e})} := \bigcap_{f \geq e} I^{(f)}$. Using J_0 one gets $I^{(0)}$, $I^{(\hat{1})}$. If $I^{(0)}, \ldots, I^{(e-1)}$ and $I^{(\hat{e})}$ are found one uses the Jacobian ideal J_e of $I^{(\hat{e})}$ to construct $I^{(e)}$ and $I^{(\hat{e+1})}$ out of $I^{(\hat{e})}$. One has $I^{(\hat{e+1})} = k[X_1, \ldots, X_n]$ iff $e = \dim I$. Hence, by checking $I^{(\hat{e+1})} = k[X_1, \ldots, X_n]$, one knows when to stop.

8 Computational requirements

In this section we are going to present a short list of fields with finitely many orderings satisfying the general assumptions in section 2. We start with an ordered field (k, α) where we assume that the arithmetic operations and the order–comparison between elements can be performed.

- k is uniquely ordered allowing factorization of univariate polynomials, e.g. $k = Q$.

  Then the decision about strict positivity of polynomials in R_α is the subject of an algorithm for elimination of quantifiers over real closed fields.

- R is the real closure of (k, α).

  By [BCR, p. 9] R allows factorization of univariate polynomials (even if k doesn't). The other assumption is fulfilled since R has elimination of quantifiers.

- l is a finite extension of the uniquely ordered field (k, α), given by an irreducible polynomial $p \in k[X]$, where k admits factorization of univariate polynomials.

  From [Se, p. 289] we obtain that l allows factorization as well. It is known that l has only finitely many orderings. Fixing a real closure R_α of (k, α) these orderings correspond to the roots of p in R_α. These roots can be coded e.g. à la Thom [CR]. Therefore we can construct all orderings of l. Having the orders of l constructed an algorithm for elimination of quantifiers will provide the required decision about positivity.

References

[AMR]   Alonso, M. E., Mora, T., Raimondo, M. (1990). **Local Decomposition Algorithms**. Proceedings of the 8th International Conference AAECC–8, Tokyo, Japan, August 1990. Lecture Notes in Computer Science 508, 208 – 221.

[B]   Becker, E. (1981). **Valuations and Real Places in the Theory of Formally Real Fields**. In: "Géométrie Algébrique Réelle et Formes Quadratiques", Proceedings Rennes 1981, Lecture Notes in Math. 959, 1 – 40.

[BCR]   Bochnak, J., Coste, M., Roy, M. F. (1987). **Géométrie algébrique réelle**. Ergebnisse der Mathematik und ihrer Grenzgebiete, Folge 3, Band 12. Springer–Verlag.

[CR]   Coste, M., Roy, M. F. (1988). **Thom's Lemma, the Coding of Real Algebraic Numbers and the Computation of the Topology of Semi–algebraic Sets**. Journal of Symbolic Computation (1988) 5, 121 – 129.

[D]   Dubois, D. W. (1969). **A nullstellensatz for ordered fields**. Arkiv for Mat. 8, 111 – 114.

[DR]   Dubois, D. W., Recio, T. (1982). **Order Extensions and Real Algebraic Geometry**. Contemporary Mathematics, Vol. 8, 265 – 288.

[EH]   Eisenbud, D., Huneke, C. (1989). **A Jacobian method for finding the radical of an ideal**. preprint.

[ELW] Elman, R., Lam, T. Y., Wadsworth, A. R. (1979). **Orderings under field extensions.** J. Reine Angewandte Mathematik 306, 6 – 27.

[GM] Gianni, P., Mora, T. (1987). **Algebraic Solution of Systems of Polynomial Equations using Gröbner Bases.** Proceedings of the 5th International Conference AAECC-5, June 1987. Lecture Notes in Computer Science 356, 247 – 257.

[GTZ] Gianni, P., Trager, B., Zacharias, G. (1989). **Gröbner Bases and Primary Decomposition of Polynomial Ideals.** In: (Robbiano, L., ed.) Computational Aspects of Commutative Algebra, Academic Press, 15 – 33.

[K] Krivine, J. L. (1964). **Anneaux préordonnés.** Journal d'analyse mathématique 12, 307 – 326.

[KL] Krick, T., Logar, A. (1991). **An Algorithm for the Computation of the Radical of an Ideal in the Ring of Polynomials.** Proceedings of the 9th International Symposium AAECC-9, New Orleans, LA, USA, October 1991. Lecture Notes in Computer Science 539, 195 – 205.

[KMH] Kobayashi, H., Moritsugu, S., Hogan, R. W. (1989). **On Radical Zero-dimensional Ideals.** Journal of Symbolic Computation (1989) 8, 545 – 552.

[KrW] Kredel, H., Weispfenning, V. (1989). **Computing Dimension and Independent Sets for Polynomial Ideals.** In: (Robbiano, L. ed.) Computational Asepects of Commutative Algebra, Academic Press, 97 – 113.

[L] Lakshman, Y. N. (1990). **On the Complexity of Computing a Gröbner Basis for the Radical of a Zero Dimensional Ideal.** In: "Proc. of 22nd ACM Symposium on Theory of Computing", May 1990.

[M] Matsumura, H. (1970). **Commutative Algebra.** W. A. Benjamin, Inc.

[P] Prestel, A. (1984). **Lectures on Formally Real Fields.** Lecture Notes in Math. 1093. Springer–Verlag.

[R] Risler, J.-J. (1970). **Une caractérisation des idéaux des variétés algébriques réelles.** C.R.A.S. Paris, série A 271, 1171 – 1173.

[Se] Seidenberg, A. (1974). **Constructions in Algebra.** Trans. Amer. Math. Soc. 197, 273 – 313.

[V]   Vasconcelos W. V. (1991). **Jacobian Matrices and Construc-
         tions in Algebra**. Proceedings of the 9th International Sympo-
         sium AAECC–9, New Orleans, LA, USA, October 1991. Lecture
         Notes in Computer Science 539, 48 – 64.

   E. Becker, R. Neuhaus
Universität Dortmund, Fachbereich Mathematik
Postfach 50 05 00, W-4600 Dortmund 50, Germany

Semialgebraic geometry of polynomial control problems

M. Briskin   Y. Yomdin

Abstract. We prove some new facts concerning metric properties of semi algebraic sets and on this base establish some geometric properties of near-critical trajectories in polynomial control problems.

1   Introduction

Consider a differential equation $\dot{x} = f(x)$, or, more generally, a control problem $\dot{x} = f(x, u)$, with a polynomial right-hand side. A well-known and difficult question related to such equations consists in establishing "algebraic finiteness" properties (like Bezout theorem) for their solution. An inherent complexity of this problem is implied by the fact that the solutions of algebraic differential equations are not algebraic, and their families (in the presence of singular points) are not analytic.

As far as control problems are considered the situation is somewhat easier from one side: usually one is interested in a behavior of solution only for a fixed time. On the other hand, the presence of an arbitrary control u makes the trajectories behavior much more complicated.

Consideration of extremal trajectories, those minimizing a certain functional, (e.g. the time required to reach the final position) reduces the freedom of a control choice. Still the behavior of extremals can be very complicated. In particular, an infinite number of control jumps in a finite time can occur in a generic way (see [7], [9]).

In [1-5] we have started investigation of some situations, where the trajectories follow algebraic patterns and thus posess finiteness properties. The results obtained suggest that there is a wide and important class of trajectories in polynomial control problems, sharing such properties namely, critical and near-critical trajectories of rank zero. These trajectories are characterized by the property that there is no state direction, by which we can move the endpoint by infinitesimal control variations both in this direction and in the opposite one. Such trajectories play an important role in the geometry of the control problem. In particular, they lead to the corners of the reachable sets. We do not discuss here these and further applications in control theory (see [4], [8-12]).

In the present paper we give a result (presumably, new) on metric properties of real semialgebraic sets, and using it establish some basic properties of near critical trajectories. As an application we give a sketch of the proof of one of the finiteness results for near-critical trajectories.

The second author would like to thank the Fields institute for research in mathematical sciences, where part of this paper was written, for its hospitality.

2   A semialgebraic result

Let $A \subseteq \mathbb{R}^n$ be a semialgebraic set. A can be constructed by a finite number of set-theoretic operations from the basic semialgebraic sets, given by a single polynomial equation for inequality. We call a set-theoretic formula of such a representation, together with the degrees of the polynomials involved and the dimension n, the diagram of A.

Now let $n = n_1 + n_2 + \ldots + n_k, \bar{r} = (r_1, \ldots, r_k), r_i \geq 0$. We denote by $B_{\bar{r}}^n \subseteq \mathbb{R}^n$ the product $B_{r_1}^{n_1} \times \ldots \times B_{r_k}^{n_k}$, where $B_{r_i}^{n_i} \subseteq \mathbb{R}^{n_i}$ is the ball of radius r_i in $\mathbb{R}^{n_i}$. Let $\pi_i : B_{\bar{r}}^n \to B_{r_i}^{n_i}$ denote the standard projections.

Theorem 2.1. Let $A \subseteq \mathbb{R}^n$ be a semialgebraic set. Then any two points, belonging to the same connected component of $A \cap B_{\bar{r}}^n$ can be joined in $A \cap B_{\bar{r}}^n$ by a semialgebraic curve s such that for each $i = 1, \ldots, k$, the length of the projection $\pi_i(s) \subseteq B_{r_i}^{n_i}$ (counted with multiplicities) does not exceed Cr_i, with the constant C depending only on the diagram of A.

Proof. As in [13], we can join these points by a semialgebraic curve s with a diagram, depending only on the diagram of A. By the projection properties of semialgebraic sets the same is true for the curves $\pi_i(s) \subseteq B_{r_i}^{n_i}$. Therefore the number of intersection points of $\pi_i(s)$ with a generic hyperplane in $\mathbb{R}^{n_i}$ is bounded in terms of the diagram of A only. Now the result follows by the basic integral-geometry formula.

Theorem 2.1 generalizes a result from [13], where $k = 1$.

3   Near-critical trajectories

We consider a control problem of the form

$$\dot{x} = f(x, u), x(0) = x^0 \tag{1}$$

where $x \in \mathbb{R}^n, u \in B_1^m \subseteq \mathbb{R}^m$- the ball of radius 1, centered at $0 \in \mathbb{R}^m$.

Let $f(x, u)$ be a polynomial of a total degree d in x and u, bounded in norm by 1 for $u \in B_1^m$ and x in the disk $\mathcal{D}$ of radius T, centered at $x^0 \in \mathbb{R}^n, T \geq 1$. In particular, this assumption guarantees that for any control $u(t) \in B_1^m, t \in [0, T]$, the corresponding solution $x(t)$ of (1) exists and belongs to $\mathcal{D}$.

Definition 3.1. For $\gamma \geq 0$, a trajectory $x(t), u(t)$ of (1) is called γ-critical of rank 0, if $\|\frac{\partial f}{\partial u}(x(t), u(t))\| \leq \gamma$ for any $t \in [0, T]$. Here $\|\ \|$ denotes the usual euclidean norm of the linear operator $\frac{\partial f}{\partial u} : \mathbb{R}^m \to \mathbb{R}^n$.

We shall consider trajectories $(x(t), u(t))$ of (1) as curves in $\mathcal{D} \times B_1^m \subseteq \mathbb{R}^n \times \mathbb{R}^m$. The projection $x(t)$ on the first factor will be called the space trajectory.

Let Ω_γ be the subset in $\mathcal{D} \times B_1^m$, defined by $\{\|\frac{\partial f}{\partial u}\| \geq \gamma\}$. Ω_γ is a semialgebraic set with the diagram, depending only on the degree d of $f(x, u)$.

By the definition 3.1, γ-critical trajectories are those, which do not enter the interior of Ω_γ i.e. they stay in $W = (\mathcal{D} \times B_1^m) \backslash \Omega_\gamma$.

Definition 3.2. For $K \geq 1$, two points $Z_1 = (x_1, u_1)$ and $Z_2 = (x_2, u_2)$ in $\mathcal{D} \times B_1^m$ are called K-visible, if they can be joined by a curve ζ in W, with $\pi_1(\zeta)$ contained in a ball of diameter $K \cdot \|x_2 - x_1\|$ in $\mathbb{R}^n$. (Compare [1]).

Thus visible are those points, which can be joined by a curve in W with the space projection "sufficiently monotone".

Lemma 3.1. If Z_1 and Z_2 can be joined W by a curve ζ, with $\pi_1(\zeta)$ contained in some ball B_r^m of diameter η, then $\|f(Z_2) - f(Z_1)\| \leq C_1 \eta + C_2 \gamma$, with C_1 and C_2 depending only on d, n and m.

Proof. Z_1 and Z_2 belong to the same connected component of $W \subseteq B_r^n \times B_1^m$, and hence by theorem 2.1 they can be joined in W by a semialgebraic curve s with lengths of projections on $\mathbb{R}^n$ and $\mathbb{R}^m$, bounded by $\tilde{C}_1 r$ and $\tilde{C}_2$, respectively. Now

$$\|f(Z_2) - f(Z_1)\| = \|\int_s \frac{df}{ds} ds\| \leq \int_s \|\frac{\partial f}{\partial x}\|\ \|dx\| + \int_s \|\frac{\partial f}{\partial u}\|\ \|du\|$$

$$\leq d^2 \int_s \|dx\| + \gamma \int_s \|du\| \leq d^2 \tilde{C}_1 r + \gamma \tilde{C}_2 = C_1 r + C_2 \gamma.$$

(By Markov's inequality (see e.g. [6]), $\|f\| \leq 1$ on D implies $\|\frac{\partial f}{\partial x}\| \leq d^2$).

Corollary 3.2. If $Z_1 = (x_1, u_1)$ and $Z_2 = (x_2, u_2)$ are K-visible, then

$$\|f(x_2, u_2) - f(x_1, u_1)\| \leq C_1 K \|x_2 - x_1\| + C_1 \gamma.$$

Now we can prove one of the main properties of γ-critical trajectories, which asserts that variation of controls (within W) cannot change the space geometry of the trajectory for more than γ, until we stay in the "visibility domain". In other words, if two γ-critical trajectories are visible, the corresponding space curves are, in fact, γ-close to one another.

Theorem 3.3. Let $(x_1(t), u_1(t))$ and $(x_2(t), u_2(t))$ be two trajectories of (1). If for any $t \in [0, T]$ the points $(x_1(t), u_1(t))$ and $(x_2(t), u_2(t))$ are K-visible then $\|x_2(t) - x_1(t)\| \leq C_2 \gamma \cdot t e^{KC_1 t}, t \in [0, T]$.

Proof. Since both the trajectories satisfy (1),
$\frac{d}{dt}\|x_2(t) - x_1(t)\| \leq \|\dot{x}_2 - \dot{x}_1\| = \|f(x_2, u_2) - f(x_1, u_1)\|$
$\leq KC_1 \|x_2(t) - x_1(t)\| + C_2 \gamma$,
with $\|x_2 - x_1\| = 0$ for $t = 0$. Solving this differential inequality, we get

$$\|x_2(t) - x_1(t)\| \leq \frac{C_2 \gamma}{KC_1}(e^{KC_1 t} - 1) \leq C_2 \gamma \cdot t e^{KC_1 t}$$

Thus to investigate the space geometry of γ-critical trajectories, we have to analyze the structure of "visibility classes". To do this we need some further properties of γ-critical trajectories. In a sense, we want to show that locally γ-critical trajectories behave as the trajectories of an ordinary differential equation: 1). If a trajectory occurs near a critical point, it stays there for a long time, 2). The "rotational velocity" of each γ-critical trajectory is bounded.

Proposition 3.4. Let $(x(t), u(t))$ be a γ-critical trajectory, $t_0 \in [0, T]$. Let $M = \|f(x(t_0), u(t_0))\|$. Then for any $t \in [0, T]$,

$$\|x(t) - x(t_0)\| \leq (C_2 \gamma + M)\Delta t \cdot e^{C_1 \Delta t},$$

where $\Delta t = |t - t_0|$.

Proof. Denote by $\rho(t)$ the radius of the minimal ball, centered at $x(t_0)$, containing the trajectory $x(\tau), \tau \in [t_0, t]$.
By lemma 3.1 $\|f(x(t), u(t) - f(x(t_0), u(t_0))\| \leq C_1 \rho(t) + C_2 \gamma$,
and hence $\|f(x(t), u(t))\| \leq M + C_1 \rho(t) + C_2 \gamma$. Clearly, $|d\rho(t)/dt|$ does not exceed $\|f(t)\|$ (and $d\rho/dt = 0$ if $\rho(t) > \|x(t) - x(t_0)\|$).

Hence $|d\rho/dt| \le C_1\rho + M + C_2\gamma$, and as above, solving this differential inequality, we get

$$\|x(t) - x(t_0)\| \le \rho(t) \le (C_2\gamma + M)(t - t_0)e^{C_1(t-t_0)}$$

Corollary 3.5. If for some $t_0 \in [0, T]$, the velocity $v(x(t_0), u(t_0))$ is of order γ, $\|f(x(t_0), u(t_0))\| \le \mathcal{L}\gamma$, then for $t \in [0, T]$ the trajectory remains in the ball of radius $(C_2 + \mathcal{L})Te^{C_1T} \cdot \gamma$, centered at $x(t_0)$.

Proposition 3.6. Let $(x(t), u(t))$ be a γ-critical trajectory, and assume that for any $t \in [0, T], \|f(x(t), u(t))\| \ge \mathcal{L}\gamma, \mathcal{L} \ge 11C_2$. Let $\Delta = 1/10\,C_1e^{C_1}$.

Then for any $t_1, t_2 \in [0, T]$ with $|t_2 - t_1| \le \Delta$, the angle φ between the velocity vectors $f(x(t_2), u(t_2))$ and $f(x(t_1), u(t_1))$ does not exceed $\pi/10$. Consequently, for any $t_1, t_2 \in [0, T]$, the angle between $f(t_2)$ and $f(t_1)$ does not exceed $([(t_2 - t_1)/\Delta] + 1) \cdot \frac{\pi}{10}$.

Proof. Let $M = \|f(x(t_1), u(t_1))\|$.

By proposition 3.4, for $|t_2 - t_1| \le \Delta, \|x(t_2) - x(t_1)\| \le (C_2\gamma + M)\Delta \cdot e^{C_1\Delta}$

In fact, the proof of this proposition shows, that the trajectory $x(\tau), \tau \in [t_1, t_2]$, is contained in a ball of this radius, centered at $x(t_1)$. Hence, by lemma 3.1, $\|\Delta f\| = \|f(t_2) - f(t_1)\| \le C_1e^{C_1\Delta}(C_2\gamma + M)\Delta + C_2\gamma$, and therefore $\sin\varphi \le \|\Delta f\| / \|f(t_1)\| \le (C_1e^{C_1\Delta}(C_2\gamma + M)\Delta + C_2\gamma)/M$
$= (C_1e^{C_1\Delta} \cdot \Delta + 1)C_2\gamma/M + C_1e^{C_1\Delta} \cdot \Delta$.

For $\Delta = 1/10\,C_1e^{C_1}$, clearly, $\Delta < 1$, and hence $e^{C_1\Delta} < e^{C_1}$ and $C_1e^{C_1\Delta} \cdot \Delta < 1/10$. Consequently, $(C_1e^{C_1\Delta} \cdot \Delta + 1)C_2\gamma/M \le 1.1\,C_2\gamma/M \le 1/10$, since $M \ge \mathcal{L}\gamma$, with $\mathcal{L} \ge 11\,C_2$. Therefore, $\sin\varphi \le 1/5$ and $\varphi < \pi/10$.

(Notice that since the control u can be chosen arbitrarily within the γ-critical set W, we cannot guarantee that the angle between $f(t_2)$ and $f(t_1)$ tends to zero as $t_2 \to t_1$).

4   A finiteness theorem

Here we illustrate the application of the general techniques developed above by sketching the proof of a finiteness result for polynomial control systems. We do not try here to give the most general form of this statement nor to indicate its relations with the problematics of the control theory.

Let $n = 2, m = 1$ in (1). We call the endpoints $x(T)$ of γ-critical trajectories the γ-critical values of (1).

Theorem 4.1. For any $\gamma \geq 0$ the set of γ-critical values of (1) in $\mathbb{R}^2$ can be covered by $\tilde{C} = C_3(d)e^{C_4(d)T}$ disks of radius γ.

Sketch of the proof. It is enough to show that each γ-critical trajectory belongs to one of $\tilde{C}$ "visibility classes", since by theorem 3.3 the endpoints of each visibility class can be covered by disk of radius $C_2Te^{KC_1T}\gamma$.

To define these classes, first of all, we can assume, that the velocity of our trajectories at each point is at least $\mathcal{L}\gamma, \mathcal{L} = 11\,C_2$. Indeed, by corollary 3.5, if at some point $\|f\| < \mathcal{L}\gamma$, then the trajectory remains in the ball of radius $(C_2+\mathcal{L})Te^{C_2T}\gamma$, centered at $x(t_0)$, for any $t \in [0,T]$. But hence the distance of the endpoint $x(T)$ from the initial point $x^0 = x(0)$, does not exceed twice this radius, and therefore all such trajectories form one class.

Now assume for a moment that x is one-dimensional. In the plane (x, u) γ-critical trajectories are those which avoid $\Omega_\gamma = \{|\frac{\partial f}{\partial u}| \geq \gamma\}$. Denote by $\Omega_1, \ldots, \Omega_N$ the components of Ω_γ. Since our trajectories always move in the same x-direction $(\|f\| \geq \mathcal{L}\gamma)$, any two trajectories, which pass on the same side of each Ω_i, are $\mathcal{L}$-visible. Therefore, the number of visibility classes does not exceed $2^N \leq 2^{Cd^2}$, by the well known bounds for the number of connected components of a semialgebraic set (See [1] for a detailed proof of a 1-dimensional result and example showing sharpness of the bound).

In higher dimensions the situation seems to differ completely, since apriori our trajectories can turn around each component of Ω_γ an arbitrary number of times, and trajectories with a different number of turns clearly can be K-invisible for any K.

However, it turns out that the algebraic nature of the problem prevents this behavior. To see this, consider one of the components $\Omega = \Omega_i$ of $\Omega_\gamma \subseteq \mathbb{R}^2 \times \mathbb{R}$. Denote $\tilde{\Omega}$ its projection on the (x_1, x_2)-plane, ω the boundary curve of $\tilde{\Omega}$ and Z- a cylinder on ω along the u-direction.

Rotation of a trajectory $(x(t), u(t))$ around Ω in a (x, u)-space corresponds to a crossing of ω by the space trajectory $x(t)$. We want to show that the number of such crossings is (essentially) bounded in terms of d. To see this consider an equation

$$f(x, u) \cdot n(x, u) = 0 \qquad (*),$$

where n is the normal vector to Z. This algebraic equation defines a curve Q on Z. The algebraic curve Q is also a trajectory of (1), since (*) implies that the velocity vectors are tangent to Z along Q.

Components of Q are projected onto segments in the boundary curve ω.

Thus ω is subdivided into segments of two types: those over which there is a component of Q and those over which $f(x, u)$ is never tangent to Z. Therefore the segments of the second type can be crossed by a space trajectory $x(t)$ only in one direction, independantly of the control $u(t)$.

The segments of the first type are the space trajectories, corresponding to the (x, u)-trajectory Q. By construction, Q is 1-visible for any γ-critical trajectory, whose space projection crosses one of these segments. Hence, by theorem 3.3, if a space trajectory crosses one of the segments of the first type, it is, in fact "γ-glued" to it.

Now we subdivide the disk $\mathcal{D} \subseteq \mathbb{R}^2$ by all the segments of the first and the second type in all the boundary curves ω of the projections $\tilde{\Omega}$ of the components of Ω_γ. The number of the elements in this semialgebraic subdivision is bounded in terms of the degree d only. Our trajectories cross the segments of the second type only in one direction, and they are "γ-glued" to the segments of the first type, if they cross them. By this property (and after some additional semialgebraic subdivision of our partition) one can show that if the trajectory crosses the same segments twice, its velocity vector rotates to at least $\pi/2$ between these two crossings. By proposition 3.6, this takes at least 5Δ time. Therefore the total number of crossings in time T does not exceed the number of segments in the partition, multiplied by $T/5\Delta$. Now one can show, that two trajectories, whose space projections cross the partition segments in the same order, and whose controls at each crossing pass on the same side of the corresponding components Ω, belong to the same visibility class.

References

[1] M. Briskin, Y. Yomdin, Critical and near critical values in polynomial control problems, I; to appear in Israel J. of Math.

[2] M. Briskin, Y. Yomdin, Vertices of reachable sets in nonlinear control problems, Proc. CDC, Hawaii, 1990, 2815-2816.

[3] M. Briskin, Y. Yomdin, Extremal and near-extremal trajectories of rank zero in nonlinear control problems, to appear in the Proc. of Control Conference, Haifa, (1990).

[4] M. Briskin, Y. Yomdin, Vertices of reachable sets in polynomial control problems, preprint, 1992.

[5] M. Briskin, Y. Yomdin, Critical and near-critical trajectories in nonlinear control problems, II, in preparation.

[6] O. Kellog, On bounded polynomials in several variables, Math. Z. 27, 1928, 55-64.

[7] I. Kupka, Generic Properties of Extremals in Optimal Control Problems, in Differential Geometric Control Theory, R.W. Brockett. R.S. Millwar, H.J. Sussmann, editors, Birkhäuser, 1983, 310-315.

[8] I. Kupka, CDC, Tampa, Dec. 1989.

[9] H.J. Sussmann, Lie Brackets, Real Analyticity and Geometric Control, in "Differential Geometric Control Theory", 1-116.

[10] H.J. Sussmann, CDC, Tampa, Dec. 1989.

[11] H.J. Sussmann, J. Diff. Eqns. 20, 292, 1979, Ibid, 31, 31.

[12] E. Sontag, Finite-dimensional open-loop control generators for non-linear systems, Int. J. Control, 1988, Vol. 47, No.2, 537-556.

[13] Y. Yomdin, The geometry of critical and near-critical values of differentiable mappings, Math. Ann.264, 1983, 495-515.

M. Briskin
Academic School for Technological Education
Givat Ram, Jerusalem 95435, Israel.

Y. Yomdin
Department of Theoritical Mathematics
The Weizmann Institute of Science
Rehovot 76100, Israel.

The Resultant via a Koszul Complex

M. Chardin

Abstract. As noticed by Jouanolou, Hurwitz proved in 1913 ([Hu])
that, in the generic case, the Koszul complex is acyclic in positive degrees
if the number of (homogeneous) polynomials is less than or equal to the
number of variables[†]. It was known around 1930 that resultants may be
calculated as a Mc Rae invariant of this complex. This expresses the resultant as an alternate product of determinants coming from the differentials
of this complex. Demazure explained in a preprint ([De]), how to recover
this formula from an easy particular case of deep results of Buchsbaum and
Eisenbud on finite free resolutions. He noticed that one only needs to add
one new variable in order to do the calculation in a non generic situation.

I have never seen any mention of this technique of calculation in recent
reports on the subject (except the quite confidential one of Demazure and
in an extensive work of Jouanolou, however from a rather different point
of view). So, I will give here elementary and short proofs of the theorems needed—except the well-known acyclicity of the Koszul complex and
the "Principal Theorem of Elimination"—and present some useful remarks
leading to the subsequent algorithm. In fact, no genericity is needed (it is
not the case for all the other techniques). Furthermore, when the resultant vanishes, some information can be given about the dimension of the
associated variety.

As an illustration of this technique, we give an arithmetical consequence on the resultant : if the polynomials have integral coefficients and
their reductions modulo a prime p defines a variety of projective dimension
zero and degree d, then p^d divides the resultant.

I The tools

Let A be a noetherian and factorial domain, k its quotient field and
$R = A[X_1, \ldots, X_n]$.

If $P_1, \ldots, P_n$ is a regular sequence of homogeneous polynomials in R,
let I be the ideal generated by the P_i's, d_i the degree of P_i and **K** the
associated Koszul complex:

[†] at least in this generic case, the Koszul complex was thus known long
before the general theory of Koszul complexes.

$$\mathbf{K} \ : \ 0 \longrightarrow \textstyle\bigwedge^n B \xrightarrow{\partial_n} \textstyle\bigwedge^{n-1} B \xrightarrow{\partial_{n-1}} \cdots \xrightarrow{\partial_2} \textstyle\bigwedge^1 B \xrightarrow{\partial_1} R \longrightarrow 0$$

where B is the free R-module R^n equipped with the base $(e_1, \ldots, e_n)$ and the differentials ∂_p are defined by

$$\partial_p(e_{i_1} \wedge \cdots \wedge e_{i_p}) = \sum_{s=1}^{p} (-1)^{s+1} P_{i_s} \, e_{i_1} \wedge \cdots \wedge \widehat{e_{i_s}} \wedge \cdots \wedge e_{i_p}.$$

If we introduce on the modules $K_p = \bigwedge^p B$ the natural graduation $\deg(e_{i_1} \wedge \cdots \wedge e_{i_p}) = d_{i_1} + \cdots + d_{i_p}$, this complex of A-modules is graduated, its differential is of degree zero, moreover, **up to the surjectivity of ∂_1 it is exact in all degrees** ([Se] IV Prop. 2, original proof in [Hu]).

We will write K_p^ν and ∂_p^ν for the degree ν parts of the modules and differentials.

Let us now recall a classical theorem from elimination theory:

Proposition 1.— *Let $A = \mathbf{Z}[U_{i,\alpha}]$ where the $U_{i,\alpha}$ are the algebraically independent coefficients of the generic homogeneous polynomials*
$P_i = \sum_{|\alpha|=d_i} U_{i,\alpha} X^\alpha \in R = A[X_1, \ldots, X_n]$ *for $i = 1, \ldots, n$.*
Then:

$$\mathrm{Ann}_A \left(R/(P_1, \ldots, P_n) \right)_\nu = \mathrm{Res}(P_1, \ldots, P_n)$$

if and only if $\nu > \sum_{i=1}^{n}(d_i - 1)$.

Proof. See [Jo2] §3.5 page 144.

We will now give the result we need from homological algebra.

If A is a noetherian factorial ring and M is a torsion A module of finite type we will denote by $\mathrm{div}(M)$ the divisor associated to M:

$$\mathrm{div}(M) = \sum_{\mathcal{P} \in \mathrm{Ass}(M), h(\mathcal{P})=1} \mathrm{length}(M_{\mathcal{P}})\mathcal{P}$$

If I is an ideal of A, and $[I]$ the principal part of I (i.e. the gcd of generators of I), then $\mathrm{div}(R/I) = \sum_i e_i \mathcal{P}_i$ if $[I] = \prod_i \mathcal{P}_i^{e_i}$ is the decomposition into irreducible factors of the principal ideal $[I]$.

A good reference for theses concepts is [Bo] Chap. 7, §4.

Proposition 2.— *Let C be a complex of finitely generated free A-modules (A factorial and noetherian):*

$$\mathbf{C} \ : \ 0 \longrightarrow C_n \xrightarrow{\partial_n} C_{n-1} \xrightarrow{\partial_{n-1}} \cdots \xrightarrow{\partial_2} C_1 \xrightarrow{\partial_1} C_0 \longrightarrow 0$$

and suppose that we have a decomposition $C_i = E_{i+1} \oplus E_i$, $E_{n+1} = 0$,
$\partial_p = \begin{pmatrix} a_p & \phi_p \\ b_p & c_p \end{pmatrix}$ *where ϕ_p is an injective endomorphism of E_p.*

Then the complex has only torsion homology, and we have:

$$\sum_i (-1)^i \operatorname{div}(H_i(C)) = \sum_i (-1)^i \operatorname{div}(\det \phi_{i+1}).$$

So if $E_0 = 0$ and $H_i(C) = 0$ for $i > 0$, the cokernel of ∂_1 being equal to $H_0(C)$, we have:

$$[\operatorname{Coker}(\partial_1)] = \prod_{i=1}^n (\det \phi_i)^{(-1)^{i+1}}.$$

Proof *(from Demazure's one).* The homology of $C \otimes_A k$ is zero because ∂_i restricted to $E_i \otimes_A k$ is an automorphism. So $H(C)$ is torsion.

Let $\partial'_i = \begin{pmatrix} 0 & I \\ 0 & 0 \end{pmatrix}$, the complex (C, ∂') has zero homology and the application $f_n = id$, $f_i = \begin{pmatrix} \phi_{i+1} & 0 \\ c_{i+1} & I \end{pmatrix}$ for $i < n$ from C_i into himself defines a morphism from (C, ∂') to (C, ∂), as we have:

$$\partial_i \circ f_i = \begin{pmatrix} a_i & \phi_i \\ b_i & c_i \end{pmatrix} \begin{pmatrix} \phi_{i+1} & 0 \\ c_{i+1} & I \end{pmatrix} = \begin{pmatrix} 0 & \phi_i \\ 0 & c_i \end{pmatrix} =$$

$$= \begin{pmatrix} \phi_i & 0 \\ c_i & I \end{pmatrix} \begin{pmatrix} 0 & I \\ 0 & 0 \end{pmatrix} = f_{i-1} \circ \partial'_i.$$

f_i is into and $\operatorname{Coker}(f_i)$ can be identified with $\operatorname{Coker}(\phi_{i+1})$, so that we get the following diagram where the vertical sequences are exact:

$$
\begin{array}{ccccccccc}
& & 0 & & & & 0 & & 0 \\
& & \uparrow & & & & \uparrow & & \uparrow \\
0 & \xrightarrow{\theta_n} & \operatorname{Coker}\phi_n & \xrightarrow{\theta_{n-1}} & \cdots & & \operatorname{Coker}\phi_2 & \xrightarrow{\theta_1} & \operatorname{Coker}\phi_1 & \to & 0 \\
& & \uparrow & & & & \uparrow & & \uparrow \\
0 \to & C_n & \xrightarrow{\partial_n} & C_{n-1} & \xrightarrow{\partial_{n-1}} & \cdots & & C_1 & \xrightarrow{\partial_1} & C_0 & \to 0 \\
& \uparrow f_n & & \uparrow f_{n-1} & & & & \uparrow f_1 & & \uparrow f_0 \\
0 \to & C_n & \xrightarrow{\partial'_n} & C_{n-1} & \xrightarrow{\partial'_{n-1}} & \cdots & & C_1 & \xrightarrow{\partial'_1} & C_0 & \to 0 \\
& \uparrow & & \uparrow & & & & \uparrow & & \uparrow \\
& 0 & & 0 & & & & 0 & & 0
\end{array}
$$

As we see, the complex of the cokernels have the same homology as (C, ∂), so we get:

$$\operatorname{div}(\operatorname{Coker} \phi_i) = \operatorname{div}(\operatorname{Im} \theta_{i-1}) + \operatorname{div}(\ker \theta_{i-1})$$
$$= \operatorname{div}(\operatorname{Im} \theta_{i-1}) + \operatorname{div}(\operatorname{Im} \theta_i) + \operatorname{div}(H_{i-1}(C)).$$

So the conclusion follows from the classical lemma ([Bo] Chap. 7, §4, n°6, corollary of prop. 13):

Lemma.—If ϕ is an injective endomorphism of a finitely generated free A-module M, we have:

$$\mathrm{div}(\mathrm{Coker}\ \phi) = \mathrm{div}(\det\ \phi)$$

Sketch of proof. Let $M' = \phi(M)$, $(e_i)_{i \in I}$ a base of M and P set of representatives of the irreducible elements of A. If $\mathcal{P} \in P$, $M'_{\mathcal{P}} \neq M_{\mathcal{P}}$ if and only if $\mathcal{P}$ divides $\det \phi$. So the two divisors have the same support. If we localize on such a $\mathcal{P}$, the ring $A_{\mathcal{P}}$ being principal, there exists two automorphisms ξ and ξ' of $M_{\mathcal{P}}$ such that $\xi \circ \phi \circ \xi'$ looks like $e_i \longmapsto \mathcal{P}^{d_i} e_i$, so $\det \phi = u\mathcal{P}^m$ with $m = \sum_i d_i$ and u a unit, and we have $M'_{\mathcal{P}} = \bigoplus_i \mathcal{P}^{d_i} A_{\mathcal{P}}$. The conclusion follows. ∎

Remarks.

1. The proposition 2 is also true if A is noetherian and integrally closed.

2. If the truncated Euler-Poincaré characteristics are positive (i.e. $\sum_{i=k}^{n-1}(-1)^{i-k}\mathrm{div}\ H_i(C)$ is effective), then $\Delta_k = \prod_{i=k}^{n-1}(\det\ \phi_{i+1})^{(-1)^{i-k}} \in A$ as

$$\sum_{i=k}^{n-1}(-1)^{i-k}\mathrm{div}\ (\det\ \phi_{i+1}) = \mathrm{div}\ (\mathrm{Im}\ \theta_k) + \sum_{i=k}^{n-1}(-1)^{i-k}\mathrm{div}\ (H_i(C)).$$

In particular, $\Delta_k \in A$ for $k \geq 0$ when $H_i(C) = 0$ for $i > 0$.

If we have an exact sequence of finitely generated free A-modules of the form

$$0 \longrightarrow C_n \xrightarrow{\partial_n} C_{n-1} \xrightarrow{\partial_{n-1}} \cdots \xrightarrow{\partial_2} C_1 \xrightarrow{\partial_1} C_0$$

with $\sum(-1)^i \mathrm{Rank}_A(C_i) = 0$, then we may construct a decomposition of the C_i's in the following way:

• fix a base B_i for each C_i,

• choose a maximal non zero minor of ∂_n (there exists one because ∂_n is into). This choice splits B_{n-1} into two parts: $B_{n-1} = B'_n \cup B''_{n-1}$ where $\mathrm{card}B'_n = \mathrm{card}B_n$,

• the restriction of ∂_{n-i} to the module $A\langle B''_{n-i}\rangle$ is into so that we can iterate the process by choosing at each step a maximal minor of ∂_{n-i} restricted to $A\langle B''_{n-i}\rangle$, and therefore a decomposition $B_{n-i-1} = B'_{n-i-1} \cup B''_{n-i-1}$,

• the matrix of ∂_1 restricted to $A\langle B''_1\rangle$ is a square matrix because of the hypotheses on the ranks.

In the case of the Koszul complex of a complete intersection, the generating function for the rank of K_p^ν is

$$G(T_p, T_\nu) = \prod_{s=1}^{n} \frac{1 + T_\nu^{d_s} \cdot T_p}{1 - T_\nu},$$

so that the alternate sum of the ranks is equal to the coefficient of the degree ν term of $\prod_{s=1}^{n} \frac{1 - T_\nu^{d_s}}{1 - T_\nu}$ and therefore vanishes if and only if $\nu > \sum_i (d_i - 1)$.

II The Algorithm

From what we have seen above, we get the following algorithm to calculate the resultant of $P_1, \ldots, P_n \in A[X_1, \ldots, X_n]$ when A is a domain (all the operations are performed in A, because we *a priori* know that the divisions are exact):

(1) Put $\nu = \sum_{i=1}^{n}(\deg P_i - 1) + 1$ and calculate the matrices M_i of ∂_i^ν for the Koszul complex corresponding to the input polynomials.

(2) Put $i := n - 1$, $\Delta_i = 1$.

(3) If M_i is not of maximal rank, $\mathrm{Res}(P_1, \ldots, P_n) = 0$, end.

(4) Take a maximal submatrix M_i' of M_i with $D_i = \det M_i' \neq 0$, put $\Delta_{i-1} = D_i / \Delta_i$ and replace M_{i-1} by the minor of M_{i-1} obtained by erasing the columns corresponding to the lines which appears in M_i'.

(5) If $i > 1$ do $i := i - 1$, and go to (3).

(6) $\Delta_0 = \pm \mathrm{Res}(P_1, \ldots, P_n)$.

III Some remarks and an improvement of the algorithm

Remarks.

1. M_n is empty in degree $\nu = \sum_i (d_i - 1) + 1$.

2. It is clear that this algorithm works over any field (or domain) because:

• if one of the matrices is not of maximal rank the complex is not exact, so that the polynomials have a non trivial common zero and the resultant vanishes,

• if a determinant is not zero in one case, it is not zero in the generic case. So if a decomposition works in a specific case it works in the generic case and rises to the resultant.

3. If, in the preceding algorithm, $M_n, \ldots, M_{i+1}$ are of maximal rank but M_i is not, there is no regular sequence of length $n-i+1$ constituted by elements of the ideal generated by the P_i's ([No] Chap. 8, Theorem 6, p. 371). So the dimension of the zero locus of the P_i's is at least i.

4. If A is any ring and the algorithm returns 0, then for every field k and every morphism $\varphi : A \longrightarrow k$, $\varphi(\text{Res}(P_1, \ldots, P_n)) = 0$; if the subring of A spanned by the coefficients of the P_i's is reduced this implies that the resultant is zero. So the algorithm gives the resultant over any reduced ring and returns 0 only if every specialization of the elements of A in any field (or domain) leads to a vanishing resultant.

5. It is easy to check that the alternate sum of the dimensions of the matrices M_i' (for $\nu = \sum_i (d_i - 1) + 1$ or bigger), is equal to the degree ν coefficient of $\frac{\partial G}{\partial T_p}(-1, T_\nu)$ and is therefore given by:

$$\frac{\partial G}{\partial T_p}(-1, T_\nu) = \sum_{i=1}^{n} \left[\frac{T_\nu^{d_i}}{1 - T_\nu} \prod_{j \neq i} \frac{1 - T_\nu^{d_j}}{1 - T_\nu} \right]$$

$$= \sum_{i=1}^{n} \left[(T_\nu^{d_i} + T_\nu^{d_i+1} + \cdots) \prod_{j \neq i} (1 + T_\nu + \cdots + T_\nu^{d_j-1}) \right].$$

From that we see that the resultant is an homogeneous polynomial in the coefficients of the generic polynomials of the expected total degree $\sum_{i=1}^{n} \prod_{j \neq i} d_j$.

6. For every i and every ν all the minors of the matrix M_k^ν of ∂_k^ν are homogeneous polynomials in the coefficients of the polynomials P_i's, because the entries of M_k^ν are either a coefficient of one of the P_i's or 0. Moreover theses polynomials are also homogeneous in the coefficients of each P_i.

Let us prove it, for example, for the coefficients of P_1. For this purpose, consider the partition of $S_k = \{e_{i_1} \wedge \cdots \wedge e_{i_k}, 1 \leq i_1 < \cdots < i_k \leq n\}$ into $S_k' \cup S_k''$ where $S_k' = \{e_1 \wedge e_{i_2} \wedge \cdots \wedge e_{i_k}, 2 \leq i_2 < \cdots i_k \leq n\}$ and $S_k'' = S_k - S_k'$.

If $e \in S_k'$, $\partial_k(e) = P_1 e'' + \sum_{k \geq 2} P_{i_k} e_k'$ with $i_k \geq 2$, $e'' \in S_{k-1}''$ and $e_k' \in S_{k-1}'$.

If $e \in S_k''$, $\partial_k(e) = \sum_{k \geq 1} P_{i_k} e_k''$ with $i_k \geq 2$ and $e_k'' \in S_{k-1}''$.

So every submatrix N_k^ν of M_k^ν splits into four blocks (eventually empty):

$$N_k^\nu = \begin{pmatrix} \text{Coef. of} & 0 \\ P_i\text{'s } i > 1 & \\ & \text{Coef. of} \\ \text{Coef. of } P_1 & P_i\text{'s } i > 1 \end{pmatrix} = \begin{pmatrix} A & 0 \\ B & C \end{pmatrix}$$

Suppose that A (resp. B and C) is an $m' \times n'$ (resp. $m'' \times n'$ and $m'' \times n''$) matrix. By a Laplace expansion we see that if $m' > n'$ we have $\det N_k^\nu = 0$

and if $m' \leq n'$ the determinant of N_k^ν is an homogeneous polynomial in the coefficients of P_1 of degree $n' - m' = m'' - n''$.

7. If the resultant is not zero, there exists a decomposition of the Koszul complex such that the determinant of the matrix M_i' does not depend of the coefficients of $P_{n-i+2}, \ldots, P_n$, because the fact that $P_1, \ldots, P_{n-i+1}$ is a regular sequence implies that $H^j(\mathbf{K}) = 0$ for $j \geq i$ independently of $P_{n-i+2}, \ldots, P_n$.

From the formula $\Delta_0 \Delta_1 = D_1$ we see that, as the resultant actually depends on the coefficients of P_n and Δ_1 does not if we make an adapted decomposition, the resultant divides Δ_0 because of its irreducibility. As Δ_0 and the resultant have the same degree they must be equal up to a constant, the case $P_i = X_i^{d_i}$ shows that this constant is ± 1.

This shows that the algorithm produces the resultant, using only the classical definition of the resultant (see e.g. [Ma] or [Gr]) regardless to proposition 1.

An improvement of the algorithm.

From the two remarks above, we deduce that we can replace in the algorithm the matrix M_i^ν by the matrix $\hat{M}_i^\nu$ obtained from M_i^ν by replacing the coefficients of $P_{n-i+2}, \ldots, P_n$ by 0. Moreover, after this replacement, the matrix split into blocks and therefore the calculation is greatly simplified ; for example the maximal minor of $\hat{\partial}_{n-1}^\nu$ may always be chosen as the product of determinants of triangular submatrices of $\hat{\partial}_{n-1}^\nu$ and maximal minors of matrices representing $(A_1, A_2) \mapsto A_1 P_1 + A_2 P_2$ in some degrees.

On the other hand, after this substitution, if the algorithm stops before the calculation of D_1 you do not get any lower bound for the dimension of the associated variety.

IV An arithmetic consequence on the resultants

Let $P_1, \ldots, P_n$ be n homogeneous polynomials of $\mathbf{Z}[X_1, \ldots, X_n]$ and $r \in \mathbf{Z}$ their resultant. If p is a prime number, p divides r if and only if the reductions modulo p of these polynomials have a non trivial common zero in an algebraic closure of $\mathbf{F}_p$.

When the algebraic variety V_p associated to the reductions modulo p of the P_i's is of dimension zero and of degree d, we will prove that $v_p(r) \geq d$, where $v_p(r)$ is the p-adic valuation of r.

In this situation, it arrives that a bigger power of p divides r. Moreover, two n-tuples of polynomials with the same reductions modulo p may have different p-adic valuation of their resultants. However, an upper bound for the p-adic valuation of the resultant of the generic lifting of n homogeneous polynomials in $\mathbf{F}_p[X_1, \ldots, X_n]$ (with $\dim V_p = 0$) can be given. This bound is d when $n = 2$ but not in general. We will not be concern with this question here.

Let us fix some notations :

- $\mathbf{Z}[X_1,\ldots,X_n] \equiv \mathbf{Z}[\mathbf{X}]$ and $\mathbf{F}_p[X_1,\ldots,X_n] \equiv \mathbf{F}_p[\mathbf{X}]$.

- p is a prime number.

- $P_1,\ldots,P_n$ are n homogeneous polynomials in $\mathbf{Z}[\mathbf{X}]$ of respective degrees $d_1,\ldots,d_n$ and we will suppose, to avoid some trivial remarks, that they are not divisible by p.

- $\overline{P}_i$ is the image of P_i by the canonical homomorphism from $\mathbf{Z}[\mathbf{X}]$ to $\mathbf{F}_p[\mathbf{X}]$, V_p is the variety associated to the homogeneous ideal spanned by the polynomials $\overline{P}_i$ and d is the degree of V_p.

- $r \in \mathbf{Z}$ is the resultant of the P_i's.

- As before K_m^ν and ∂_m^ν are the homogeneous part of degree ν of K_m and ∂_m for the polynomials P_i with $A = \mathbf{Q}$ and, in a similar fashion, $\overline{K}_m^\nu$ and $\overline{\partial}_m^\nu$ with $A = \mathbf{F}_p$.

Our result is the following :

Proposition 3. *With the above notations, p^d divides r if $\dim V_p = 0$ and $\deg V_p = d$.*

As in remark 3 of section III, we notice that $\dim V_p = 0$ implies that $H_i(\overline{\mathbf{K}}) = 0$ for $i > 1$. And the proposition follows from two easy lemmas :

Lemma 1. *If $H_i(\overline{\mathbf{K}}) = 0$ for $i > 1$, the p-adic valuation of r is the minimum of the p-adic valuations of the maximal minors of ∂_1^ν for all ν strictly bigger than $\delta = \sum_{i=1}^n (d_i - 1)$.*

Proof. For all $\nu > \delta$, as $H_i(\overline{\mathbf{K}}^\nu) = 0$ for $i > 1$, the construction made in the algorithm (for the $\overline{P}_i$'s) gives for $m > 1$ some square submatrix M_m' of M_m (the matrix of ∂_m^ν) whose determinant is not divisible by p (because the matrix associated to $\overline{\partial}_m^\nu$ is the reduction mod p of M_m) and a square submatrix of maximal size M_1' of M_1 such that $r = \pm \prod_{i=1}^n (\det M_i')^{(-1)^{i+1}}$.

So the p-adic valuation of the resultant is the one of a maximal minor of ∂_1^ν, and this for all $\nu > \delta$. As the resultant divides all the maximal minors, the lemma follows. ∎

The corank of $\overline{\partial}_1^\nu$ is the Hilbert function of V_p in degree ν and therefore, for ν big enough, equal to the degree of V_p. It remains to show the elementary following lemma :

Lemma 2. *Let M be a square matrix with entries in $\mathbf{Z}$ whose reduction modulo p is a matrix of corank d, then p^d divides $\det M$.*

Proof. If $\overline{M}$ is the reduction modulo p of M, we can find two invertible matrices with entries in $\mathbf{F}_p$ such that : $A\overline{M}B = \begin{pmatrix} I & 0 \\ 0 & 0 \end{pmatrix}$ where I is the identity of size $(n - d) \times (n - d)$. If we choose some liftings A' and B' of A and B in $\mathbf{Z}$, the determinant of the matrix $A'MB'$ has the same p-adic valuation as the one of M and is divisible by p^d as the last d lines of $A'MB'$ are divisible by p. ∎

Remark 1. We have just seen that, if V_p is of dimension zero, the p-adic valuation of r is the one of the gcd of the maximal non zero minors of the so called "Generalized Sylvester matrix", that is the matrix of the application $\partial_1^{\nu} : (A_1, \ldots, A_n) \longmapsto \sum_{i=1}^{n} A_i P_i$, written on the monomial bases, and in degree ν strictly bigger than $\sum_{i=1}^{n}(d_i - 1)$.

In particular, if, for all p, V_p is of dimension zero, the resultant is the gcd of the maximal non zero minors of this matrix, as in the generic case.

Remark 2. If $(P_1, \ldots, P_n)$ is a regular sequence on $\mathbf{Z}[\mathbf{X}]$ the homology modules $H_i(\mathbf{K})$ $(i > 0)$ are vanishing (they may have torsion even if the P_i's defines an empty variety). The classical exact sequence of ([Se] IV-2 prop. 1) then implies that, for all prime p, $H_i(\overline{\mathbf{K}}) = 0$ for $i > 1$.

So, from the construction above, the p-adic valuation of r is the one of a maximal minor of ∂_1^{ν} for all ν big enough. Lemma 2 implies that the corank of $\overline{\partial}_1^{\nu}$, that is the Hilbert function of V_p, is bounded by $v_p(r)$ for all ν big enough.

So, for all p, V_p is either empty or of dimension zero.

V The Decomposition of Demazure

We will now give an explicit decomposition of the modules K_p due to Demazure that gives a method to calculate the resultant adding only one parameter. First a definition:

Definition.—*Given an n-tuple $(d_1, \ldots, d_n)$, a monomial $X_1^{i_1} \cdots X_n^{i_n}$ is called k reduced if $i_s < d_s$ for all $s < k$. The set of k-reduced monomials will be denoted by $\Re_k$.*

In our situation k-reduced will mean k-reduced with respect to the n-tuple of the degrees of the P_i's. We will denote by V_i the free A-module $V_i = \bigoplus_{X^{\alpha} \in \Re_i} AX^{\alpha} \subseteq A[X_1, \ldots, X_n]$.

With this definition K_p splits into $K_p = K'_{p+1} \oplus K''_p$ with:

$$K''_p = \bigoplus_{i_1 < \cdots < i_p} V_{i_1} \, e_{i_1} \wedge \cdots \wedge e_{i_p}$$

$$K'_{p+1} = \bigoplus_{i_1 < \cdots < i_p} V^c_{i_1} \, e_{i_1} \wedge \cdots \wedge e_{i_p}$$

where V_i^c is the free A-module spanned by the non-reduced monomials.

As the resultant is an universal object, the following lemma enables us to calculate the resultant with one added parameter:

Lemma.—*If* $1 \leq p \leq n$ *the composed application*

$$K_p'' \hookrightarrow K_p \xrightarrow{\partial_p} K_{p-1} \xrightarrow{s} K_p'$$

where s *is the canonical surjection, is a bijection for the regular sequence* $P_i = X_i^{d_i}$.

Proof. See [De] p. 12 .

This lemma shows that, in this case, all the determinants coming from this decomposition equal ± 1; so, replacing every polynomials P_i by $P_i + \lambda X_i^{d_i}$, the associated determinants are monic polynomials in λ.

So the resultant is the constant term of a quotient two products of monic polynomials in λ that can be calculated by exact divisions which remain in $A[\lambda]$.

The idea of using a deformation parameter this way was also used by others, for example by Canny *et al.* (to compute the resultant using Macaulay formulas) or by Chistov & Grigoriev (among others) to determine the solutions of a zero dimensional variety.

Remark. With this decomposition, the matrix M_i' does not depend on the coefficients of $P_n, \ldots, P_{n-i+2}$ (see [De] p. 15), it gives an effective version of what we have said at the remark 7 of part III.

References

[Bo] N. BOURBAKI, *Algèbre Commutative Chapitres 1 à 9*, Masson 1983 et 1985.

[De] M. DEMAZURE, *Une définition constructive du résultant*, Notes Informelles de Calcul Formel **2**, prépublications du Centre de Mathématiques de l'École Polytechnique, 1984.

[Gr] W. GRÖBNER, *Modern Algebraische Geometrie*, Springer-Verlag Wien und Innsbruck, 1949.

[Hu] A. HURWITZ, *Über die Trägheitsformen eines algebraischen Moduls*, Annali di Mathematica pura ed applicata (3) 20, 1913, pp. 113–151.

[Jo1] J.-P. JOUANOLOU, *Le formalisme du résultant*, Publication de l'IRMA 417/P-234, Université de Strasbourg, 1990.

[Jo2] J.-P. JOUANOLOU, *Aspects invariants de l'élimination*, Publication de l'IRMA 457/P-263, Université de Strasbourg, 1991.

[Ma] F. S. MACAULAY, *The Algebraic Theory of Modular Systems*, Stechert-Hafner Service Agency, New-York and London, 1964, *(Originally published in 1916 by Cambridge University Press)*.

[No] D.G. NORTHCOTT, *Lessons on Rings Modules and Multiplicities*, Cambridge University Press, 1968.

[Se] J.-P. SERRE, *Algèbre locale et multiplicités*, Lectures Notes in Mathematics **11**, 1965.

Marc Chardin

Équipe de Calcul Formel
Centre de Mathématiques
École Polytechnique
F–91128 Palaiseau cedex
e-mail : chardin@polytechnique.fr

Gröbner Bases and
Standard Monomial Theory

A.M. Cohen   R.H. Cushman

Abstract.   For rings of polynomials on varieties corresponding to minuscule weight representations of Lie groups, we show how the standard monomial theory of Seshadri et al. can be used to compute Gröbner bases. This generalizes results of Sturmfels and White in which straightening has been interpreted as a normal form computation with respect to a Gröbner basis.

keywords: Lie groups, minuscule weights, straightening, Gröbner bases

1. Introduction

The Grassmann varieties have been studied in various algorithmic ways. For instance, rectangular Young tableaux have proved to be useful for describing monomials in the ring of polynomial functions on a Grassmannian. The associated straightening procedure is an effective way of expressing such a monomial as a linear combination of certain monomials, called standard monomials. Moreover, the standard monomials form a vector space basis of the ring of polynomial functions on the Grassmannian.

The theory of Gröbner bases deals with the effective study of rings of polynomial functions on algebraic varieties in much greater generality. It entails the existence of a vector space basis consisting of special monomials, which – by fortunate coincidence? – are also called standard!

One would expect from the very nature of the Grassmannians that at least certain Gröbner bases for the ring of polynomial functions on them would be obtainable from the much older procedure of straightening. This in fact is true. The correspondence between Gröbner bases and straightening has been discussed in [StWh1]. Here we show that this correspondence naturally extends to the standard monomial theory for minuscule weight varieties as developed by Seshadri (see [Sesh]). Thereby we answer part of the question raised in [StWh1]. The combinatorics we need to obtain the reduction ordering, an ingredient to the Gröbner basis, is that of the Bruhat order on cosets of the Weyl group with respect to parabolic subgroups. This is dealt with in §4. The main result is to be found in §5, at the end of which an explicit example, related to E_6 in a 27-dimensional representation, is worked out. An effective version of the main result is described in §6. There, it is shown how the Casimir operator can be used to find the reduced Gröbner basis. But first we survey the two main ingredients to this paper, Gröbner bases in §2, and straightening in §3.

2. Gröbner bases

This section contains the necessary background material from Gröbner basis theory. See [Buch] for further details and references.

Let $\mathbf{F}$ be a field. Let $\mathbf{F}[x]$ be the ring of polynomials in the variables $x = (x_1, \ldots, x_n)$. A *monomial* is an expression of the form $x_1^{\alpha_1} \cdots x_n^{\alpha_n}$; it is frequently abbreviated to x^α, where $\alpha = (\alpha_1, \ldots, \alpha_n) \in \mathbf{N}^n$ is the *exponent* of the monomial. We can identify $\mathbf{N}^n$ with the monoid M of all monomials of $\mathbf{F}[x]$ via the map $\alpha \mapsto x^\alpha$. Let I be an ideal of $\mathbf{F}[x]$. Then the quotient $A = \mathbf{F}[x]/I$ is a finitely generated commutative $\mathbf{F}$-algebra. A key procedure in many computational problems regarding A consists of producing a unique representative in $\mathbf{F}[x]$ of $f + I \in A$ when given a polynomial $f \in \mathbf{F}[x]$.

To describe this procedure, we first order the variables $x_1, \ldots, x_n$ in $\mathbf{F}[x]$ as follows: $x_1 < x_2 < \ldots < x_n$. Then we extend it to a linear ordering $<$ on the set M of all monomials of $\mathbf{F}[x]$ by setting

$$x_1^{\alpha_1} \cdots x_n^{\alpha_n} < x_1^{\beta_1} \cdots x_n^{\beta_n}$$

if and only if either $\sum_{i=1}^n \alpha_i < \sum_{i=1}^n \beta_i$ or $\sum_{i=1}^n \alpha_i = \sum_{i=1}^n \beta_i$ and there is $j \in \{1, \ldots, n\}$ with $\beta_k = \alpha_k$ for all $k \in \{1, \ldots, j-1\}$ and $\beta_j > \alpha_j$. This ordering $<$ on M is called the *total degree lexicographic order*. When viewed as an order on the exponents, it has the following property:

$$\left. \begin{array}{l} \gamma \geq 0, \quad \text{and} \\ \text{if } \alpha < \beta \text{ then } \alpha + \gamma < \beta + \gamma \end{array} \right\} \text{ for all exponents } \alpha, \beta, \gamma \in \mathbf{N}^n. \quad (1)$$

An ordering $<$ on M with this property is called a *reduction ordering*. An important feature of a reduction ordering is that it is *Noetherian*, that is, every strictly descending chain is finite. For nonzero $f \in \mathbf{F}[x]$, we denote by lead (f) the *leading monomial* of f, that is the highest monomial which occurs in f (with nonzero coefficient).

Suppose that $F = \{f_1, \ldots, f_m\}$ is a finite set of polynomials in $\mathbf{F}[x]$ generating the ideal I (such a finite set exists as $\mathbf{F}[x]$ is Noetherian). If $g \in \mathbf{F}[x]$ is given, then $h \in \mathbf{F}[x]$ is called a *reduct of g with respect to F* if lead $(h) <$ lead (g) and there is an $f \in F$ such that, for some $c \in \mathbf{F}$ and $u \in M$, we have $h = g - c\,u\,f$. (Note that then lead (g) is divisible by lead (f) with quotient u.) If we continue with h instead of g and form successive reducts with respect to F until no further reduction is possible, then we arrive at a polynomial which is called a *normal form of g with respect to F*. The normal form need not be unique. Nevertheless, we shall write $NormalForm_F(g)$ to denote a normal form of g with respect to F. Each reduct h of g represents the same element $h + (F) = g + (F)$ of $\mathcal{A}$, and so $NormalForm_F(g) \in g + I$.

Define the S-polynomial of two polynomials $f_1, f_2 \in \mathbf{F}[x]$ as follows:

$$S(f_1, f_2) = c_2 u_1 f_1 - c_1 u_2 f_2, \qquad (2)$$

where, for $i = 1, 2$, c_i is the coefficient of the term in f_i containing $\operatorname{lead}(f_i)$ and $u_i \operatorname{lead}(f_i)$ is the least common multiple of $\operatorname{lead}(f_1)$ and $\operatorname{lead}(f_2)$.

Let F be a finite subset of $\mathbf{F}[x]$. For $I = (F)$, denote by $\operatorname{lead}(I)$ the $\mathbf{F}$-linear subspace of $\mathbf{F}[x]$ spanned by all leading monomials of polynomials in I. It is an ideal of $\mathbf{F}[x]$. For $m \in M$ define m to be *non-standard* if it belongs to $\operatorname{lead}(I)$, and *standard* otherwise. If J is the $\mathbf{F}$-linear span of all standard monomials, then

$$\mathbf{F}[x] = I \oplus J = \operatorname{lead}(I) \oplus J. \qquad (3)$$

2.1 Effective characterizations of Gröbner bases. *Let $<$ be a reduction ordering on the collection of monomials in the polynomial ring $\mathbf{F}[x]$. Then, for a fixed finite subset F of $\mathbf{F}[x]$ generating $I = (F)$, the following statements are equivalent.*

(i)   *if $f, g \in F$, then $NormalForm_F(S(f,g)) = 0$ (for some computation of $NormalForm$);*

(ii) *for each $f \in I$, we have $NormalForm_F(f) = 0$ (for any computation of $NormalForm$);*

(iii) *$\operatorname{lead}(I)$ is spanned by $\{\operatorname{lead}(f) \mid f \in F\}$ as an ideal, or, equivalently, by $\{m \operatorname{lead}(f) \mid f \in F, m \in M\}$ as a vector space;*

(iv) *for each $f \in \mathbf{F}[x]$, we have $NormalForm_F(f) \in J$.*

If F satisfies these properties, it is called a *Gröbner basis*. An ideal I may have many Gröbner bases. Given one, say F, we first normalize each $f \in F$ so that the coefficients of the terms containing $\operatorname{lead}(f)$ are 1. We say that f is *monic* if it is normalized in this way. F is a *reduced Gröbner basis* for the ideal $I = (F)$ if the normal form of any $f \in F$ with respect to $F \setminus \{f\}$ is f and each $f \in F$ is monic. It is not hard to show that I has a *unique* reduced Gröbner basis. Some caution is in order though: the uniqueness only holds for a fixed ordering $<$ (because the ideal $\operatorname{lead}(I)$ may change if the ordering is changed).

In algebraic geometry, the quotient ring $\mathcal{A} = \mathbf{F}[x]/I$ is the ring of all polynomial functions on the subset of $\mathbf{F}^n$ consisting of all common zeros of the polynomials in I. If I is a homogeneous ideal, the zero set can be interpreted as a subset of the projective space $\mathcal{P}\mathbf{F}^n$ on $\mathbf{F}^n$. The projective zero set X of I is then called a *projective variety* and the polynomial ring $\mathcal{A}$, denoted by $\mathbf{F}[X]$, will then be viewed as a graded ring. In the next section, we shall consider a special kind of projective variety, the Grassmann variety.

3. Straightening

We begin by reviewing some basic facts on Grassmann varieties. Viewing an $n \times d$-matrix as an array of d vectors $v_1, \ldots, v_d$ of length n we obtain the mapping $\phi : \mathcal{M}_{n \times d} \longrightarrow \bigwedge^d \mathbf{F}^n$ sending $(v_1, \ldots, v_d)$ to $v_1 \wedge \ldots \wedge v_d$. The image under ϕ of the subset of $\mathcal{M}_{n \times d}$ consisting of all matrices of rank d in the projective space $\mathcal{P}(\bigwedge^d(\mathbf{F}^n))$ is called the *Grassmann variety* $G(n, d)$. Its embedding in $\mathcal{P}(\bigwedge^d(\mathbf{F}^n))$ is called the Plücker embedding. The projective points of this variety correspond bijectively to the d-dimensional subspaces of $\mathbf{F}^n$: the projective point (= 1-dimensional subspace) corresponding to $v_1 \wedge \ldots \wedge v_d \in \bigwedge^d \mathbf{F}^n$ does not depend on the choice of basis $v_1, \ldots, v_d$ in the d-dimensional subspace of $\mathbf{F}^n$ that it spans.

We shall now justify the use of the word variety by exhibiting a set of homogeneous polynomials $F_{n,d}$ whose projective zero set coincides with $G(n, d)$. For each $1 \leq i_1, \ldots, i_d \leq n$ define the *bracket* $[i_1, \ldots, i_d]$ to be the coordinate function $x_{i_1} \wedge \ldots \wedge x_{i_d}$, where $x_1, \ldots, x_n$ are the standard coordinate functions on $\mathbf{F}^n$. (Note, here we use the standard identification of $\bigwedge^d(\mathbf{F}^n)^*$ with $(\bigwedge^d \mathbf{F}^n)^*$ by means of the pairing determined by $(\mu_1 \wedge \ldots \wedge \mu_d, \nu_1 \wedge \ldots \wedge \nu_d) = \det (\mu_i(\nu_j))_{1 \leq i,j \leq d}$ for $\mu_i \in (\mathbf{F}^n)^*$ and $\nu_i \in \mathbf{F}^n$.) Using the map ϕ the bracket $[i_1, \ldots, i_d]$ can be interpreted as the determinant of the $d \times d$-matrix formed by the rows $i_1, \ldots, i_d$ of the $n \times d$-matrix $(v_1, \ldots, v_d)$. Thus $\mathbf{F}[x]$, where $x = ([i_1, \ldots, i_d])_{1 \leq i_1 < \ldots < i_d \leq n}$, is the polynomial ring of the vector space $\bigwedge^d(\mathbf{F}^n)$. Usually the bracket $[i_1, \ldots, i_d]$ is defined for an arbitrary sequence $i_1, \ldots, i_d$ of elements of $\{1, \ldots, n\}$ by use of the conventions $[i_1, \ldots, i_\ell, i_{\ell+1}, \ldots, i_d] = -[i_1, \ldots, i_{\ell+1}, i_\ell, \ldots, i_d]$ and $[i_1, \ldots, i_{m-1}, i_m, i_m, i_{m+1} \ldots, i_{d-1}] = 0$ for all ℓ, m. We shall say that the bracket $[i_1, \ldots, i_d]$ is *normalized* if it satisfies $i_1 < \ldots < i_d$.

It is customary to write the bracket monomial $[i_{11}, \ldots, i_{1d}][i_{21}, \ldots, i_{2d}] \cdots [i_{\ell 1}, \ldots, i_{\ell d}]$ as a tableau

$$T = \begin{bmatrix} i_{11} & \cdots & i_{1d} \\ i_{21} & \cdots & i_{2d} \\ \vdots & \vdots & \vdots \\ i_{\ell 1} & \cdots & i_{\ell d} \end{bmatrix} \tag{4}$$

of length ℓ. Since rows may be interchanged at will, and since a transposition of entries from the same row results in a sign change of the corresponding term, we can always normalize a tableau so that it either vanishes or has the following properties:

- each row corresponds to a normalized bracket, and
- the rows are weakly increasing (with respect to the lexicographical ordering).

We shall refer to a tableau satisfying these two conditions as a *normalized tableau*. Furthermore, we shall say that the t-th column of T is weakly increasing if

$$i_{1t} \leq i_{2t} \leq \cdots \leq i_{tt}.$$

If X is a set, we write $\mathrm{Sym}X$ for the group of all permutations of X. For each pair of brackets $[i_1, \ldots, i_d]$ and $[j_1, \ldots, j_d]$, and each $\ell \in \{1, \ldots, d\}$, consider the following polynomial

$$\sum_\sigma \mathrm{sign}(\sigma) \begin{bmatrix} i_1 & \cdots & i_{\ell-1} & \sigma(i_\ell) & \sigma(i_{\ell+1}) & \cdots & \sigma(i_d) \\ \sigma(j_1) & \cdots & \sigma(j_{\ell-1}) & \sigma(j_\ell) & j_{\ell+1} & \cdots & j_d \end{bmatrix} \quad (5)$$

where the sum runs over a system of coset representatives in $\mathrm{Sym}\{i_\ell, \ldots, i_d, j_1, \ldots, j_\ell\}$ with respect to $\mathrm{Sym}\{i_\ell, \ldots, i_d\} \times \mathrm{Sym}\{j_1, \ldots, j_\ell\}$. For some choices, e.g., $j_1 \in \{i_1, \ldots, i_d\}$ and $\ell = 1$, the corresponding polynomial vanishes identically. Otherwise, (5) is a quadratic polynomial and vanishes on $G(n, d)$. Let $F_{n,d}$ denote the set of all nonzero quadratic polynomials occurring in (5). Its members, when equated to zero, are (special instances of) the well known *Plücker relations*.

Example The set $F_{n,1}$ is empty, whereas $F_{n,2}$ consists of all $[i, j][k, \ell] + [i, k][\ell, j] + [i, \ell][j, k]$ for i, j, k, ℓ distinct integers in $\{1, \ldots, n\}$.

3.1 The second fundamental theorem of invariant theory. *The Grassmann variety $G(n, d)$ is the zero set of the ideal $(F_{n,d})$. Conversely, any polynomial vanishing on $G(n, d)$ belongs to $(F_{n,d})$.*

PROOF. See [ACGH].   QED

Set $R = \mathbf{F}[([i_1, \ldots, i_d])_{1 \leq i_1 < \ldots < i_d \leq n}]$. We introduce the total degree lexicographical ordering $<$ on M, the set of monomials of R, which is determined by ordering the variables so that $[i_1, \ldots, i_d] < [j_1, \ldots, j_d]$ if, for some ℓ we have $i_s = j_s$ ($1 \leq s < \ell$) and $i_\ell < j_\ell$. The theory of §2 applied to $F_{n,d}$ now gives:

3.2 Proposition. *A normalized tableau is a standard monomial (with respect to the total degree lexicographic ordering $<$) if and only if all of its columns are non-decreasing.*

PROOF. Let T be a tableau and suppose it is normalized but non-standard. Then there are two consecutive rows in T:

$$\begin{bmatrix} i_1 & \cdots & i_d \\ j_1 & \cdots & j_d \end{bmatrix}$$

and a column $\ell > 1$ such that $i_r \leq j_r$ for all $r < \ell$ and $i_\ell > j_\ell$. The Plücker relation (5) shows that T can be replaced with a linear combination of tableaux each of which is smaller than T with respect to $<$ on M. For, if σ

is a nontrivial coset representative as in (5), then $\sigma(i_m) \in \{j_1, \ldots, j_s\}$ for at least one $m \in \{\ell, \ldots, d\}$, yielding

$$[i_1 \quad \ldots \quad i_{\ell-1} \quad i_\ell \quad \ldots \quad i_d] >$$

$$[i_1 \quad \ldots \quad i_{\ell-1} \quad \sigma(i_\ell) \quad \ldots \quad \sigma(i_m) \quad \ldots \quad \sigma(i_d)]$$

as $\sigma(i_m) \leq j_\ell < i_\ell$. Thus, by use of (5), every nonstandard quadratic bracket monomial can be written as a sum of two smaller quadratic bracket monomials. Consequently, every nonstandard tableau may be *straightened*, that is, by use of $F_{n,d}$, it may be rewritten as a linear combination of standard tableaux modulo $(F_{n,d})$. Since this operation is nothing but the linear projection onto the **F**-span of the standard monomials with kernel $(F_{n,d})$, straightening of a monomial m produces the normal form $NormalForm_{F_{n,d}}(m)$. QED

Observe that, in general, the Plücker relations do not give a reduced basis. But, for $d = 2$, each polynomial in $F_{n,2}$ has a single non-standard term and each nonsquare quadratic monomial occurs in precisely one member of $F_{n,2}$, from which it is easy to see that we do obtain a reduced Gröbner basis. Putting all this together, we obtain the result covered by Sturmfels and White in [StWh1]:

3.3 Corollary. *The set $F_{n,d}$ is a Gröbner basis for the ideal with quotient ring $\mathbf{F}[G(n,d)]$. If $d = 2$, it is even a reduced Gröbner basis.*

4. The Bruhat order

Consider the set $P_{n,d}$ whose elements are the normalized brackets and supply it with the partial ordering $\prec$ in which $[i_1, \ldots, i_d] \preceq [j_1, \ldots, j_d]$ if $i_s \leq j_s$ for all $s \in \{1, \ldots, d\}$. Then $<$ is a refinement of $\prec$, and the tableau $[i_1, \ldots, i_d] \cdot [j_1, \ldots, j_d]$ is nonstandard if and only if the two brackets involved are incomparable (i.e., they are related by neither $\prec$ nor $\succ$).

Example The Hasse diagram of $(P_{6,2}, \prec)$ is given in Figure 1.

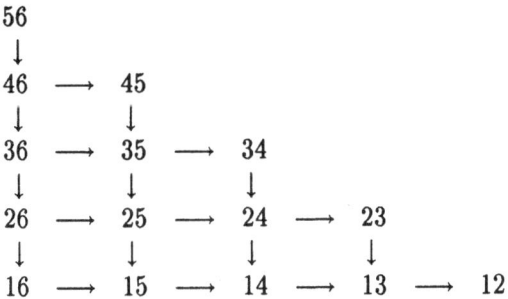

Figure 1. Hasse diagram of $(P_{6,2}, \prec)$.

The node ij represents the bracket $[i, j]$ and the arrow $b \longrightarrow a$ connects the two nodes a and b related by $a \prec b$ in the poset $(P_{6,2}, \prec)$ for which there is no node $c \in P_{6,2}$ such that $a \prec c \prec b$. In other words, b covers a. A straightforward check shows that there are 15 incomparable pairs in the Hasse diagram, each corresponding to a unique non-standard, square-free quadratic monomial in lead $F_{6,2}$.

The poset $(P_{n,d}, \prec)$ can be obtained from the theory of Coxeter groups. We first recall the definition of a Coxeter group. See [Bour], [Hum] or [Coh] for introductions.

A *Coxeter matrix* of rank n is an $n \times n$ matrix $M = (m_{i,j})_{1 \leq i, j \leq n}$ with $m_{i,i} = 1$ and $m_{i,j} = m_{j,i} > 1$ (possibly ∞) for all $i, j = 1, \ldots, n$. The *Coxeter group* associated with the Coxeter matrix M is the group generated by elements ρ_i $(i = 1, \ldots, n)$ subject to the relations

$$(\rho_i \rho_j)^{m_{i,j}} = 1.$$

It is denoted by $W(M)$ or just W. Furthermore, we set $I = \{1, \ldots, n\}$ and $R = \{\rho_i \mid i \in I\}$. The pair (W, R) is called the *Coxeter system* of type M. The number n is called the *rank* of the system (or group).

If $n = 1$ then $W = \{1\} \cup R \cong \mathbf{Z}/(2)$, the group of order 2. If $n = 2$, then $W \cong \langle r, s \mid r^2 = s^2 = (r\,s)^{m_{12}} = 1 \rangle$, the dihedral group of order $2m_{1\,2}$.

It is common practice to provide a pictorial presentation of M by means of the labeled graph (I, M) with vertex set I, no loops, and (undirected) edges $\{i, j\}$, labeled $m_{i,j}$, if the latter number exceeds 2. Also, if $m_{i,j} = 3$, the label may be omitted.

Let I be an index set and W any group generated by a set $R = \{\rho_i \mid i \in I\}$. The free monoid on the alphabet I with unit (usually denoted by ϵ) is denoted by I^* and $\rho : I^* \to W(R)$ stands for the monoid morphism determined by $\rho(i) = \rho_i$ $(i \in I)$. There is a natural notion of length for an element of I^*; the length of the empty element is 0, the length of an element of the alphabet I equals 1, and so on. A typical element of I^* will be written as i and its length as $\ell(i)$. Thus, if $\ell(i) = q$, there are $i_j \in I$ $(1 \le j \le q)$ such that $i = i_1 \cdots i_q$. The *length* of an element $w \in W$, denoted by $\ell(w)$, or $\ell_R(w)$ if more precision is required, is $\min\{\ell(i) \mid \rho(i) = w\}$. For each element $i = i_1 \cdots i_q \in I^*$ with $\rho(i) = w$, we call the product $\rho(i_1) \cdots \rho(i_q)$ an *expression* of w. If $q = \ell(w)$, the expression is called *reduced*.

Let (W, R) be a Coxeter system with $R = \{\rho_1, \ldots, \rho_n\}$ of cardinality n. We shall write $\prec$ for the relation on W defined by $x \prec w$ if there is a reduced expression $s_1 \cdots s_q$ $(s_j \in R)$ for w such that $x = s_{i_1} \cdots s_{i_m}$ (where $1 \le i_1 < i_2 < \ldots < i_m \le q$). It is well known (cf. [Bour]) that $\succ$ defines an ordering on W; it is called the *Bruhat order*. Clearly, 1 is the smallest element of W; in case W is finite, the group W has a unique longest element w_0 which is the largest element with respect to $\succ$.

Let I be a subset of R. The subgroup $\langle I \rangle$ of W, also denoted by W_I, is the subgroup generated by I. It is again a Coxeter group. In particular, $W_\emptyset = \{1\}$ and $W_R = W$. Set

$$D_I = \{w \in W \mid \ell(ws) > \ell(w) \text{ for all } s \in I \}.$$

The set D_I is a natural system of W_I-coset representatives. As W-sets $D_I \cong W/W_I$.

4.1 Proposition. *Let (W, R) be a Coxeter system, and let I be a subset of R. The map $D_I \to W/W_I$ sending $w \in D_I$ to wW_I is a bijection. Each $w \in W$ has a reduced expression $w = dv$ with $d \in D_I$, $v \in W_I$; in particular, w is the unique shortest element of wW_I.*   QED

PROOF. See [Bour].   QED

The order $\succ$ on W induces a partial order, also denoted by $\succ$, on W/W_I. It can be characterized by $xW_I \succ yW_I$ if and only if $x \succ y$ for $x, y \in D_I$ and has the following useful property ([Deodh] Cor. 3.5]):

> if $x, y \in D_I$ and $x \succ y$ then there is a chain $x = x_0 \succ x_1 \succ \ldots \succ x_t = y \in W$ of elements of D_I such that $\ell(x_{i-1}) = \ell(x_i) - 1$ for all $i \in \{1, \ldots, t\}$.

Example The symmetric group Sym_{n+1} on $n + 1$ letters is the Coxeter group $W(A_n)$ where

$$A_n = \begin{array}{ccccccc} & 1 & 2 & 3 & n-1 & n \\ & \circ\!\!-\!\!\!-\!\!\!-\!\!\circ\!\!-\!\!\!-\!\!\!-\!\!\circ\!\!-\!\!\cdots\cdots\!\!-\!\!\circ\!\!-\!\!\!-\!\!\!-\!\!\circ \end{array}.$$

The evident morphism $W \to \mathrm{Sym}_{n+1}$ sending ρ_i to $(i, i+1)$ for each $i \in I$ is an isomorphism. Set $W = W(A_{n-1}) = \mathrm{Sym}_n$, where we identify ρ_i with the transposition $(i, i+1)$. It is easy to check that the longest element is the permutation $w_0 = (1, n)(2, n-1)(3, n-2) \cdots$ which is the mapping $[[n, n-1, \ldots, 1]]$ sending $j \in \{1, \ldots, n\}$ to $n+1-j$. (The double square bracket notation for a sequence of length n indicates the permutation sending $i \in \{1, \ldots, n\}$ to the i-th element of the sequence.)

We now look at the coset space W/W_J, where $J = \{\rho_1, \ldots, \rho_{d-1}, \rho_{d+1}, \ldots, \rho_{n-1}\}$. It has $\binom{n}{d}$ elements. The subgroup W_J is isomorphic to $\mathrm{Sym}_d \times \mathrm{Sym}_{n-d}$, and there is a bijection between $D_{\emptyset,J}$ and the set of normalized tableaux $t = [i_1, \ldots, i_d]$ with $1 \leq i_1 < \ldots < i_d \leq n$. The element in $D_{\emptyset,J}$ corresponding to t is the permutation

$$w_t = (\rho_{i_1-1} \cdots \rho_1)(\rho_{i_2-1} \cdots \rho_2) \cdots (\rho_{i_d-1} \cdots \rho_d), \qquad (6)$$

that is, the permutation $(i_1, \ldots, 1)(i_2, \ldots, 2) \cdots (i_d, \ldots, d)$, or, written in yet another notation, $[[i_1, \ldots, i_d, i_{d+1}, \ldots, i_n]]$ with $i_{d+1} < i_{d+2} < \cdots < i_n$ and $i_\ell \in \{1, \ldots, n\} \setminus \{i_1, \ldots, i_d\}$ for $\ell = d+1, \ldots, n$. For example, if $[i, j] = [2, 5]$ then

$$w_{2,5} = [[2, 5, 1, 3, 4]] = (1, 2)(4, 5)(3, 4)(2, 3) = \rho_1 \rho_4 \rho_3 \rho_2.$$

From (6) we see that if t and t' are normalized tableaux with $t' = [i_1, \ldots, i_{j-1}, i_j - 1, i_{j+1}, \ldots, i_d]$ and $t = [i_1, \ldots, i_{j-1}, i_j, i_{j+1}, \ldots, i_d]$, then t' is covered by t. This follows because $w_{t'}$ is obtained from w_t by removing ρ_{i_j-1}.

4.2 Proposition. *If* $W = W(A_{n-1}) = \mathrm{Sym}_n$ *and* $J = \{\rho_1, \ldots, \rho_{d-1}, \rho_{d+1}, \ldots, \rho_{n-1}\}$, *the Bruhat poset* $(D_J, \prec)$ *is isomorphic to the poset* $(P_{n,k}, \prec)$; *the isomorphism is given by*

$$w\langle J \rangle \mapsto w[[i_1, \ldots, i_d]] \quad (w \in D_J), \qquad \text{and its inverse by} \quad t \mapsto w_t \ (t \in P_{n,k}).$$

5. The highest weight orbit

Let $\mathbf{F}$ be a field of characteristic 0. Consider a simple split algebraic group G defined over $\mathbf{F}$ and fix a Tits system (B, N, W, R) in G, where B is a Borel subgroup of G (cf. [Bourb]). Denote by T the maximal torus $B \cap N$ of G in B. Then $N = N_G(T)$ is the normalizer in G of T and $W = N/T$ is a Weyl group with Coxeter system (W, R). For example, if G is the special linear group $SL(n, \mathbf{C})$ consisting of all $n \times n$ matrices with determinant 1, then B can be taken to be the subgroup of all upper triangular matrices in G, and the subgroup N can be taken to be the subgroup of all monomial matrices (one nonzero entry in each row and each column). The group W, which arises as the quotient of N by the diagonal subgroup $T = B \cap N$ is then isomorphic to Sym_n, and the Coxeter system (W, R) is of type A_{n-1}.

If $\mathbf{F} = \mathbf{C}$, the Bruhat order is closely related to the topological structure of G (viewed as a Lie group). For each $I \subset R$, the subgroup $P_I = B\langle I \rangle B$ has the property that G/P_I inherits the structure of a variety. The subset BwP_I/P_I for $w \in W$ is a cell in G/P_I, whose closure is $\{bxP_I \mid x \preceq w, b \in B\}$. We shall exploit the Bruhat order in a slightly different way.

Suppose λ is a dominant weight. Then $\lambda \in \mathbf{N}^n$, where n is the Lie rank of G (i.e., the dimension of T), and the highest weight module $V(\lambda)$ has a highest weight vector v_λ (unique up to scalar multiples). The stabilizer of the projective point $\langle v_\lambda \rangle$ is the parabolic subgroup $P = B\langle I \rangle B$ containing B of type I, where $\langle I \rangle = W_\lambda$ ($=$ the stabilizer in W of the weight λ). Thus the projective variety G/P embeds in the projective space $\mathcal{P}V(\lambda)$ with image $G\langle v_\lambda \rangle$. If V is any G-module, then, as a T-module, it has a basis of eigenvectors whose corresponding projective points are permuted by W. We shall call such a basis of V a T-frame.

We shall be primarily interested in the easiest cases, namely those where λ is *minuscule*, that is, $V(\lambda)$ has a T-frame consisting of a single W-orbit. (See [Proc] or [Hill] for many equivalent definitions.) Fix such a T-frame $(v_\mu)_{\mu \in W\lambda}$ in $V(\lambda)$. Accordingly, for $\mu \in W\lambda$ we denote by x_μ the element of the dual space $V(\lambda)^*$ that is dual to v_μ with respect to the selected basis of $V(\lambda)$. Write $x_\mu \prec x_\nu$ whenever $w \prec v$ for $w, v \in W$ with $\mu = w\lambda$ and $\nu = v\lambda$. Thus we transport the poset structure from W/W_λ, where W_λ is the stabilizer of λ, to $\{x_\mu\}_{\mu \in W\lambda}$. Since $W_\lambda = W_I$ for some subset I of R, we can apply Proposition 4.1. By the way, we shall write λ^* for the dominant weight satisfying $V(\lambda^*) \cong V(\lambda)^*$.

For $G = SL(n, \mathbf{C})$ the representation $V(\lambda)$ is the one obtained from the standard n-dimensional one by taking the symmetrized power (or plethysm) of the standard representation V with respect to the partition $(\lambda_1 + \cdots + \lambda_{n-1}, \lambda_2 + \cdots + \lambda_{n-1}, \ldots, \lambda_{n-1})$. Thus $V((0, 0, 1, 0, 0))$ and $V((3, 0, 0, 0, 0))$ are the third exterior and the third symmetric power of V, respectively. In particular, the minuscule weights of $SL(n, \mathbf{C})$ are the fundamental ones, i.e., those of the form $V((0, \ldots, 0, 1, 0, \ldots, 0))$, the T-frames of which are the Plücker coordinates.

The straightening phenomena in §2 have the following generalizations:

5.1 Theorem. *Let F be a field of characteristic 0. Suppose G is a simple split algebraic group over F with Tits system (B, N, W, R), and set $T = B \cap N$. Let λ be a dominant weight, and let $P = BW_\lambda B$ be the corresponding parabolic subgroup of G.*

(i) (cf. [Brion], [Lich]) *The G-module $S^2 V(\lambda)$ contains the highest weight module $V(2\lambda^*)$ with multiplicity 1 and has a G-invariant complement M. The ideal I in $\mathbf{F}[V(\lambda)]$ of the highest weight orbit $G\langle v_\lambda \rangle \cong G/P$ of G in $\mathcal{P}V(\lambda)$ is generated by the polynomial quadratic maps forming a T-frame of M.*

(ii) (cf. [Sesh]) *If λ is minuscule, then $\{x_\alpha x_\beta \mid \alpha, \beta \in W\lambda, \ \alpha \geq \beta\}$ is a basis of a complement of M in $S^2 V(\lambda)$.*

(iii) *If λ is minuscule, then there is a set $F_{G,\lambda}$ of polynomials $f_{\phi,\tau}$ indexed by the incomparable (unordered) pairs $\{\phi, \tau\}$ from the poset $(W/W_J, \prec)$ in such a way that, for each indexing pair $\{\phi, \tau\}$, the polynomial $f_{\phi,\tau}$ has shape*

$$x_\phi x_\tau - \sum_{\substack{\alpha, \beta \in W\lambda \\ x_\alpha \succeq x_\beta}} c_{\alpha,\beta} x_\alpha x_\beta \tag{7}$$

for certain coefficients $c_{\alpha,\beta}$.

(iv) (cf. [SeSh]) *In (7), any pair α, β with $c_{\alpha,\beta} \neq 0$ satisfies*

$$x_\alpha \succeq x_\phi, \quad x_\alpha \succeq x_\tau, \quad \text{and} \quad x_\beta \preceq x_\phi, \quad x_\beta \preceq x_\tau.$$

Consequently, $F_{G,\lambda}$ is a reduced Gröbner basis (with respect to any total degree lexicographical ordering $<$ extending $\prec$ on the variables) for the projective variety G/P embedded in $\mathcal{P}V(\lambda)$.

PROOF. For (i) and (ii), we refer to the cited papers.

(iii) The standard monomials form a complement J_2 of M in $S^2 V(\lambda)$. But the non-standard monomials span a complementary space I_2 to J_2. Thus, modulo J_2, the vector spaces M and I_2 are isomorphic. This means that there is a map $M \to I_2$ and a basis

$$\{f_{\phi,\tau} \mid \phi, \tau \in W\lambda, \ \phi, \tau \text{ incomparable}\}$$

of M such that each $f_{\phi,\tau}$ is the sum of $x_\phi x_\tau \in I_2$ and a linear combination of standard monomials.

(iv) The first statement is due to [Sesh], see also [LaSh]. It implies that, for $f \in I_2$, we have $NormalForm_F(f) \in J_2$, where $F = F_{G,\lambda}$. Together with (i), this readily implies that, for any $f \in \mathbf{F}[x]$, we have $NormalForm_F(f) \in J$, the $\mathbf{F}$-linear span of all monomials in which no product $x_\phi x_\tau$ with incomparable ϕ, $\tau \in W\lambda$ occurs. Hence F is a Gröbner basis. Since removal of an element g from F would make the non-standard monomial $lead(g)$ reduced with respect to $F \setminus \{g\}$, the Gröbner basis is reduced.   QED

To give an impression of the scope of the theorem, we list all miniscule weights of simple Lie groups. We use the labeling of Dynkin diagrams of [Bour]. If ω_j is the miniscule weight in the case of type Y_n, we write $Y_{n,j}$.

$$A_{n,j}, \; B_{n,n}, \; C_{n,1}, \; D_{n,1}, D_{n,n-1}, \; D_{n,n}, \; E_{6,1}, \; E_{6,6}, \; E_{7,7}.$$

We finish this section by giving examples of Gröbner bases in some of these cases.

Example $A_{3,2}$. Here $V(\lambda)$ has dimension 6, so $S^2V(\lambda)$ has dimension 21. As $V(2\lambda^*)$ has dimension 20, the Gröbner basis consists of a single relation. This is the well known one, given in the $G(4, 2)$ example of §3.

Example $C_{n,1}$. We consider Sp_{2n} in its natural $2n$-dimensional representation. Every nonzero vector is in the high weight vector orbit, so the set of quadratic equations is the empty set. Correspondingly, the Bruhat poset is a chain (of length $2n$). Thus the Gröbner basis is empty.

Example $D_{n,1}$. Another easy example is furnished by the standard representation of the orthogonal group in $2n$ dimensions of maximal Witt index (the split group of type D_n). Here the Hasse diagram looks like

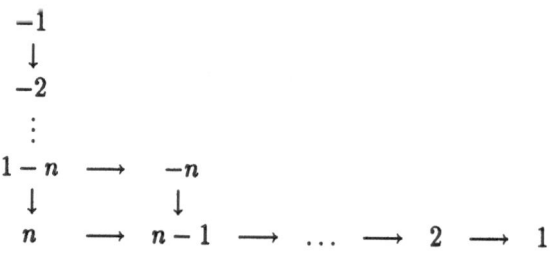

Figure 2. Hasse diagram of $D_{n,1}$.

This is in accordance with the well-known fact that the high weight orbit is a quadric in the natural representation space. Thus, the Gröbner basis is nothing but the quadric.

Example $E_{6,1}$. Let $\mathbf{K}$ denote the 27-dimensional vector space over the field $\mathbf{F}$ consisting of all ordered triples $x = [x^{(1)}, x^{(2)}, x^{(3)}]$ of 3×3-matrices $x^{(i)}$

$(1 \leq i \leq 3)$ (addition and scalar multiplication are entrywise.) The vector space $\mathbf{K}$ is supplied with the symmetric cubic form $D : \mathbf{K} \rightarrow \mathbf{F}$ given by :

$$D(x) = \det x^{(1)} + \det x^{(2)} + \det x^{(3)} - \text{trace} \ x^{(1)}x^{(2)}x^{(3)} \quad (x \in \mathbf{K}) \quad (8)$$

The group $G = GL(\mathbf{K})_D$ of (invertible) linear transformations g of $\mathbf{K}$ such that $D(g(x)) = D(x)$ for all $x \in \mathbf{K}$ is the simply connected Lie group G of type E_6. Moreover, $\mathbf{K}$ is a highest weight representation space of G with highest weight ω_1.

Next, we describe the ordering of the variables which we exploit. We shall write $\mathbf{F}[\mathbf{K}] = \mathbf{F}[x]$ with $x = (x_i)_{1 \leq i \leq 27}$ such that x_i is the coordinate function attached to the standard basis element labeled i the following scheme:

$$\left[\begin{pmatrix} 6 & 4 & 11 \\ 7 & 5 & 12 \\ 25 & 26 & 27 \end{pmatrix} , \begin{pmatrix} 14 & 13 & 15 \\ 17 & 16 & 18 \\ 9 & 8 & 10 \end{pmatrix} , \begin{pmatrix} 23 & 20 & 2 \\ 24 & 21 & 3 \\ 22 & 19 & 1 \end{pmatrix} \right].$$

Thus, for example $x_6 = x_{1,1}^{(1)}$.

The Bruhat ordering of W/W_J, where $J = \{2, 3, 4, 5, 6\}$ is depicted in the Hasse diagram of Figure 3. It is refined to a total ordering on the variables by putting $1 < x_1 < x_2 < \ldots < x_{27}$.

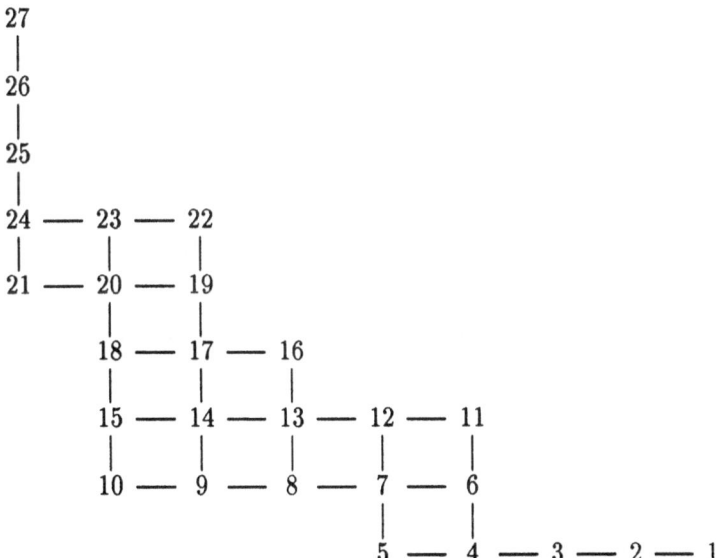

Figure 3. Hasse diagram of $(W/W_J, \prec)$, where $W = W(E_6)$ and $J = \{2, 3, 4, 5, 6\}$.

There are 27 incomparable pairs in the poset $(W/W_{\{2,3,4,5,6\}}, \prec)$. On the other hand, the variety G/P is well known (cf. [CoCo]) to be the zero set of the 27 partial derivatives $\partial_{x_i} D(x)$ $(i \in \{1, \ldots, 27\})$ of the cubic form D of (8). These defining equations are

$$x_{10}x_{19} - x_{25}x_4 + x_{26}x_6 + x_9x_{20} + x_8x_{21},$$

$$x_{15}x_{22} - x_5x_{27} + x_{12}x_{26} + x_{14}x_{23} + x_{13}x_{24},$$

$$x_{22}x_{20} - x_{19}x_{23} - x_{25}x_{13} - x_{26}x_{16} - x_{27}x_8,$$

$$x_{11}x_9 - x_{21}x_1 + x_3x_{19} + x_6x_{14} + x_4x_{17},$$

$$x_9x_{13} - x_8x_{14} - x_{22}x_4 - x_{19}x_5 - x_1x_{26},$$

$$x_{16}x_{10} - x_{18}x_8 - x_{23}x_6 - x_{20}x_7 - x_2x_{25},$$

$$x_{11}x_5 - x_4x_{12} + x_{14}x_2 + x_{13}x_3 + x_{15}x_1,$$

$$x_{18}x_{22} - x_{12}x_{25} + x_7x_{27} + x_{17}x_{23} + x_{16}x_{24},$$

$$x_{11}x_{10} - x_{20}x_3 + x_2x_{21} + x_6x_{15} + x_4x_{18},$$

$$x_{12}x_9 - x_3x_{22} + x_{24}x_1 + x_7x_{14} + x_5x_{17},$$

$$x_{15}x_{16} - x_{13}x_{18} + x_{23}x_{11} + x_{20}x_{12} + x_2x_{27},$$

$$x_{17}x_{10} - x_{18}x_9 + x_{24}x_6 + x_{21}x_7 + x_3x_{25},$$

$$x_{11}x_7 - x_6x_{12} - x_{17}x_2 - x_{16}x_3 - x_{18}x_1,$$

$$x_{10}x_{22} - x_7x_{26} + x_5x_{25} + x_9x_{23} + x_8x_{24},$$

$$x_{12}x_{10} - x_2x_{24} + x_{23}x_3 + x_7x_{15} + x_5x_{18},$$

$$x_{21}x_{22} - x_{24}x_{19} + x_{25}x_{14} + x_{26}x_{17} + x_{27}x_9,$$

$$x_{15}x_{17} - x_{14}x_{18} - x_{24}x_{11} - x_{21}x_{12} - x_3x_{27},$$

$$x_{16}x_9 - x_{17}x_8 + x_{22}x_6 + x_{19}x_7 + x_1x_{25},$$

$$x_6x_5 - x_4x_7 - x_9x_2 - x_8x_3 - x_{10}x_1,$$

$$x_{15}x_{19} - x_{26}x_{11} + x_{27}x_4 + x_{14}x_{20} + x_{13}x_{21},$$

$$x_{23}x_{21} - x_{20}x_{24} - x_{25}x_{15} - x_{26}x_{18} - x_{27}x_{10},$$

$$x_{11}x_8 - x_{19}x_2 + x_1x_{20} + x_6x_{13} + x_4x_{16},$$

$$x_{14}x_{16} - x_{13}x_{17} - x_{22}x_{11} - x_{19}x_{12} - x_1x_{27},$$

$$x_{10}x_{13} - x_8x_{15} + x_{23}x_4 + x_{20}x_5 + x_2x_{26},$$

$$x_{18}x_{19} - x_{27}x_6 + x_{25}x_{11} + x_{17}x_{20} + x_{16}x_{21},$$

$$x_{12}x_8 - x_1x_{23} + x_{22}x_2 + x_7x_{13} + x_5x_{16},$$

$$x_{10}x_{14} - x_9x_{15} - x_{24}x_4 - x_{21}x_5 - x_3x_{26}.$$

Each polynomial contains a unique monomial indexed by an incomparable pair. Since $S^2 V(\omega_1) \cong V(2\omega_6) \oplus V(\omega_1)$, Theorem §5.1 yields that these 27 polynomials form a reduced Gröbner basis for the highest weight orbit of E_6 in $\mathcal{P}V(\omega_1)$.

6. The algorithm

The Gröbner bases in the examples of §5 have been obtained by ad hoc arguments. In order to present an effective version of Theorem 5.1, we need the Casimir operator.

Retain the notation of §5 for G, B, T, W and J. Let Φ denote the root system of G with respect to T, let $\{\alpha_1, \ldots, \alpha_n\}$ be a set of fundamental roots of T, defining B. The dual basis is the set of fundamental weights, $\{\omega_1, \ldots, \omega_n\}$. The numbering of the roots is as in [Bour]. Consider the Lie algebra L corresponding to G, with Chevalley basis $(X_\alpha)_{\alpha \in \Phi}$, $(H_i)_{1 \le i \le n}$, where $H_i = H_{\alpha_i} = [X_{\alpha_i}, X_{-\alpha_i}]$. Let $Z_\alpha \in \mathbf{F}X_\alpha$ be a vector with $\kappa(X_\alpha, Z_\alpha) = 1$, where κ is the Killing form on L. In the Cartan subalgebra $\sum_i \mathbf{F}H_i$, take $\{K_i\}_i$ to be a dual basis of $\{H_i\}_i$ with respect to κ. Then

$$\Omega = \sum_{\alpha \in \Phi} X_\alpha Z_\alpha + \sum_{i=1}^{\ell} H_i K_i$$

is the *Casimir element* of L. It is a central element of the universal enveloping algebra of L. As such, Ω is an operator on each G-module V, in particular on SV, the ring of polynomial functions on V. The crucial property of the Casimir operator is that, on the G-module with highest weight μ, we have

$$\Omega|_{V(\mu)} \text{ is scalar multiplication by } (\mu + 2\delta, \mu),$$

where $\delta = \omega_1 + \cdots + \omega_n$ and $(\cdot, \cdot)$ is defined on weights by $(\mu, \nu) = \sum_{i=1}^{n} \mu(H_i)\nu(K_i)$.

6.1 Lemma. *Let G, B, λ, P be as in Theorem 5.1. If λ is minuscule, then, for every dominant weight μ occurring in $W\lambda + W\lambda$ distinct from 2λ,*

$$(\mu + 2\delta, \mu)(2\lambda + 2\delta, 2\lambda).$$

In particular, M as defined in Theorem 5.1 is generated by all $(\Omega - (\lambda + 2\delta, \lambda))v$ for v running through a T-frame of $S^2 V(\lambda)$.

PROOF. Using the standard inner product on the root space to identify the weights with vectors in the root space, the inner product $(\cdot, \cdot)$ on the weight space becomes a positive multiple of the standard inner product on the root space. Thus, $(\omega_i, \alpha_j) > 0$ if $i = j$ and equal to 0 otherwise. Set $\nu = 2\lambda - \mu$. By assumption $\mu \in W\lambda + W\lambda$, so there are $w_1, w_2 \in W$ such that $\mu = w_1\lambda + w_2\lambda$. Now $\lambda - w_i\lambda$ is a sum of positive roots for both $i \in \{1, 2\}$ (cf. [Proc, Lemma 3.1]), so the same holds for $\nu = (\lambda - w_1\lambda) + (\lambda - w_2\lambda)$. Writing $\nu = \sum_{i=1}^{n} \nu_i \alpha_i$ with $\nu_i \in \mathbf{N}$, and $\lambda = \omega_k$ for some k (recall that λ is fundamental hence of this form), we obtain $(\nu, \lambda + \delta) = \nu_k(\omega_k, \alpha_k) + \sum_{i=1}^{n} \nu_i(\omega_i, \alpha_i)$. Hence

$$(\nu, \lambda + \delta) \ge 0, \quad \text{with equality if and only if } \nu = 0. \tag{9}$$

Next note, that, since μ is dominant (that is, of the form $\sum_{i=1}^{n} \mu_i \omega_i$ with $\mu_i \geq 0$ for all i) we have $(\mu, \nu) = \sum_{i=1}^{n} \mu_i \nu_i (\omega_i, \alpha_i) \geq 0$. In view of (1), this yields

$$(2\lambda + 2\delta, 2\lambda) - (\mu + 2\delta, \mu) = 2(\nu, \lambda + \delta) + (\nu, \mu) \geq 0,$$

with a strict inequality whenever $\nu \neq 0$, that is, whenever $\mu \neq 2\lambda$. Hence the first statement.

For the second statement, observe that $\delta^* = \delta$ and $(\mu + 2\delta, \mu) = (\mu^* + 2\delta, \mu^*)$ for each dominant weight μ.   QED

For computations, it is convenient to rewrite Ω. Let Φ^+ be the set of positive roots in Φ and put

$$C = \sum_{\alpha \in \Phi^+} (\alpha, \alpha) X_{-\alpha} X_\alpha. \qquad (10)$$

Then, on a vector v in a weight space of some G-module with weight μ, we have

$$\Omega v = Cv + (\mu + 2\delta, \mu)v.$$

From this we obtain

$$\begin{aligned}
\Omega(v\,w) &= C(v\,w) + (\mu + \nu + 2\delta, \mu + \nu)v\,w \\
&= (\mu + \nu + 2\delta, \mu + \nu)v\,w + \\
&\quad (Cv)\,w + v\,(Cw) + \sum_{\alpha \in \Phi^+} (\alpha, \alpha)\left((X_\alpha v)(X_{-\alpha} w) + (X_{-\alpha} v)(X_\alpha w)\right)
\end{aligned}$$

for v a vector of weight μ^* and w a vector of weight ν^* in $S^2 V(\lambda)$. Thus, the computation of the Casimir operator on $S^2 V(\lambda)$ is brought back to a computation of X_α on monomials from a T-frame.

Now, by Lemma 6.1, the operator $\Omega - (2\lambda + \delta, 2\lambda)$ projects $S^2 V(\lambda)$ onto M. Thus, we come to the following algorithm for finding the basis in each of the minuscule cases. The partitioning of the Gröbner basis by means of weights is a good tool to cut down the size of the computations.

MinusculeWeightStandardBase Algorithm   *Given a minuscule weight λ of a simple Lie group G, compute a Gröbner basis of the ideal of polynomials vanishing on the highest weight orbit.*

$MWStandardBase(G, \lambda) =$

   $\Lambda := W\lambda; \ \Lambda_2 := \Lambda + \Lambda;$

   $C := \sum_{\alpha \in \Phi^+}(\alpha, \alpha)X_{-\alpha}X_\alpha;$ # the operator of (10)

   for each $\mu \in \Lambda_2$ **do**

      $S_\mu := \{x_\alpha x_\beta \mid \alpha, \beta \in \Lambda, \ \alpha + \beta = -\mu\};$

      # a $-$ sign as x_α has weight $-\alpha$

      $E_\mu := \{(C - (2\lambda + 2\delta, 2\lambda) + (\mu + 2\delta, \mu))p \mid p \in S_\mu\};$

      $E_\mu := GaussElim(B_\mu);$

         # to get the identity submatrix on incomparable pairs

      $E := E \cup E_\mu$

   od ;

   return E.

Here, the procedure *GaussElim* sees to it that the principal minor of E_μ, viewed as a matrix with columns indexed by monomials, the non-standard ones coming first, becomes the identity (of size the number of incomparable pairs from $W\lambda$ whose sums equal μ) by means of elementary row operations.

   In general, the computation of Ω (or, to be more precise, of X_α on a weight vector) requires knowledge of the structure constants of the Lie algebra L. Apart from these data, all necessary ingredients are implemented in LiE (cf. [LiE]).

Example In the case of A_n, the knowledge of the structure constants is hidden in the tableaux calculus. We have

$$\Phi^+ = \{\epsilon_i - \epsilon_j \mid 1 \le i < j \le n+1\}$$
$$X_\alpha = X_{ij} \text{ if } \alpha = \epsilon_i - \epsilon_j \in \Phi$$

$$X_{ij}[k_1, \ldots, k_t] = \begin{cases} [k_1, \ldots, k_{\ell-1}, i, k_{\ell+1} \ldots, k_t] & \text{if } j = k_\ell \text{ for a} \\ & \text{unique } \ell \in \{1, \ldots, t\} \\ 0 & \text{otherwise} \end{cases}$$

$$X_{ji}X_{ij}[k_1, \ldots, k_t] = \begin{cases} [k_1, \ldots, k_t] & \text{if } j = k_\ell \text{ for a unique} \\ & \ell \in \{1, \ldots, t\} \text{ and } i \\ & \text{does not occur,} \\ 0 & \text{otherwise} \end{cases}$$

$$\kappa(\alpha, \beta) = 2(n+1)\alpha^\top C\beta \quad (\alpha, \beta \in \Phi),$$

where C is the Cartan matrix of A_n

$$(\mu, \nu) = \mu^\top C^{-1}\nu \quad (\mu, \nu \text{ weights})$$

We finish by illustrating the algorithm for $A_{5,3}$, focusing on the most difficult weight occurring in $S^2(V(\lambda))$, namely the zero weight. The computations have been done in LiE. In the lexicographic total ordering, there are
10 quadratic monomials of weight 0. They are

$$[136][245], [145][236], [146][235], [156][234], [126][345],$$
$$[123][456], [124][356], [125][346], [134][256], [135][246].$$

Their order of appearance is such that the non-standard ones come first.
Applying $\Omega - (2\lambda + \delta, 2\lambda) = \Omega - 24$ to each of these monomials gives
polynomials which are presented as rows of the following coefficient matrix.
Here, the j-th entry of the i-th row represents the coefficient of the j-th
monomial in the image of the i-th monomial under $\Omega - 24$.

$$2 \cdot \begin{pmatrix}
-3 & 1 & 1 & -1 & 1 & -1 & -1 & 1 & -1 & 1 \\
1 & -3 & 1 & -1 & -1 & 1 & 1 & -1 & -1 & 1 \\
1 & 1 & -3 & 1 & -1 & -1 & -1 & -1 & 1 & 1 \\
-1 & -1 & 1 & -3 & 1 & 1 & -1 & -1 & 1 & 1 \\
1 & -1 & -1 & 1 & -3 & 1 & -1 & 1 & -1 & 1 \\
-1 & 1 & -1 & 1 & 1 & -3 & 1 & -1 & -1 & 1 \\
-1 & 1 & -1 & -1 & -1 & 1 & -3 & 1 & 1 & 1 \\
1 & -1 & -1 & -1 & 1 & -1 & 1 & -3 & 1 & 1 \\
-1 & -1 & 1 & 1 & -1 & -1 & 1 & 1 & -3 & 1 \\
1 & 1 & 1 & 1 & 1 & 1 & 1 & 1 & 1 & -3
\end{pmatrix}.$$

The rank of this matrix is 5. Now we can apply classic Gauss elimination
to obtain reduced Gröbner basis elements. The result is:

$$\begin{pmatrix}
1 & 0 & 0 & 0 & 0 & 1 & 0 & 0 & 1 & -1 \\
0 & 1 & 0 & 0 & 0 & 1 & -1 & 1 & 1 & -1 \\
0 & 0 & 1 & 0 & 0 & 1 & 0 & 1 & 0 & -1 \\
0 & 0 & 0 & 1 & 0 & -1 & 1 & 0 & -1 & 0 \\
0 & 0 & 0 & 0 & 1 & -1 & 1 & -1 & 0 & 0
\end{pmatrix}$$

which tells us that the contribution E_0 to E is the following set of five
quadratic polynomials:

$$E_0 = \begin{cases}
[136][245] + [123][456] + [134][256] - [135][246], \\
[145][236] + [123][456] - [124][356] + [125][346] + \\
\quad + [134][256] - [135][246], \\
[146][235] + [123][456] + [125][346] - [135][246], \\
[156][234] - [123][456] + [124][356] - [134][256], \\
[126][345] - [123][456] + [124][356] - [125][346]
\end{cases}.$$

Together with the 30 Plücker relations of the form

$$[i\,j\,k]\,[i\,\ell\,m] + [i\,j\,\ell]\,[i\,m\,k] + [i\,j\,m]\,[i\,k\,\ell]$$

$(\{i, j, k, \ell, m\}$ a 5-set in $\{1, 2, 3, 4, 5, 6\})$,

the set E_0 yields the reduced Gröbner basis E (of size 35) of the Grassmannian $G(6, 3)$ with respect to any lexicographic ordering $<$ in which $x_\alpha > x_\beta$ whenever $x_\alpha \succ x_\beta$.

7. Concluding remarks

Theorem 5.1(i) gives a clear indication that in the non-minuscule weights case, all hope for finding a Gröbner basis need not be lost. Also, standard monomial theory has been generalized to the case of quasi-minuscule weights (see [LaSh]). It is our intention to study how far the correspondence with Gröbner basis theory can be pushed.

As we have indicated above, the Plücker relations are in general different from the reduced Gröbner basis elements obtained by use of the Casimir operator. To derive the actual Plücker relations, we have reason to believe that Casimir operators relative to Levi subalgebras are the appropriate tool.

In [StWh2], it is indicated how the poset $(P_{n,d}, \succ)$ can be used to find a Stanley decomposition of the ring $\mathcal{A}$. By means of a lexicographical shelling of the poset (cf. [BjWa]), this construction can also be generalized to one for all minuscule weights.

8. References

[ACGH] E. Arbarello, M. Cornalba, P.A. Griffiths, J. Harris, *Geometry of Algebraic Curves, Volume I*, Springer Verlag, Berlin, 1984.

[BjWa] A. Björner and M. Wachs, "Bruhat order of Coxeter groups and shellability", *Adv. in Math.* **43**, (1982), 87–100.

[Bour] N. Bourbaki, *Groupes et algèbres de Lie, Chap. IV, V, VI*, Hermann, Paris, 1968.

[Brion] M. Brion, "Représentation exceptionnelles des groupes semi simples", *Ann. Scient. Éc. Norm. Sup.* **18**, (1985), 345–387.

[Buch] B. Buchberger, *Gröbner bases – an algorithmic method in polynomial ideal theory*, in "Multidimensional Systems Theory" (N.K. Bose, ed.), D. Reidel, Dordrecht, 1985.

[Coh] A.M. Cohen, *Coxeter Groups and three Related Topics,*, pp. 235–278 in "Generators and Relations in Groups and Geometries, Castelvecchio Pascoli, Italy (April 1990)" (A. Barlotti, E.W. Ellers, P. Plaumann, K. Strambach, eds.), NATO ASI Series C: Math. and Phys. Sciences 333, Kluwer Acad. Publ., Dordrecht, 1991.

[CoCo] A.M. Cohen and B.N. Cooperstein, "The 2-spaces of the standard $E_6(q)$-module", *Geometriae Dedicata* **24**, (1988), 467–480.

[Deo] V.V. Deodhar, "Some characterizations of Bruhat ordering on a Coxeter group and determination of the relative Möbius Function", *Inventiones Math.* **39**, (1977), 187–198.

[Hill] H. Hiller, *Geometry of Coxeter Groups*, Research Notes in Math.,
Pitman, Boston, 1982.

[Hum] J.E. Humphreys, *Reflection groups and Coxeter groups*, Cambridge
University Press, 1990.

[LaSh] V. Lakshmibai and C.S. Seshadri, "Geometry of $G/P - V$", *J. Algebra* **100**, (1986), 462–557.

[Lich] W. Lichtenstein, "A system of quadrics describing the orbit of the
highest weight vector", *Proc. Amer. Math, Soc.* **84**, (1982), 605–608.

[LiE] M.A.A. van Leeuwen, A.M. Cohen, B. Lisser, *LiE manual, describing
version 2.0*, CAN, Amsterdam, 1992.

[Proc] R.A. Proctor, "Bruhat lattices, plane partition generating functions, and minuscule representations", *European J. Combinatorics*
5, (1984), 331–350.

[Sesh] C.S. Seshadri, *Geometry of $G/P - I$, Theory of standard monomials
for minuscule representations*, pp. 207–239 in "C.P. Ramanujam –
A tribute", (for Tata Institute) Springer-Verlag, Berlin, 1978.

[StWh1] B. Sturmfels and N. White, "Gröbner bases and invariant theory",
Adv. in Math. **76**, (1989), 245–259.

[StWh2] B. Sturmfels and N. White, "Stanley decompositions of the
bracket ring", *Math. Scandinavia* **87**, (1990), 183–189.

Arjeh M. Cohen & Richard H. Cushman
Mathematics Institute
Rijksuniversiteit Utrecht,
3508 TA Utrecht,
The Netherlands

A continuous and rational solution
to Hilbert's 17th problem
and several cases of the Positivstellensatz

C.N. Delzell[1]   L. González-Vega[2]   H. Lombardi

Abstract: From the Positivstellensatz we construct a continuous and rational solution for Hilbert's 17th problem and for several cases of the Positivstellensatz. The solutions are obtained using an especially simple method.

I. Introduction.

Let $\mathbf{K}$ be an ordered field, and $\mathbf{R}$ its real closure. Hilbert's 17th problem asks if an everywhere nonnegative polynomial $f \in \mathbf{K}[\mathbf{X}] := \mathbf{K}[X_1, \ldots, X_n]$ can be expressed as a sum of squares of rational functions in $\mathbf{K}(\mathbf{X})$ with positive weights in $\mathbf{K}$. Since the answer is well known to be 'Yes,' we now seek more information, in particular, on the way the coefficients of the solution can vary in terms of the coefficents of f. The main tools we shall use to obtain this extra information are the notion of semipolynomial function and the Positivstellensatz for $\mathbf{K}[\mathbf{X}]$.

A semipolynomial function with coefficients in $\mathbf{K}$ (a $\mathbf{K}$-semipolynomial) from $\mathbf{R}^n$ to $\mathbf{R}$ is a function obtained by a finite iteration of composition of polynomials in $\mathbf{K}[\mathbf{X}]$ and the function absolute value. A well-known proposition, not used here, assures that the set of $\mathbf{K}$-semipolynomials agrees with the minimal max-min stable set of functions containing polynomials in $\mathbf{K}[\mathbf{X}]$ (see, for example, [Del₄]).

By the name Positivstellensatz we shall refer to the more general version of this theorem, i.e., the one assuring that it is possible to associate to every incompatible finite conjunction of generalized sign conditions on a list of polynomials in $\mathbf{K}[\mathbf{X}]$ an algebraic identity in $\mathbf{K}[\mathbf{X}]$ making this incompatibility evident (see the beginning of §II).

---

[1] Supported by NSF, the Louisiana Board of Regents Research and Development Program (Education Quality Support Fund), and the Alexander von Humboldt Foundation.

[2] Partially supported by CICyT PB 89/0379/C02/01 and Esprit/Bra 6846 (Posso).

Using this Positivstellensatz we shall provide, in a constructive way, a continuous and rational solution for Hilbert's 17th problem. More precisely, let $f_{n,d}(\mathbf{c}, \mathbf{X})$ be the general polynomial of degree d in the n variables $\mathbf{X} := (X_1, \ldots, X_n)$ with coefficients $\mathbf{c} := (c_1, \ldots, c_m)$, and consider the semialgebraic set

$$\mathbf{F}_{n,d} = \{\mathbf{c} \in \mathbf{R}^m : \forall \mathbf{x} \in \mathbf{R}^n \quad f_{n,d}(\mathbf{c}, \mathbf{x}) \geq 0\}.$$

Theorem II.5. (Main Theorem) *The general polynomial $f_{n,d}$ of degree d in n variables can be written as a weighted sum of squares of rational functions*

$$f_{n,d}(\mathbf{c}, \mathbf{X}) = \sum_j p_j(\mathbf{c}) \left(\frac{q_j(\mathbf{c}, \mathbf{X})}{k(\mathbf{c}, \mathbf{X})} \right)^2$$

(for all $\mathbf{c} \in \mathbf{R}^m$), where

- $k(\mathbf{c}, \mathbf{X})$ *and the* $q_j(\mathbf{c}, \mathbf{X})$ *are polynomials in the variables* $\mathbf{X}$ *whose coefficients are $\mathbf{Q}$-semipolynomials in the coefficients* $\mathbf{c}$. *Moreover, if* $\mathbf{c} \in \mathbf{F}_{n,d}$, *then $k(\mathbf{c}, \mathbf{X})$ vanishes only on the zeros of $f_{n,d}(\mathbf{c}, \mathbf{X})$, and*

- *each $p_j(\mathbf{c})$ is a $\mathbf{Q}$-semipolynomial which is nonnegative on $\mathbf{F}_{n,d}$. Moreover, under the hypothesis $\mathbf{c} \in \mathbf{F}_{n,d}$, the nonnegativity of $p_j(\mathbf{c})$ is 'clearly' evident.*

A. Prehistory of Theorem II.5

Hilbert posed his 17th problem in 1900 [Hil]; E. Artin solved it in 1927 [Art] by a non-constructive method. In 1955 Artin asked Kreisel whether, from *Artin's own* proof, one could extract bounds on the number and degrees of the rational functions, in terms of suitable elements of the data (n, d, and possibly $\mathbf{c}$); Kreisel gave a sketch of such an 'unwinding' in [Kre2]; the bounds so constructed were in terms of n and d (not $\mathbf{c}$), and primitive recursive; in 1961 Daykin [Day] worked out Kreisel's sketch, showing that, roughly, the bounds were obtained by applying primitive recursion at least twice to exponential functions of n and d.[3] Independently, A. Robinson (see [Rob1] and [Rob2]) got (by definition total) general recursive bounds. These authors also expressed the weights and coefficients of the rational functions as $\mathbb{Z}$-piecewise-polynomial functions g of $\mathbf{c}$. All this handles the case in which $\mathbf{K}$ is given with computable arithmetic operations and sign test (e.g., when $\mathbf{K} = \mathbf{Q}$)—and it was more than enough for Artin himself.

But the fact that those g were, *a priori*, discontinuous in $\mathbf{c}$ for the usual, order topology on $\mathbf{R}$, meant that they were computationally inadequate for the case where $\mathbf{K} = \mathbf{R}$, since if we want to have computable arithmetic operations, elements of $\mathbf{R}$ must be given by, say, rational approximations; this makes equality and, *a fortiori*, the order relation, undecidable. Intuitionistic logic gives small changes in the logical laws which do ensure

---

[3] The statement of Daykin's bounds was oversimplified in [Del1] and [Del2].

continuity of functions constructed (and for a wide range of topologies). So after the above contributions by *classical* logic, Kreisel asked in 1962 [Kre₃] whether intuitionistic logic could also contribute, by determining whether the g could be chosen to be continuous, or the rational functions $\in \mathbf{K(X)}$ could be chosen to be continuously extendible to $\mathbf{R}^n$. So far, intuitionistic logic has contributed little, and real algebraic geometry much:

(a) In 1978 Kreisel noticed [Kre₄] that Stengle's 1974 "Positivstellensatz" [Ste] (which others have called a "Nichtnegativstellensatz") easily represents a positive semidefinite ('psd') $f \in \mathbf{K[X]}$ in the form $f = \sum_j p_j (q_j/k)^2$ with $0 \le p_j \in \mathbf{K}$ and $k, q_j \in \mathbf{K[X]}$ and, most importantly here, such that the functions q_j/k extend continuously to $\mathbf{R}^n$ (see the end of §II below for details).

(b) In 1980 the first author showed [Del₂] that for even $d \ge 4$, the weights $p_j(\mathbf{c})$ in (II.5)($\star$) cannot be chosen to be rational functions ($\in \mathbf{R(c)}$); and in 1990 ([Del₆] and [Del₇]) he excluded even (germs of) real analytic functions when $\mathbf{R} = \mathbb{R}$.

(c) Also in 1980 he found his first positive result on this ([Del₃]): for all $d \ge 0$ the p_j and the coefficients of k and the q_j can be chosen to be (locally uniformly) continuous $\mathbb{Q}$-semialgebraic functions of $\mathbf{c}$; functions with these properties *can* be effectively evaluated even over $\mathbb{R}$, in the—only reasonable—sense that they are computable on, say, the rationals, and they take approximations to approximations. And from the mere existence of this representation of f (and the completeness of the theory of real closed fields), the semialgebraic descriptions of these functions are given by general recursive functions of n and d.

There were two shortcomings in (c): (1) As Scowcroft [Sco₁] and the third author [Lom₂] remarked, it contained (only) one nonconstructive step, namely, the use of Stengle's theorem, proved up until then by means of Zorn's lemma. In 1956 Kreisel had observed [Kre₁] that by relativising a proof to the constructible universe, the axiom of choice (and even the generalized continuum hypothesis) can be eliminated from any proof of an *arithmetical* theorem, a fact which has been used to sanitize proofs of results on the order of homotopy groups (by Serre) and on p-adic fields (by Ax and Kochen). Thus, being arithmetic (at least if, say, $\mathbf{K} = \mathbb{Q}$, which was the only case used in (c)), Stengle's theorem follows from ZF without further ado. It is clear that this 'trick' would not satisfy constructivists, however: often there is more to constructivity than purging the axiom of choice, namely purging the principle of the excluded middle. Note also that ZF with intuitionistic logic seems not constructively meaningful. Scowcroft [Sco₁] offered a sketch of a proof of Stengle's theorem without AC or any other constructively dubious principle; if successful it would yield bounds which belong not only to the set of general recursive functions, but to the proper subset of those which are provably total in formal arithmetic (intuitionistic or classical, both having the same provably total recursive functions). The third author gave a direct, constructive proof of the Positivstellensatz [Lom$_1$], with explicit (primitive recursive) bounds [Lom$_3$]. Thus, by changing only one entry in the bibliography of [Del$_3$] (namely, from [Ste] to [Lom$_1$] and [Lom$_3$]), the proof in [Del$_3$] becomes constructive, and the semialgebraic descriptions of those functions also become primitive recursive. Thus, if all that we had wanted in a solution to Hilbert's 17th problem over **R** were brutal constructivity, then we could have stopped here; in particular, there would have been no need for (II.5) (though the luxury of a simpler proof is welcome).

(2) The other shortcoming of (c), which was not to be overcome by the above kind of tinkering with the proof in [Del$_3$], was that the functions which it produced are 'only' semialgebraic, and therefore do not necessarily take **K**-rational values at **K**-rational arguments c, as Hilbert asked for (unless **K** is real-closed). So in the first author's thesis, and in [Del$_1$], [Del$_3$], and elsewhere, he proposed to construct continuous, **Z**-piecewise-polynomial weights and coefficients; such functions obviously do take values in **K** at **K**-rational arguments c, uniformly for all **K**. And in [Del$_4$] he conjectured the full statement of Theorem II.5, that the functions could even be chosen to be **Z**-semipolynomials, which would make their continuity 'evident'. This last conjecture/theorem appears to be 'strongest possible' by (b), and it contains the results in (a) and (c) above, as explained in more detail at the end of §II below.

B. History of Theorem II.5

Theorem II.5 is a special case of the Positivstellensatz for semipoly-nomials ('Pfs'). The first author proved II.5 and the Pfs in 1988 (see the abstracts [Del$_5$] and [Del$_7$]). The second and third authors, jointly, re-discovered these results in 1991, independently of the first author, and by a different method. This paper presents their proof of (II.5), along with their view (§IV) of it as being primarily a contribution to constructive mathe-matics (à la Bishop). Their proof of the Pfs will appear in [GV-L]. The first author's proof of both results will appear in [Del$_8$], along with his view of them as being primarily a contribution to (classical) topological algebra and to the *examination* of assumptions of the logical tradition, specifically, the assumption that logic contributes to this part of mathematics.

Perhaps the main difference between the two proofs of the Pfs is that the first author reduces it to the abstract Stellensätze for the real spec-trum of an arbitrary commutative ring, while the second and third authors reduce it to the Positivstellensatz for polynomials. An advantage of the former proof is that it has some surprising facts about abstract semial-gebraic (in particular, abstract 'semipolynomial') functions in f-rings as by-products. For example, while the absolute value function $c_1 \mapsto |c_1|$ is obviously positive semidefinite on **R**, it is not psd on the real spectrum of the ring of semipolynomials. This example motivated a large part of N. Schwartz's recent investigation [Sch] into abstract piecewise-polynomial functions. An advantage of the latter proof of (II.5) is that it avoids even the appearance of reliance on Zorn's lemma, simply by avoiding the abstract Stellensätze. This relies, instead, on—the third author's direct, constructive proof of—the Positivstellensatz for polynomials [Lom₁]; as a result, the functions produced by this proof are primitive recursive in n and d; as in (A.c.1) and in Kreisel's and Daykin's results at the beginning of subsection A, these primitive recursive functions are of iterated exponential complexity in n and d,[4] while those of [Del₈] are only general recursive. Thus the latter proof provides another constructive solution to Hilbert's 17th problem over $\mathbb{R}$, simpler and more informative than that described in (A.c.1) above (see §IV for details).

The first author's proof of (II.5) was, at first, more complicated than necessary; A. Prestel simplified the proof, and incidentally rearranged it along the lines of the method below by the second and third authors. He also asked in [BP] whether the result extended to higher even powers; cf. the abstract in [Pre₂] when $n = 1$.

The underlying ideas used to prove (II.5) and the Pfs led all of us (again independently) to improve the continuous and semialgebraic variation in Scowcroft's Positivstellensatz [Sco₂] to semipolynomial variation.

Finally, unlike the proofs of (II.5) presented in this paper and in [Del₈], the proof presented in [GV-L] reduces the result to the Pfs.

II. A rational and continuous solution to Hilbert's 17th problem.

First we recall the definitions of strong incompatibility and the general form for the Real Nullstellensatz in the polynomial case (see [Lom₁] and [Lom₃]). We consider an ordered field $\mathbf{K}$, and $\mathbf{X}$ denotes a list of variables $X_1, X_2, \ldots, X_n$. We then denote by $\mathbf{K[X]}$ the ring $\mathbf{K}[X_1, X_2, \ldots, X_n]$. If F is a finite subset of $\mathbf{K[X]}$, we let F^{*2} be the set of squares of elements in F, and $\mathcal{M}(F)$ be the *multiplicative monoid generated by* $F \cup \{1\}$. $\mathcal{C}p(F)$ will be the *positive cone generated by* F (= the additive monoid generated by elements of type pPQ^2, where $0 \le p \in \mathbf{K}$, $P \in \mathcal{M}(F)$, and $Q \in \mathbf{K[X]}$). Finally, let $I(F)$ be the ideal generated by F.

Definition II.1. *Consider 4 finite subsets of* $\mathbf{K[X]}$: $F_>, F_\ge, F_=, F_{\ne}$, *containing polynomials for which we want respectively the sign conditions* > 0, ≥ 0, $= 0$, *and* $\ne 0$: *we say that* $\mathbf{F} := [F_>, F_\ge, F_=, F_{\ne}]$ *is strongly incompatible in* $\mathbf{K}$ *if we have in* $\mathbf{K[X]}$ *an equality of the following type:*

$$S + P + Z = 0 \quad \text{with} \quad S \in \mathcal{M}(F_> \cup F_{\ne}^{*2}), \ P \in \mathcal{C}p(F_\ge \cup F_>), \ Z \in I(F_=).$$

It is clear that a strong incompatibility is a very strong form of incompatibility. In particular, it implies that it is impossible to give the indicated signs to the polynomials considered, in any ordered extension of $\mathbf{K}$. If one

---

[4] However, they involve fewer iterations of the exponential function than Daykin's bounds.

considers the real closure $\mathbf{R}$ of $\mathbf{K}$, the previous impossibility is testable by Hörmander's algorithm, for example (see [BCR], chapter 1).

The different variants of the Nullstellensatz in the real case are a consequence of the following general theorem:

Theorem II.2. *Let $\mathbf{K}$ be an ordered field and $\mathbf{R}$ a real closed extension of $\mathbf{K}$. The three following conditions, concerning a generalized system of sign conditions on polynomials of $\mathbf{K}[\mathbf{X}]$, are equivalent:*
- *strong incompatibility in $\mathbf{K}$;*
- *impossibility in $\mathbf{R}$; and*
- *impossibility in all the ordered extensions of $\mathbf{K}$.*

This Nullstellensatz was first proved in 1974 [Ste]. Less general variants were given by Krivine [Kri], Dubois [Du], Prestel [Pre₁], Risler [Ris] and Efroymson [Efr]. All the proofs until [Lom₁] and the sketch in [Sco₁] 'used' the axiom of choice (recall (I.A.c.1)).

II.3 Parameterizing Hilbert's 17th problem.

Let $f_{n,d}(\mathbf{c}, \mathbf{X})$ be the general polynomial of degree d in n variables ($\mathbf{c}$ denotes the list of coefficients $c_1, \ldots, c_m$ and $\mathbf{X}$ the list of variables $X_1, \ldots, X_n$). It is a standard fact in real algebraic geometry that the set

$$\mathbf{F}_{n,d} = \{\mathbf{c} \in \mathbf{R}^m : \forall \mathbf{x} \in \mathbf{R}^n \quad f_{n,d}(\mathbf{c}, \mathbf{x}) \geq 0\}$$

is a closed $\mathbf{Q}$-semialgebraic set. So, applying the Finiteness Theorem, we have that $\mathbf{F}_{n,d}$ is a finite union of 'basic' closed $\mathbf{Q}$-semialgebraic sets:

$$\mathbf{F}_{n,d} = \bigcup_{i=1}^{k} \bigcap_{j=1}^{n_i} \{\mathbf{c} : R_{n,d,i,j}(\mathbf{c}) \geq 0\}.$$

Here the $R_{n,d,i,j}$ are polynomials in $\mathbf{Z}[\mathbf{c}]$.[5] The last equation allows us to describe the set $\mathbf{F}_{n,d}$ in the following way:

$$\mathbf{F}_{n,d} = \left\{\mathbf{c} : \max_{i=1,\ldots,k} \left\{ \min\{R_{n,d,i,j}(\mathbf{c}) : j = 1, \ldots, n_i\}\right\} \geq 0\right\}.$$

So, if for every i in $\{1, \ldots, k\}$ we define

$$H_{n,d,i}(\mathbf{c}) = \min_{j=1,\ldots,n_i} \{R_{n,d,i,j}(\mathbf{c})\}$$

---

[5] Some of the published proofs of the Finiteness Theorem (e.g., [Del₁]) explicitly mention the fact that the coefficients of the $R_{n,d,i,j}$ may be chosen to be rational numbers (even integers, after clearing denominators), while others (e.g., [BCR]) assert only that they can be chosen in $\mathbf{R}$; but even these other proofs actually do yield the integrality of the coefficients, if one merely pays attention to it. Likewise, in most of the proofs, the authors do not pay attention to the constructive character of their proofs; if one does, one sees that in fact most of the proofs are constructive; [Del₁] does pay attention, and [Sol] goes even further, by giving a careful complexity analysis.

and

$$H_{n,d}(c) = \max_{i=1,\ldots,k} \{H_{n,d,i}(c)\},$$

we have obtained the following description for the set $\mathbf{F}_{n,d}$:

$$\mathbf{F}_{n,d} = \{c : H_{n,d}(c) \geq 0\},$$

where $H_{n,d}(c)$ is a $\mathbf{Q}$-semipolynomial. Therefore we have shown the equivalence

$$c \in \mathbf{F}_{n,d} \iff H_{n,d}(c) \geq 0 \iff \forall x \in \mathbf{R}^n \quad f_{n,d}(c,x) \geq 0.$$

II.4 The proof of the parameterized theorem.

The last equivalence allows us to conclude

$$\forall c \in \mathbf{R}^m \; \forall x \in \mathbf{R}^n \; \{H_{n,d}(c) \geq 0 \implies f_{n,d}(c,x) \geq 0\},$$

or, what is the same, the incompatibility of the system of generalized sign conditions

$$H_{n,d}(c) \geq 0, \quad f_{n,d}(c, X) < 0. \tag{1}$$

If $H_{n,d}(c)$ were a polynomial and not a semipolynomial, then, applying the classical Positivstellensatz (theorem II.2) to the incompatibility (1), we would get an algebraic identity in c and X making this incompatibility evident. This would give, for the polynomial $f_{n,d}(c, X)$, a solution to Hilbert's 17th problem parameterized by polynomials in c and so, a rational and continuous solution.

As this way of attacking the problem is not feasible, we shall try to translate our incompatible system to another one using only polynomials in order to be able to apply the Positivstellensatz (Theorem II.2). To achieve this goal we introduce new variables $z_1, \ldots, z_k$, and for every $i \in \{1, \ldots, k\}$ we consider the following polynomial system of generalized sign conditions:

$$\mathbf{H}_i = \begin{cases} (z_i - R_{n,d,i,1}(c))(z_i - R_{n,d,i,2}(c)) \cdots (z_i - R_{n,d,i,n_i}(c)) = 0, \\ z_i - R_{n,d,i,1}(c) \leq 0, \\ z_i - R_{n,d,i,2}(c) \leq 0, \\ \quad\quad \vdots \\ z_i - R_{n,d,i,n_i}(c) \leq 0. \end{cases}$$

It is clear from the definitions of $H_{n,d,i}$ and $\mathbf{H}_i$ that, for fixed $c \in \mathbf{R}^m$, if the system $\mathbf{H}_i$ is verified then $z_i = H_{n,d,i}(c)$.

Next we consider a new variable z and the following polynomial systems of generalized sign conditions:

$$\mathbf{H} = \begin{cases} (z - z_1)(z - z_2) \cdots (z - z_k) = 0, \\ z - z_1 \geq 0, \\ z - z_2 \geq 0, \\ \quad\quad \vdots \\ z - z_k \geq 0. \end{cases} \quad\quad \mathbf{K} = \{\mathbf{H}_1, \ldots, \mathbf{H}_k, \mathbf{H}\}.$$

Clearly we have that, for fixed $c \in R^m$, if the system $\mathbb{K}$ is verified, then $z = H_{n,d}(c)$.

After introducing in this way the variables $z, z_1, \ldots, z_k$, what we have obtained is the following incompatible system of generalized sign conditions on polynomials in $\mathbb{K}[X, z, z_1, \ldots, z_k]$:

$$\mathbb{K}, \ z \geq 0, \ f_{n,d}(c, X) < 0.$$

Applying to this system the Positivstellensatz (theorem II.2) and replacing in the equality obtained every z_i by $H_{n,d,i}(c)$ and z by $H_{n,d}(c)$, we obtain an algebraic identity concerning polynomials in X whose coefficients are $\mathbb{Q}$-semipolynomials in c.

Next we study the different parts appearing in this algebraic identity:

- The strictly positive part in the initial identity (before the z_i's replacement) was $f_{n,d}(c, X)^{2r}$ and remains unchanged.
- The null part in the initial identity was a polynomial in the ideal of $\mathbb{K}[X, z, z_1, \ldots, z_k]$ generated by the polynomials with '$= 0$' in $\mathbb{K}$, i.e.,

$$(z_i - R_{n,d,i,1}(c))(z_i - R_{n,d,i,2}(c)) \cdots (z_i - R_{n,d,i,n_i}(c)) \quad 1 \leq i \leq k$$

$$(z - z_1)(z - z_2) \cdots (z - z_k).$$

After replacing z by $H_{n,d}(c)$ and every z_i by $H_{n,d,i}(c)$, this part becomes a function of c identically 0 (zero for all $c \in R$), and this is the reason why it will not appear in the final identity which concerns polynomials in X with coefficients $\mathbb{Q}$-semipolynomials in c.

- The nonnegative part in the initial identity was a polynomial in the positive cone generated by the polynomials:

$$-f_{n,d}(c, X), z, z - z_1, \ldots, z - z_k,$$

$$R_{n,d,i,1}(c) - z_i, \ldots, R_{n,d,i,n_i}(c) - z_i \quad i \in \{1, \ldots, k\}.$$

After the replacement only $-f_{n,d}(c, X)$ remains unchanged, and the other generators of the cone become $\mathbb{Q}$-semipolynomials in c which are clearly nonnegative for every c (by the definition of the functions $H_{n,d,i}(c)$ and $H_{n,d}(c)$ in terms of max and min) or the $\mathbb{Q}$-semipolynomial $H_{n,d}(c)$ that is nonnegative if $f_{n,d}(c, x)$ is nonnegative for all $x \in R^n$.

Summarizing, we have found an algebraic identity with the following structure:

$$f_{n,d}(c, X)g(c, X) = f_{n,d}(c, X)^{2r} + h(c, X), \tag{2}$$

where $g(c, X)$ and $h(c, X)$ are polynomials in X whose coefficients are $\mathbb{Q}$-semipolynomials in c. More precisely, $g(c, X)$ and $h(c, X)$ are sum of terms

$$p_j(\mathbf{c})q_j(\mathbf{c}, \mathbf{X})^2,$$

where the $q_j(\mathbf{c}, \mathbf{X})$ are polynomials in $\mathbf{X}$ with coefficients $\mathbb{Q}$-semipolynomials in $\mathbf{c}$ and the $p_j(\mathbf{c})$ are $\mathbb{Q}$-semipolynomials nonnegative under the hypothesis $H_{n,d}(\mathbf{c}) \geq 0$. More precisely we have that every $p_j(\mathbf{c})$ is a product whose factors have one of the following type:

- the $\mathbb{Q}$-semipolynomial $H_{n,d}(\mathbf{c})$,
- a $\mathbb{Q}$-semipolynomial $H_{n,d}(\mathbf{c}) - H_{n,d,i}(\mathbf{c})$,
- a $\mathbb{Q}$-semipolynomial $R_{n,d,i,j}(\mathbf{c}) - H_{n,d,i}(\mathbf{c})$, and
- a positive rational or the square of a $\mathbb{Q}$-semipolynomial in $\mathbf{c}$.

If we multiply by $f_{n,d}(\mathbf{c}, \mathbf{X})$ every member of equation (2), we get

$$f_{n,d}(\mathbf{c}, \mathbf{X}) = \frac{f_{n,d}(\mathbf{c}, \mathbf{X})^2 g(\mathbf{c}, \mathbf{X})}{f_{n,d}(\mathbf{c}, \mathbf{X})^{2r} + h(\mathbf{c}, \mathbf{X})},$$

and denoting by $k(\mathbf{c}, \mathbf{X})$ the denominator of this fraction, we obtain finally

$$f_{n,d}(\mathbf{c}, \mathbf{X}) = \frac{f_{n,d}(\mathbf{c}, \mathbf{X})^2 g(\mathbf{c}, \mathbf{X}) k(\mathbf{c}, \mathbf{X})}{k(\mathbf{c}, \mathbf{X})^2} = \frac{g_1(\mathbf{c}, \mathbf{X})}{k(\mathbf{c}, \mathbf{X})^2},$$

where $g_1(\mathbf{c}, \mathbf{X})$ is of the same type as $g(\mathbf{c}, \mathbf{X})$ and $h(\mathbf{c}, \mathbf{X})$. Moreover, $k(\mathbf{c}, \mathbf{X})$ vanishes only at the zeros of $f_{n,d}(\mathbf{c}, \mathbf{X})$ if $H_{n,d}(\mathbf{c}) \geq 0$, because then $h(\mathbf{c}, \mathbf{x})$ is positive for all $\mathbf{x} \in \mathbb{R}^n$.

Most of the following theorem has now been proved:

Theorem II.5. *The general polynomial $f_{n,d}$ of degree d in n variables can be written as a weighted sum of squares of rational functions*

$$f_{n,d}(\mathbf{c}, \mathbf{X}) = \sum_j p_j(\mathbf{c}) \left(\frac{q_j(\mathbf{c}, \mathbf{X})}{k(\mathbf{c}, \mathbf{X})} \right)^2, \tag{$\star$}$$

where

- *the $q_j(\mathbf{c}, \mathbf{X})$ and $k(\mathbf{c}, \mathbf{X})$ are polynomials in the variables $\mathbf{X}$ whose coefficients are $\mathbb{Q}$-semipolynomials in the coefficients $\mathbf{c}$. Moreover, if $\mathbf{c} \in \mathbb{F}_{n,d}$, then $k(\mathbf{c}, \mathbf{X})$ vanishes only on the zeros of $f_{n,d}(\mathbf{c}, \mathbf{X})$;*
- *each $p_j(\mathbf{c})$ is a product whose factors are $H_{n,d}(\mathbf{c})$, or one of the $\mathbb{Q}$-semipolynomials $H_{n,d}(\mathbf{c}) - H_{n,d,i}(\mathbf{c})$, or one of the $\mathbb{Q}$-semipolynomials $R_{n,d,i,j}(\mathbf{c}) - H_{n,d,i}(\mathbf{c})$, or a positive rational, or the square of a $\mathbb{Q}$-semipolynomial in $\mathbf{c}$. So, under the hypothesis $H_{n,d}(\mathbf{c}) \geq 0$, the non-negativity of $p_j(\mathbf{c})$ is 'clearly' evident; and*
- *the equation*

$$f_{n,d}(\mathbf{c}, \mathbf{X}) k(\mathbf{c}, \mathbf{X})^2 - \sum_j p_j(\mathbf{c}) q_j(\mathbf{c}, \mathbf{X})^2 = 0$$

is especially evident in the following sense: the first member of the equality, as polynomial in $\mathbf{X}$, *has as coefficients* $\mathbf{Q}$-*semipolynomials in* **c** *which are identically* 0 *(without assuming* $\mathbf{c} \in \mathbf{F}_{n,d}$).

Equation $(\star)$ *provides a continuous, rational-valued solution to Hilbert's* 17th *problem, because*

- *all the coefficients (the* $p_j(\mathbf{c})$ *and the* $\mathbf{X}$-*coefficients of the* $q_j(\mathbf{c}, \mathbf{X})$ *and* $k(\mathbf{c}, \mathbf{X})$) *appearing in equation* $(\star)$ *are continuous, rational-valued functions of* **c**; *more precisely, they are* $\mathbf{Q}$-*semipolynomials in* **c**; *and*
- *every summand in* $(\star)$

$$p_j(\mathbf{c})\left(\frac{q_j(\mathbf{c}, \mathbf{X})}{k(\mathbf{c}, \mathbf{X})}\right)^2$$

is a function which is rational in $\mathbf{X}$, *and which can be continuously and semialgebraically extended to the closed semialgebraic set* $\mathbf{F}_{n,d} \times \mathbf{R}^n$.

Proof:

The only statement not yet proved is the last: the semialgebraicity of the extension of $p_j(q_j/k)^2$ is obvious. To see its continuity, we use an argument of Kreisel [Kre₃] (with parameters **c**): note that $k = f^{2r} + h$ (with h nonnegative over $\mathbf{F}_{n,d} \times \mathbf{R}^n$), which can vanish at some point $(\mathbf{c}; \mathbf{x}) \in \mathbf{F}_{n,d} \times \mathbf{R}^n$ only if f vanishes there, forcing each $p_j(q_j/k)^2$ to tend to 0 near $(\mathbf{c}; \mathbf{x})$, by $(\star)$. The fact that this pointwise continuity is actually locally uniform follows from the corresponding property of f; we leave the ϵ's and δ's to the reader (see [GV-L] for details). ∎

III. Rational and continuous solution to other cases of the classical Real Positivstellensatz.

The solution for Hilbert's 17th problem can be seen as a particular case of the Real Positivstellensatz, and for this case we have just proved, in the previous section, the existence of a solution depending on the parameters of the problem in a semipolynomial way. So what we shall do in this section is to generalize this result to other cases.

Let $\mathbb{H}(\mathbf{c}, \mathbf{X})$ be a system of generalized sign conditions on polynomials in $K[\mathbf{c}, \mathbf{X}]$ where the X_i's are considered as variables and the c_j's as parameters. We denote by $\mathbf{S_H}$ the semialgebraic set defined by

$$\mathbf{S_H} = \{\mathbf{c} : \forall \mathbf{x} \in \mathbf{R}^n \quad \mathbb{H}(\mathbf{c}, \mathbf{x}) \text{ is incompatible}\}.$$

If $\mathbf{S_H}$ is locally closed (i.e., intersection of a closed and an open semialgebraic set), then, applying the Finiteness Theorem (see [BCR] or elsewhere) and the strategy followed in §II when dealing with the set $\mathbf{F}_{n,d}$, it is possible to construct two $\mathbf{K}$-semipolynomials $H_1(\mathbf{c})$ and $H_2(\mathbf{c})$ satisfying

$$\mathbf{c} \in \mathbf{S_H} \iff \left[H_1(\mathbf{c}) \geq 0, \ H_2(\mathbf{c}) > 0\right] \iff$$

$$\Longrightarrow \forall \mathbf{x} \in \mathbf{R}^n \quad \mathbb{H}(\mathbf{c}, \mathbf{x}) \text{ is incompatible.}$$

We have obtained the incompatible system of sign conditions

$$\left[H_1(\mathbf{c}) \geq 0, \ H_2(\mathbf{c}) > 0, \ \mathbb{H}(\mathbf{c}, \mathbf{X}) \right],$$

but with $H_1(\mathbf{c})$ and $H_2(\mathbf{c})$ **K**-semipolynomials. Now we proceed in the same way as in §II: we consider new variables z_1, $z_{1,i}$ $(i \in \{1, \ldots, k_1\})$, z_2 and $z_{2,i}$ $(i \in \{1, \ldots, k_2\})$, which are used to construct a system of polynomial sign conditions $\mathbb{K}(\mathbf{c})$ translating the definition of $H_1(\mathbf{c})$ and $H_2(\mathbf{c})$ as semipolynomials.

As the incompatible system of generalized sign conditions

$$\{\mathbb{K}(\mathbf{c}), \ z_1 \geq 0, \ z_2 > 0, \ \mathbb{H}(\mathbf{c}, \mathbf{X})\}$$

involves only polynomials, we can apply the Positivstellensatz (theorem II.2), obtaining an algebraic identity making this incompatibility evident. Finally, we replace in the algebraic identity obtained the variables z_1, $z_{1,i}$ $(i \in \{1, \ldots, k_1\})$, z_2 and $z_{2,i}$ $(i \in \{1, \ldots, k_2\})$ by the **K**-semipolynomials they are representing. So we have obtained an algebraic identity concerning polynomials in the variables **X** with coefficients **K**-semipolynomials in **c**.

The next theorem summarizes the results obtained in this section and provides a rational and continuous solution for some cases of the Real Positivstellensatz.

Theorem III.1. *Let* $\mathbb{H}(\mathbf{c}, \mathbf{X})$ *be a system of generalized sign conditions on polynomials in* $\mathbf{K}[\mathbf{c}, \mathbf{X}]$, *where the* X_i's *are considered as variables and the* c_j's *as parameters. If* $\mathbf{S_H}$ *is the semialgebraic set defined by*

$$\mathbf{c} \in \mathbf{S_H} \iff \forall \mathbf{x} \in \mathbf{R}^n \quad \mathbb{H}(\mathbf{c}, \mathbf{x}) \text{ is incompatible,}$$

and if $\mathbf{S_H}$ *is locally closed, then (Finiteness Theorem) there exist* **K**-*semipolynomials* $H_1(\mathbf{c})$ *and* $H_2(\mathbf{c})$ *such that*

$$\mathbf{c} \in \mathbf{S_H} \iff \left[H_1(\mathbf{c}) \geq 0, \ H_2(\mathbf{c}) > 0 \right].$$

If $\mathbf{c} \in \mathbf{S_H}$, *then the incompatibility of* $\mathbb{H}(\mathbf{X}) := \mathbb{H}(\mathbf{c}, \mathbf{X})$ *inside* $\mathbf{R}^n$ *is made obvious by a strong incompatibility of fixed type (independent of* **c***) and with coefficients given by* **K**-*semipolynomials in* **c**. *Moreover,*

- *the algebraic identity obtained, seen as a polynomial in* **X**, *has an especially simple structure. More precisely, every* **X**-*coefficient of this identity is a* **K**-*semipolynomial in* **c** *identically 0 (without assuming* $H_1(\mathbf{c}) \geq 0$ *and* $H_2(\mathbf{c}) > 0$), *and*

- *every coefficient* $p(\mathbf{c})$ *in the algebraic identity which must be nonnegative (resp. positive) is given by a* **K**-*semipolynomial showing such*

*character in an especially clear way under the hypothesis $H_1(\mathbf{c}) \geq 0$
and $H_2(\mathbf{c}) > 0$.*

In the same way that our rational and continuous solution for Hilbert's
17th problem (§II) improves the first author's result [Del₃], (III.1) improves
Scowcroft's results [Sco₂] in four respects:

a-. for us, the semialgebraic set $\mathbf{S_H}$ need not be closed (Scowcroft knew
   this, but chose to use hypotheses involving the logical form of the
   implications implicit in $\mathbb{H}$, rather than topological hypotheses on $\mathbf{S_H}$);

b-. the coefficients of our solution are continuous, rational-valued functions
   (more precisely, $\mathbf{K}$-semipolynomials) in the parameters $\mathbf{c}$ and not only
   continuous semialgebraic as in [Sco₂];

c-. the algebraic identity obtained, seen as a polynomial in $\mathbf{X}$, has an
   especially simple structure: its coefficents are $\mathbf{K}$-semipolynomials in $\mathbf{c}$
   identically 0 (without assuming $H_1(\mathbf{c}) \geq 0$ and $H_2(\mathbf{c}) > 0$); and

d-. the nonnegativity or positivity of those coefficients in the solution
   which must satisfy such conditions, is clearly evident under the hypothesis $H_1(\mathbf{c}) \geq 0$ and $H_2(\mathbf{c}) > 0$.

Finally, we note a strong converse of (III.1): the hypothesis that $\mathbf{S_H}$
be locally closed is also *necessary* for the existence of a semipolynomially
(or even continuously) varying Positivstellensatz; the proof is in [Del₈].

IV. Conclusion: the constructive content of the results.

In constructive mathematics (see [BB] and [MRR]), the theorems presented
in §§II and III are valid when the parameters $\mathbf{c}$ take values in an ordered
discrete field [LR], because in this setting we have a constructive proof of
the Positivstellensatz ([Lom₁]).

An interpretation of the results admissible for everybody is the following one: all our proofs are effective, in particular without using the axiom
of choice and, more precisely, providing uniformly primitive recursive algorithms if the structure of the field of parameters is given by an oracle giving
the sign of every polynomial with integer coefficients on the parameters of
the problem.

The only thing remaining to be mentioned is the constructive content of
the results when dealing with the field $\mathbb{R}$ of the real numbers in constructive
analysis [BB], i.e., the real numbers defined as—equivalence classes of—
Cauchy sequences of rational numbers. From an algorithmic point of view
this means that the real parameters $\mathbf{c}$ are given by oracles providing suitable
rational approximations (depending on the request made to the oracle) of
the real numbers involved, and that we are looking for a uniformly primitive
recursive algorithm. More details on this question can be found in [GV-L].

The answer to Hilbert's 17th problem provided by theorem II.5 uses
polynomials and semipolynomials with coefficients in $\mathbb{Q}$ that can be computed explicitly. The nonnegativity of the weights is clear from a constructive point of view when dealing with real numbers "à la Cauchy" under

the hypothesis $H_{n,d}(c) \geq 0$. This implies that, if the parameters c satisfy the condition $H_{n,d}(c) \geq 0$, then the polynomial $f_{n,d}(c, X)$ is everywhere nonnegative. So, for the polynomial $f_{n,d}(c, X)$, Hilbert's 17th problem is solved in a continuous, rational-valued way with respect to its coefficients. Moreover, since we can constructively prove (see [Lom$_2$] or [GV-L]) the converse

$$\forall x \in \mathbb{R}^n \quad f_{n,d}(c, x) \geq 0 \implies H_{n,d}(c) \geq 0,$$

we can conclude that also this continuous, rational-valued solution for Hilbert's 17th problem is complete and constructive for the field $\mathbb{R}$.

References.

[Art] Artin E.: *Uber die Zerlegung definiter Funktionen in Quadrate*. Abh. Math. Sem. Hamburg, **5** (1927), 100–115.

[BB] Bishop E., Bridges D.: *Constructive Analysis*. Springer-Verlag (1985).

[BCR] Bochnak J., Coste M. and Roy M.-F.: *Géométrie Algébrique Réelle*. Ergebnisse vol. **12**, Springer-Verlag (1987).

[BP] Bradley M., Prestel A.: *Representation of a real polynomial $f(X)$ as a sum of 2m-th powers of rational functions*. Ordered Algebraic Structures, J. Martinez, ed., Kluwer (1989), 197–207.

[Day] Daykin D.: *Hilbert's 17th problem*. Ph.D. Thesis, unpublished (1961).

[Del$_1$] Delzell C.N.: *A finiteness theorem for open semialgebraic sets, with applications to Hilbert's 17th problem*. Ordered Fields and Real Algebraic Geometry, Contemp. Math. **8**, AMS, D. Dubois and T. Recio, (eds.), Providence 1982, 79–97.

[Del$_2$] Delzell C.N.: *Case distinctions are necessary for representing polynomials as sums of squares*. Proc. Herbrand Symp., Logic Coll. 1981, J. Stern (ed.), Amsterdam-Oxford-New York 1982, 87–103.

[Del$_3$] Delzell C.N.: *A continuous, constructive solution to Hilbert's 17th problem*. Inventiones Mathematicae **76** (1984), 365–384.

[Del$_4$] Delzell C.N.: *On the Pierce-Birkhoff conjecture over ordered fields*. Rocky Mountain Journal of Mathematics **19**(3) (Summer 1989), 651–668.

[Del$_5$] Delzell C.N.: *A sup-inf-polynomially varying solution to Hilbert's 17th problem*. AMS Abstracts **10**(3), (Issue 63, April 1989), 208–209, #849-14-160.

[Del$_6$] Delzell C.N.: *On analytically varying solutions to Hilbert's 17th problem*. Submitted to Proc. Special Year in Real Algebraic Geometry and Quadratic Forms at UC Berkeley, 1990–1991, (W. Jacob, T.-Y. Lam, R. Robson, eds.), Contemporary Mathematics.

[Del$_7$] Delzell C.N.: *On analytically varying solutions to Hilbert's 17th problem*. AMS Abstracts **12**(1), (Issue 73, January 1991), page 47, #863-14-743.

[Del$_8$] Delzell C.N.: *Continuous, piecewise-polynomial functions which solve Hilbert's 17th problem.* Submitted for publication (1992).

[Du] Dubois D.W.: *A nullstellensatz for ordered fields.* Arkiv for Mat. **8** (1969), 111–114.

[Efr] Efroymson G.: *Local reality on algebraic varieties.* Journal of Algebra **29** (1974), 113–142.

[GV-L] González-Vega L., Lombardi H.: *A Real Nullstellensatz and Positivstellensatz for the Semipolynomials over an Ordered Field.* Submitted to the Journal of Pure and Applied Algebra (1992).

[Hil] Hilbert D.: *Mathematische Probleme.* Göttinger Nachrichten (1900), 253–297, and Archiv der Math. u. Physik (3rd ser.) **1** (1901), 44–53, 213–237.

[Kre$_1$] Kreisel G.: *Some uses of metamathematics.* British J. Phil. Sci. **7**(26) (August 1956), 161–173.

[Kre$_2$] Kreisel G.: *Sums of squares.* Summaries of Talks Presented at the Summer Institute in Symbolic Logic in 1957 at Cornell Univ., Institute Defense Analyses, Princeton (1960), 313–320.

[Kre$_3$] Kreisel G.: *Review of Goodstein,* MR **24A**, 336–7, #A1821 (1962).

[Kre$_4$] Kreisel G.: *Review of Ershov,* Zbl. **374**, 02027 (1978).

[Kri] Krivine J.L.: *Anneaux préordonnés.* Journal d'Analyse Mathématique **12** (1964), 307–326.

[Lom$_1$] Lombardi H.: *Effective real Nullstellensatz and variants.* Effective Methods in Algebraic Geometry. Editors T. Mora and C. Traverso. Progress in Mathematics 94 (1991), 263–288, Birkhauser. Detailed French version in *Théorème effectif des zéros réel et variantes (avec une majoration explicite des degrés),* Memoire d'habilitation (1990).

[Lom$_2$] Lombardi H.: *Une étude historique sur les problèmes d'effectivité en algèbre réelle.* Memoire d'habilitation (1990).

[Lom$_3$] Lombardi H.: *Une borne sur les degrés pour le théorème des zéros réel effectif.* Proceedings of the conference on Real Algebraic Geometry held in La Turballe (1991). To appear as Lecture Notes in Mathematics.

[LR] Lombardi H., Roy M.-F.: *Théorie constructive élémentaire des corps ordonnés.* Publications Mathematiques de Besançon, Théorie de Nombres, 1990–1991. English version in: *Constructive elementary theory of ordered fields,* Effective Methods in Algebraic Geometry. Editors T. Mora and C. Traverso. Progress in Mathematics **94** (1991), 249–262, Birkhauser.

[MRR] Mines R., Richman F., Ruitenburg W.: *A Course in Constructive Algebra.* Universitext, Springer-Verlag (1988).

[Pre$_1$] Prestel A.: *Lectures on Formally Real Fields.* IMPA Lecture Notes **22**, Rio de Janeiro 1975; reprinted in Lecture Notes in Mathematics **1093**

(1983), Springer-Verlag.

[Pre₂] Prestel A.: *Continuous representations of real polynomials as sums of 2^m-th powers.* Math. Forschungsinst. Oberwolfach Tagungsbericht (Reelle algebraische Geometrie, June 10–16, 1990) **25** (1990), 18.

[Ris] Risler J.-J.: *Une caractérisation des idéaux des variétés algébriques réelles.* C.R.A.S. Paris, Série A, **271** (1970), 1171–1173.

[Rob₁] Robinson A.: *On ordered fields and definite forms.* Math. Ann. **130** (1955), 257–271.

[Rob₂] Robinson A.: *Further remarks on ordered fields and definite forms.* Math. Ann. **130** (1956), 405–409.

[Sch] Schwartz N.: *Piecewise-polynomial functions.* Submitted.

[Sco₁] Scowcroft P.: *A transfer theorem in constructive real algebra.* Ann. Pure and Appl. Logic **40**(1), 29–87 (1988).

[Sco₂] Scowcroft P.: *Some continuous Positivstellensatze.* Journal of Algebra **124** (1989), 521–532.

[Sol] Solernó P.: *Effective Lojasiewicz inequalities in semialgebraic geometry.* Applicable Algebra in Engineering, Communication and Computing **2**(1) (1991), 1–14.

[Ste] Stengle G.: *A Nullstellensatz and a Positivstellensatz in semialgebraic geometry.* Math. Ann. **207** (1974), 87–97.

Charles N. Delzell   Laureano González-Vega
Dept. Mathematics   Dept. Matemáticas
Louisiana State University   Universidad de Cantabria
Baton Rouge, LA 70803, U.S.A.   Santander 39071, Spain
(delzell@lsuvax.sncc.lsu.edu)   (g_vega@ccucvx.unican.es)

Henri Lombardi
Lab. de Mathématiques
Université de Franche-Comté
Besançon 25030, France
(tdnbesac@frgren81.bitnet)

The analytic spread of the ideal of a monomial curve in projective 3-space

P. Gimenez   M. Morales   A. Simis [*]

Introduction

Let k be an algebraically closed field and let $C \subset \mathbb{A}^n$ stand for a quasi-homogeneous surface admitting a rational parametrization given by monomials. Consider the blowing-up $\widetilde{\mathbb{A}^n}$ of $\mathbb{A}^n$ with center C and look at the fibre of the structural morphism

$$\widetilde{\mathbb{A}^n} \to \mathbb{A}^n$$

over the origin 0 (the *special fibre* of the blowing-up).

A question of some interest regards the possible values of $\dim \widetilde{\mathbb{A}^n}$.

The question is certainly trivial for $n \leq 3$. This paper is concerned with the case where $n = 4$. We will give a complete answer for the homogeneous case.

More precisely, switching to an algebraic version of the problem, one is given a polynomial ring $R = k[x, y, z, w]$ (k an arbitrary field) and monomials S^a, $S^{a-c}T^c$, $S^{a-b}T^b$, T^a in new variables S, T, where a, b, c are positive integers such that $a > b > c$ and $(a, b, c) = 1$. Let

$$I = \ker R \to k[S, T]$$
$$x \mapsto S^a, y \mapsto S^{a-c}T^c, z \mapsto S^{a-b}T^b, w \mapsto T^a.$$

Clearly, I is a homogeneous ideal.

Let $R[It] = \oplus I^n t^n$ be the Rees algebra of I and let $\mathfrak{m} = (x, y, z, w) \subset R$. Then one is looking for the value of the so-called *analytic spread*

$$\ell(I) = \dim R[It]/\mathfrak{m}R[It].$$

It is well known that $2 \leq \ell(I) \leq 4$. We prove:

Theorem *If I is not a perfect ideal then $\ell(I) = 3$.*

The proof of this result uses effective methods from Gröbner bases, as will made more explicit at the end of this introduction.

[*]Partially supported by CNPq, Brazil

The result itself has implications for the structure of the Rees algebra itself. As a sample, using a criterion established by Huckaba and Huneke [4], [5], one obtains:

Theorem *If the two affine curves at $x = 1$ and $w = 1$ are ideal-theoretically complete intersections then:*

1. *$R[It]$ is a Gorenstein normal domain*

2. *$R[It]$ is presented by equations of degree at most two.*

Actually, the generators of the presentation ideal of $R[It]$ can be obtained in a structured fashion, as Plücker equations of a certain matrix. This line of thought will be more fully pursued in a forthcoming paper in which the quasi-homogeneous case will be treated.

As it will then turn out, a fundamental tool for understanding the above structural result lies on the following:

Theorem *The ideal I is minimally generated by a subset of the set of 2×2 minors of a $2 \times \tau$ matrix with monomials entries, for some $\tau \geq 3$.*

After [2], one has learned that I is generated by binomial equations of a special format. Obviously, any of these can be individually expressed as a 2×2 minor. To our knowledge, however, the precise *quasi-determinantal* structure of I has never been fully registered in the earlier literature, except for a par-enthetical approach taken by Huneke and Bresinsky for a diverse purpose in [1].

Special cases of this *quasi-determinantal* structure have long been known to the last two of the present authors – the case where the surface lies on the quadric $xw - yz$ has appeared in [10]. The general argument given in [9] stemmed from intense interchange between the same authors based on preliminary calculations by the second author.

Due both to its intrinsically algorithmic nature and some of its features, we give here the precise subset of minors that minimally generate the ideal of all minors of the matrix. A good reason to do this stems from the fact that, as pointed out earlier, in order to obtain the above estimate for the analytic spread, one needs a detailed consideration of the format of the Plücker relations coming from this quasi-determinantal structure of I.

In the present work, we will mainly consider such equations after cancelling superfluous factors and reducing modulo the ideal (x, y, z, w). Let $\mathcal{F}$ denote the list of the polynomials in $k[\underline{T}]$ obtained in this way. One of our main effective results reads as follows:

Theorem *Let $\mathcal{A} \subset k[\underline{T}]$ denote the ideal generated by $\mathcal{F}$.*

1. *$\mathcal{F}$ is a Gröbner basis of $\mathcal{A}$ for the diagonal order of the $\underline{T}$.*

2. dim $k[\underline{T}]/\mathcal{A} = 3$

An easy corollary of this effective result is the aforementioned value of the analytic spread of I.

It seemed suitable to this conference to produce a non-trivial example where most of the effective features of our results would be present. The arithmetic characters of this example, theoretically given in the general approach of next section, were computed with the aid of a *PASCAL* program written by the second author using the construction of [9].

1   The quasi-determinantal structure

In this section, we discuss the explicit form of the minimal generators of the ideal I, heavily relying on the construction of [9]. In [7], a similar but not complete approach has been undertaken.[1]

The defining matrix

We introduce the relevant notation. As above, let a, b, c be integers such that $a > b > c > 0$ and $(a, b, c) = 1$.

Set $d = (a, b)$, $\alpha = a/d$ and $\beta = b/d$. Let s_0 denote the unique integer satisfying $s_0\beta \equiv c \pmod{\alpha}$ and $0 < s_0 < \alpha$. Set $s_{-1} = \alpha$ and apply the euclidian algorithm (with negative remainders) in order to determine $\gcd(s_{-1}, s_0)$:

$$
\begin{aligned}
s_{-1} &= q_1 s_0 - s_1 \\
s_0 &= q_2 s_1 - s_2 \\
&\ \ \vdots \\
s_{m-1} &= q_{m+1} s_m \\
s_{m+1} &-\ \ 0
\end{aligned}
$$

where $q_i \geq 2$ and $s_i \geq 0$, $-1 \leq i \leq m + 1$.

We next define three numerical sequences p_i, r_i, r'_i $(-1 \leq i \leq m + 1)$ as follows:

$$
\begin{aligned}
p_{-1} = 0 \ ; \ p_0 = d \ ; \ p_{i+1} = q_{i+1}p_i - p_{i-1} \ , \ 0 \leq i \leq m \\
r_i = (s_i b - p_i c)/a \ , \ -1 \leq i \leq m + 1 \\
r'_i = s_i - p_i - r_i \ , \ -1 \leq i \leq m + 1.
\end{aligned}
$$

Lemma 1.1 *In the above notation, the following holds:*

1. The numbers s_i, p_i, r_i, r'_i, $-1 \leq i \leq m + 1$ are integers

---

[1] We thank the referee for pointing out this work of which we had no knowledge before this paper was finished.

2. *The sequence of $\{p_i\}$ is strictly increasing, while the sequences $\{s_i\}$, $\{r_i\}$, $\{r_i'\}$ are strictly decreasing*

3. *There is a unique μ in the integer interval $-1 \leq \mu \leq m$ such that $r_{\mu+1} \leq 0 < r_\mu$*

4. *There is a unique ν in the integer interval $-1 \leq \nu \leq m$ such that $r_{\nu+1}' < 0 \leq r_\nu'$*

Example 1.2 *Consider the curve defined by the integers*

$$a = 885, b = 884, c = 559.$$

With the aid of the aforementioned *PASCAL* program, one obtains the following values:

i	s_i	p_i	r_i	r_i'	q_i
-1	885	0	884	1	
0	326	1	325	0	
1	93	3	91	-1	3
2	46	11	39	-4	4
3	45	30	26	-11	3
4	44	49	13	-18	2
5	43	68	0	-25	2

from which one sees that $\mu = 4$ and $\nu = 0$.

It is possible to see that the case where $\mu = \nu$ corresponds exactly to the case where the ideal of the corresponding monomial curve in $\mathbb{P}^3$ is perfect. Since we are mainly interested in the non-perfect case (for structural results concerning the perfect case, cf. [11]), we will henceforth assume that $\mu > \nu$, which can always be attained by exchanging the roles of the variables x and w, respectively y and z.

We need yet to consider the set of integers

$$\{i_1, \ldots, i_t\} = \{i|\ \nu + 2 \leq i \leq \mu + 1 \text{ and } q_i \neq 2\},$$

where, say, $i_1 > \ldots > i_t$. Set $i_0 = \mu + 2$ and $i_{t+1} = \nu + 1$.

We also consider the complementary integers $j_u := \mu + 2 - i_u$, for $0 \leq u \leq t + 1$. Clearly then

$$j_0 = 0 < 1 \leq j_1 < j_2 < \ldots < j_t \leq \mu - \nu < \mu - \nu + 1 = j_{t+1}.$$

We are now ready to introduce the fundamental matrix, namely

$$M := A\tilde{P}_1 \ldots \tilde{P}_{j_1} \tilde{I}_{j_1}^1 \ldots \tilde{I}_{j_1}^{q_{i_1}-2} \tilde{P}_{j_1+1} \ldots \tilde{P}_{j_2} \tilde{I}_{j_2}^1 \ldots \quad \ldots \tilde{I}_{j_t}^{q_{i_t}-2} \tilde{P}_{j_t+1} \ldots \tilde{P}_{j_{t+1}} \tilde{I}_{j_{t+1}}^1$$

obtained by concatenation from the following one-column matrices:

$$A = \begin{pmatrix} x^{-r'_{\nu+1}} \\ y^{p_{\nu+1}-p_\nu} \end{pmatrix}, \quad \tilde{P}_k = \begin{pmatrix} y^{p_m-(p_{\nu+1}-p_\nu)}w^{\sigma r_m} \\ x^{r'_{\nu+1}-r'_m}z^{s_m}w^{(\sigma-1)r_m} \end{pmatrix},$$

where, for a given $1 \le k \le \mu - \nu + 1$, we have set $m = \mu + 2 - k$ and where

$$\sigma = \begin{cases} 0 & \text{if } k=1 \\ 1 & \text{otherwise} \end{cases},$$

and

$$\tilde{I}^\alpha_{j_*} = \begin{pmatrix} y^{(q_m-\alpha)p_{m-1}-p_{m-2}-(p_{\nu+1}-p_\nu)}z^{s_{m-2}-(q_m-\alpha)s_{m-1}} \\ x^{r'_{\nu+1}+r'_{m-2}-(q_m-\alpha)r'_{m-1}}w^{r_{m-2}-(q_m-\alpha)r_{m-1}} \end{pmatrix},$$

where α is an integer in the interval $1 \le \alpha \le q_{i_*} - 2$, for $1 \le u \le t$ (with $\alpha = 1$ if $u = t+1$) and where we have set $m = i_u$.

So far for the definition of the matrix.

A closer inspection at the nature of M will allow for a more convenient notation, as follows. After the first column, the matrix decomposes into $2t + 2$ blocks of $\tilde{P}$'s (*piles*) and $\tilde{I}$'s (*interiors*). Denoting $\varepsilon_i + 1$ the length of the ith block, for $0 \le i \le 2t + 1$, one sees that

$$\begin{aligned} \varepsilon_0 + 1 &= j_1 = \mu + 2 - i_1 \\ \varepsilon_{2u} + 1 &= j_{u+1} - j_u = i_u - i_{u+1}, \ 1 \le u \le t - 1 \\ \varepsilon_{2u-1} + 1 &= q_{i_*} - 2, \ 1 \le u \le t \\ \varepsilon_{2t} + 1 &= \mu - \nu - j_t + 1 = i_t - (\nu + 1) \\ \varepsilon_{2t+1} + 1 &= 1 \end{aligned}$$

We will henceforth denote

$$P^\alpha_{2u} := \tilde{P}_{j_*+\alpha+1}, \quad \text{for } 0 \le u \le t \text{ and } 0 \le \alpha \le \varepsilon_{2u}$$

and, similarly,

$$I^\alpha_{2u+1} := \tilde{I}^{\alpha+1}_{j_*+1}, \quad \text{for } 0 \le u \le t \text{ and } 0 \le \alpha \le \varepsilon_{2u+1}.$$

In this modified notation, the matrix becomes

$$M = AP^0_0 P^1_0 \dots P^{\varepsilon_0}_0 I^0_1 I^1_1 \dots I^{\varepsilon_1}_1 P^0_2 \dots \dots I^{\varepsilon_{2t-1}}_{2t-1} P^0_{2t} \dots P^{\varepsilon_{2t}}_{2t} I^0_{2t+1}.$$

Remark 1.3 *The following equality follows suit easily for every integer u such that $0 \le u \le t + 1$:*

$$\varepsilon_0 + \varepsilon_2 + \dots + \varepsilon_{2u-2} + u = j_u.$$

In particular, $\varepsilon_0 + \varepsilon_2 + \dots + \varepsilon_{2t} + t = \mu - \nu$.

Definition 1.4 *We call the matrix M the defining matrix of the monomial curve defined by the integers a, b, c and the above decomposition its canonical form.*

Example 1.5 *The defining matrix corresponding to the data of Example 1.2 is*

$$
\begin{pmatrix}
x & y^{66} & y^{47}w^{13} & y^{28}w^{26} & y^{17}z & y^9 w^{39} & y^6 z^{47} & y^3 z^{140} & yw^{91} & z^{233} \\
y^2 & x^{24}z^{43} & x^{17}z^{44} & x^{10}z^{45} & x^6 w^{13} & x^3 z^{46} & x^2 w^{52} & xw^{143} & z^{93} & w^{234}
\end{pmatrix}
$$

that is $M = A P_0^0 P_0^1 P_0^2 I_1^0 P_2^0 I_3^0 I_3^1 P_4^0 I_5^0$.

The 2×2 minors that generate

We now make explicit the minors of the defining matrix M that constitute a minimal set of generators of the ideal I.

The minor with columns C_i, C_j will be denoted $\Delta_{C_i C_j}$. Pairs $(u, \alpha) \in \mathbb{N}^2$ will be ordered lexicographically. For convenience, we will set up:

$$
\sigma = \begin{cases} 0 & \text{if } u = \alpha = 0 \\ 1 & \text{otherwise} \end{cases}
$$

and while dealing with the pile P_{2u}^α (resp. with the interior I_{2u+1}^α), we will set $k := j_u + \alpha + 1$ and $m := \mu + 2 - k$ (resp. $k := j_{u+1}$ and $m := \mu + 2 - k = i_{u+1}$). If the pile is denoted P_{2v}^β (resp. the interior I_{2v+1}^β), the corresponding integers will be denoted l and n instead of k and m.

Lemma 1.6 *With the above notation, the following holds:*

(i)
$$
\Delta_{P_{2u}^\alpha P_{2v}^\beta} = -x^{r'_{\nu+1} - r'_n} y^{p_n - (p_{\nu+1} - p_\nu)} z^{s_m} w^{\sigma r_m} \bar{\Delta}_{P_{2u}^\alpha P_{2v}^\beta}
$$

whenever $(u, \alpha) < (v, \beta)$ and where

$$
\bar{\Delta}_{P_{2u}^\alpha P_{2v}^\beta} = \begin{cases} \Delta_{A I_{2u+1}^0} & \text{if } u = v, \beta = \alpha + 1 \\ & \text{or if } v = u + 1, \alpha = \varepsilon_{2u}, \beta = 0 \\ 0 \pmod{\mathfrak{m}I} & \text{otherwise.} \end{cases}
$$

(ii)
$$
\Delta_{P_{2u}^\alpha I_{2v+1}^\beta} = -x^{r'_{\nu+1} - r'_n + \bar{\beta}r'_{n-1}} y^{p_n - \bar{\beta}p_{n-1} - (p_{\nu+1} - p_\nu)} w^{(\sigma - 1)r_m} \bar{\Delta}_{P_{2u}^\alpha I_{2v+1}^\beta}
$$

for $0 \le u \le v \le t$, with $\bar{\beta} = \beta + 1$ and where

$$
\bar{\Delta}_{P_{2u}^\alpha I_{2v+1}^\beta} = \begin{cases} \Delta_{A P_{2u}^{\alpha+1}} & \text{if } \beta = 0, v = u, \alpha \ne \varepsilon_{2u} \\ \Delta_{A P_{2u+2}^0} & \text{if } \beta = 0, v = u < t, \alpha = \varepsilon_{2u} \\ \Delta_{I_{2t+1}^0 P_{2t}^{\varepsilon 2t}} & \text{if } \beta = 0, v = u = t, \alpha = \varepsilon_{2t} \\ 0 \pmod{\mathfrak{m}I} & \text{otherwise} \end{cases}
$$

(iii)
$$
\Delta_{I_{2u+1}^\alpha P_{2v}^\beta} = -x^{r'_{\nu+1} - r'_n} y^{p_n - (p_{\nu+1} - p_\nu)} \bar{\Delta}_{I_{2u+1}^\alpha P_{2v}^\beta}
$$

for $0 \leq u < v \leq t$, *where*

$$\bar{\Delta}_{P^\alpha_{2u+1} I^\beta_{2v}} = \begin{cases} \Delta_{AI^{\alpha+1}_{2u+1}} & \text{if } \beta = 0, v = u+1, \alpha \neq \varepsilon_{2u+1} \\ \Delta_{AI^0_{2u+3}} & \text{if } \beta = 0, v = u+1, \alpha = \varepsilon_{2u+1} \\ 0 \pmod{\mathfrak{m}I} & \text{otherwise} \end{cases}$$

(iv) $\quad \Delta_{I^\alpha_{2u+1} I^\beta_{2v+1}} =$

$$- x^{r'_{v+1} - r'_n + \bar{\beta} r'_{n-1}} y^{p_n - \bar{\beta} p_{n-1} - (p_{v+1} - p_v)} z^{\bar{\alpha} s_{m-1} - s_m} w^{\bar{\alpha} r_{m-1} - r_m} \bar{\Delta}_{I^\alpha_{2u+1} I^\beta_{2v+1}}$$

whenever $(u, \alpha) < (v, \beta)$ *with* $\bar{\alpha} = \alpha + 1$, $\bar{\beta} = \beta + 1$ *and where*

$$\bar{\Delta}_{I^\alpha_{2u+1} I^\beta_{2v+1}} = \begin{cases} \Delta_{AP^0_{2u+2}} & \text{if } u = v, \beta = \alpha + 1 \\ & \text{or if } v = u+1, \alpha = \varepsilon_{2u+1}, \beta = 0 \\ 0 \pmod{\mathfrak{m}I} & \text{otherwise.} \end{cases}$$

One further bit of notation, namely set:

$$\begin{aligned} F_{2u,\alpha} &:= \Delta_{AP^\alpha_{2u}}, \quad 0 \leq u \leq t, \ 0 \leq \alpha \leq \varepsilon_{2u} \\ F_{2u+1,\alpha} &:= \Delta_{AI^\alpha_{2u+1}}, \quad 0 \leq u \leq t, \ 0 \leq \alpha \leq \varepsilon_{2u+1} \\ F_{2t+2,0} &:= \Delta_{I^0_{2t+1} P^{\varepsilon_{2t}}_{2t}}. \end{aligned}$$

We can now state:

Proposition 1.7 *The ideal I of the monomial curve in $\mathbb{P}^3$ with defining matrix M is generated by the 2×2 minors of M of the form*

$$F_{ij}, \quad 0 \leq i \leq 2t+2, \ 0 \leq j \leq \varepsilon_i$$

with the convention that $\varepsilon_{2t+2} = 0$.

Proof. The proof is a long but straightforward calculation, checking all cases - refer to [9] for details.   □

Example 1.8 *Going back to Example 1.2, the generators of I in that case are*

$$\begin{aligned} F_{00} &= x^{25} z^{43} - y^{68} & F_{01} &= x^{18} z^{44} - y^{49} w^{13} & F_{02} &= x^{11} z^{45} - y^{30} w^{26} \\ F_{10} &= x^7 w^{13} - y^{19} z & F_{20} &= x^4 z^{46} - y^{11} w^{39} & F_{30} &= x^3 w^{52} - y^8 z^{47} \\ F_{31} &= x^2 w^{143} - y^5 z^{140} & F_{40} &= x z^{93} - y^3 w^{91} & F_{50} &= x w^{234} - y^2 z^{233} \\ & & F_{60} &= z^{326} - y w^{325} \end{aligned}$$

Actually, these generators are minimal and the remaining minors are contained in I. Thus, one has, for example:

$$\Delta_{P_{0,0}P_{0,1}} = -x^{17}y^{47}z^{43}F_{10}$$

$$\Delta_{P_{0,0}I_{1,0}} = -x^6 y^{17} F_{01}$$

$$\Delta_{P_{0,1}I_{1,0}} = -x^6 y^{17} F_{02}$$

$$\Delta_{P_{0,0}P_{4,0}} = -yz^{43}(x^3 w^{52}(x^{14}w^{26} + x^7 y^{19} zw^{13} + y^{38} z^2)F_{10} + y^{57} z^3 F_{30})$$

2   The Gröbner basis

The essential fiber ideal

Let $\underline{T} = \{\, T_{ij} \mid 0 \le i \le 2t + 2,\ 0 \le j \le \varepsilon_i \,\}$ be independent variables over R and consider the surjective R-homomorphism

$$\Phi : R[\underline{T}] \to R[It], \quad T_{ij} \mapsto F_{ij}t,$$

where $I \subset R$ is the ideal of the monomial curve and F_{ij} its minimal generators as found in Proposition 1.7. Set $\tilde{J} = ker\,\Phi$.

For any 2×4 matrix, one has the following relation between its 2×2 minors:
$$\Delta_{1,2}\Delta_{3,4} - \Delta_{1,3}\Delta_{2,4} + \Delta_{1,4}\Delta_{2,3} = 0$$
where $\Delta_{i,j}$ is the minor with columns i and j.
The so-called Plücker equations coming from the 2×4 submatrices of the defining matrix of I obviously yield elements of the presentation ideal $\tilde{J}$ because of Lemma 1.6. We will be interested in a distinguished subset of such submatrices, namely, those corresponding to columns $AC_iC_jC_{k_j}$, where

- C_i is any column following P_0^0 and preceding the column before $P_{2t}^{\varepsilon_{2t}}$

- C_j is any column following C_i and preceding $P_{2t}^{\varepsilon_{2t}}$

- The last column has the following values:

$$C_{k_j} = \begin{cases} I_{2u+1}^0 & \text{if } C_j = P_{2u}^\alpha \\ P_{2u+2}^0 & \text{if } C_j = I_{2u+1}^\alpha \end{cases}$$

The corresponding relations are not minimal generators of $\tilde{J}$. Nevertheless, $\tilde{J}$ being a prime ideal generated in positive degree in the $\underline{T}$, we can cancel any common factor in each one of these relations that involve solely monomials in x, y, z, w.

The so-obtained relations form a subset of a minimal set of generators of $\tilde{J}$ and will be called *essential reduced relations*. We consider the ideal of $k[\underline{T}]$ generated by the residues modulo $\mathfrak{m}$ of the essential reduced relations. Clearly, this ideal is contained in the ideal $\mathcal{J} \subset k[\underline{T}]$ generated by the residues modulo $\mathfrak{m}$ of the generators of $\tilde{\mathcal{J}}$ – in other words, $\mathcal{J}$ is a presentation ideal of the special fiber $R[It]/\mathfrak{m}R[It]$.

Set

$$\mathcal{H} = \{ (r,\alpha) \in \mathbb{N}^2 \mid 0 \le r \le 2t, \ 0 \le \alpha \le \varepsilon_r \}$$

Definition 2.1 *With the above notation, the essential fiber ideal is the ideal $\mathcal{A} \subset k[\underline{T}]$ generated by the polynomials*

$$G_{r,\alpha}^{s,\beta} = T_{r,\alpha}T_{s,\beta+1} - (\sigma\delta_{r,s})\, T_{s,\beta}T_{r,\alpha+1},$$

for all values of $((r,\alpha),(s,\beta)) \in \mathcal{H} \times \mathcal{H} \mid (r,\alpha) < (s,\beta)$, *with the conventions*

$$\delta_{r,s} = \begin{cases} 1 & \text{if } r = s \\ 0 & \text{otherwise} \end{cases}, \sigma = \begin{cases} 0 & \text{if } r = \alpha = 0, r_{\mu+1} \ne 0 \\ 1 & \text{otherwise} \end{cases}$$

and $T_{l,\varepsilon_l+1} = T_{l+2,0}$ *for* $0 \le l \le 2t$.

Remark 2.2 *It can actually be shown that the essential fiber ideal coincides with the ideal generated by the residues of the essential reduced relations. We can also prove that this ideal coincides with the presentation ideal of the special fiber $\mathcal{J}$. This result won't have any special effect in the arguments to follow and will be proved in a forthcoming paper.*

It is at least allectionating to look back at Example 1.2. With the data found in Example 1.5 and Example 1.8, the Plücker relation coming from the submatrix $AP_0^0 P_0^1 I_1^0$ is

$$-x^6 y^{17}(F_{0,0}F_{0,2} - F_{0,1}{}^2 + x_{11}y^{30}z^{43}F_{1,0}{}^2) = 0$$

whose corresponding essential reduced relation yields the generator $T_{0,0}T_{0,2} - T_{0,1}{}^2 \in k[\underline{T}]$ of the essential fiber ideal $\mathcal{A}$.

Actually, the essential fiber ideal is naturally associated to the following block matrix

$$\begin{pmatrix} T_{00} & T_{01} & T_{02} & T_{10} & T_{20} & T_{30} & T_{31} & T_{40} \\ T_{01} & T_{02} & T_{20} & T_{30} & T_{40} & T_{31} & T_{50} & T_{60} \end{pmatrix}.$$

Namely, indexing the columns of this matrix by the indices of their top entries $(r,\alpha) \in \mathbb{N}^2$, the generators of the ideal are the polynomials

$$G_{r,\alpha}^{s,\beta} = \begin{cases} \Delta_{(r,\alpha),(s,\beta)} & \text{if } (r,\alpha),(s,\beta) \text{ belong to the same block} \\ M_{(r,\alpha)} & \text{otherwise} \end{cases},$$

where $M_{(r,\alpha)}$ denotes the monomial of $\Delta_{(r,\alpha),(s,\beta)}$ divisible by $T_{r,\alpha}$.

The initial ideal of the essential fiber ideal

We will need some results on Gröbner bases, for which standard references are [3], [6], [14]. We thank L.Robbiano and W.Vasconcelos for their precious help on Gröbner basis.

Let $k[\underline{T}]$ be a polynomial ring over the field k in n variables. Recall that the *diagonal order* on the monomials of $k[\underline{T}]$ is defined in terms of the linear form $L = \sum_{i=1}^{n} x_i$, to wit, if $\underline{T}^\alpha, \underline{T}^\beta$ are monomials , with $\alpha = (\alpha_1, \ldots, \alpha_n)$ and $\beta = (\beta_1, \ldots, \beta_n)$, then

$$\underline{T}^\alpha < \underline{T}^\beta \Leftrightarrow \begin{cases} L(\alpha) < L(\beta) & \text{or} \\ L(\alpha) = L(\beta) & \text{and } \exists\, 1 \leq i \leq n|\ \alpha_i < \beta_i \text{ and } \alpha_j = \beta_j\ \forall j < i \end{cases}$$

Following widespread notation, we denote by in (P) the leading term of the polynomial $P \in k[\underline{T}]$, $P \neq 0$. If $\mathcal{F} \subset k[\underline{T}] \setminus \{0\}$ is a finite set, an *admissible combination* on $\mathcal{F}$ is an expression of the form

$$G := \sum_{P \in \mathcal{F}} G_P P, G_P \in k[\underline{T}] \mid \text{in } (G) = \max\ \{\text{in } (G_P P), P \in \mathcal{F}, G_P \neq 0\ \}.$$

The following result is the property on which Buchberger's algorithm is based.

Lemma 2.3 *The following conditions are equivalent for a finite subset $\mathcal{F} \subset k[\underline{T}] \setminus \{0\}$:*

(i) $\mathcal{F}$ is a Gröbner basis, for the diagonal order, of the ideal $(\mathcal{F})$

(ii) For every pair P, Q of elements of $\mathcal{F}$, the s-polynomial $S(P, Q)$ is an admissible combination on $\mathcal{F}$.

Proof. It follows from [13, 3.5]. $\square$

We will also use the well-known

Lemma 2.4 *Let $J \subset k[\underline{T}]$ be an ideal and let $\mathcal{F}$ be a Gröbner basis of J for the diagonal order. Then*

$$\dim k[\underline{T}]/J = \dim k[\underline{T}]/(\text{in } (\mathcal{F})),$$

where in $(\mathcal{F}) = \{\text{in } (P) \mid P \in \mathcal{F}\}$.

Proof. See, e.g., [6, 7.6]. $\square$

We next apply these results to our original set-up. Thus, $\underline{T} = \{T_{00}, T_{01}, \ldots, T_{2t,\epsilon_{2t}}\}$, ordered by means of the lexicographic order of their indices. Consider the essential fiber ideal $\mathcal{A} \subset k[\underline{T}]$ and let $\mathcal{G}$ denote its set of generators as introduced in Definition 2.1.

Lemma 2.5 *For $\mathcal{G}$ as above, one has*

$$in\ (\mathcal{G}) = \{T_{r,\alpha}T_{s,\beta+1} | ((r,\alpha),(s,\beta)) \in \mathcal{H} \times \mathcal{H}\}$$

with respect to the diagonal order, the double indices running as in Definition 2.1.

Proof. It follows immediately from the definitions that

$$T_{r,\alpha}T_{s,\beta+1} > T_{s,\beta}T_{r,\alpha+1}$$

for the diagonal order. □

Proposition 2.6 $\mathcal{G}$ *is a Gröbner basis of the essential fiber ideal $\mathcal{A}$.*

Proof. According to Lemma 2.3, we have to show that the s-polynomials of elements of $\mathcal{G}$ are admissible combinations on $\mathcal{G}$. Thus, let $G = G_{r,\alpha}^{s,\beta}$, $G' = G_{r',\alpha'}^{s',\beta'} \in \mathcal{G}$. By Lemma 2.5, in (G) and in (G') have a common factor exactly in one of the following cases:

1. $(r,\alpha) = (r',\alpha')$

2. $(s,\beta) = (s',\beta')$

3. $(r',\alpha') = (s,\beta+1)$ with $\beta \neq \varepsilon_s$ or $(r',\alpha') = (s+2,0)$ with $\beta = \varepsilon_s$

4. $(r,\alpha) = (s',\beta'+1)$ with $\beta' \neq \varepsilon_{s'}$ or $(r,\alpha) = (s'+2,0)$ with $\beta' = \varepsilon_{s'}$

The restrictions on the set of double in the definition of $\mathcal{G}$ are easily seen to rule out case (4); for the remaining cases, the s-polynomial has the following form:

1. $(r,\alpha) = (r',\alpha') < (s,\beta) < (s',\beta')$: then $S(G,G') = \delta_{r,s}T_{r,\alpha+1}G_{s,\beta}^{s',\beta'}$.

2. $(r,\alpha) < (r',\alpha') < (s,\beta) = (s',\beta')$: then $S(G,G') = \delta_{r',s}T_{s,\beta}G_{r,\alpha}^{s,\beta}$.

3. $(r,\alpha) < (s,\beta) < (s,\beta+1) = (r',\alpha') < (s',\beta')$ with $\beta \neq \varepsilon_s$, or $(r,\alpha) < (s,\beta) < (s+2,0) = (r',\alpha') < (s',\beta')$ with $\beta = \varepsilon_s$: then $S(G,G') = \delta_{s,s'}T_{s',\beta'}G_{r,\alpha}^{r',\alpha'} - \delta_{r,s}T_{r,\alpha+1}G_{s,\beta}^{s',\beta'}$.

Thus, in all cases, one obtains an admissible combination on $\mathcal{G}$. □

The dimension of the special fiber

We now inspect closely the initial ideal $\mathcal{I} := (in\ (\mathcal{G})) \subset k[\underline{T}]$. Its basic properties are given in the following result.

Proposition 2.7 *(i)* $\mathcal{I}$ *is a radical ideal, whose minimal primes are* $\mathcal{I}_{r,\alpha}$, *one for each* $(r,\alpha) \in \mathcal{H}$, $r \le 2t$, *with generators* $\{T_{s,\beta}\}$ *satisfying the following conditions:*

$$(s,\beta) \ne (r,\alpha),\ (s,\beta) \ne (r+1,0),\ (s,\beta) \ne \begin{cases} (r,\alpha+1) & \text{if } \alpha \ne \varepsilon_r \\ (r+2,0) & \text{if } \alpha = \varepsilon_r \end{cases}$$

(ii) $\dim k[\underline{T}]/\mathcal{I} = 3$.

Proof. (i) First, by Lemma 2.5, a typical generator of $\mathcal{I}$ is of the form $T_{r,\alpha}T_{s,\beta+1}$, with

$$((r,\alpha),(s,\beta)) \in \mathcal{H} \times \mathcal{H},\ (r,\alpha) < (s,\beta) \le (2t, \varepsilon_{2t}).$$

If now $(r',\alpha') \in \mathcal{H}$ is arbitrarily given, then one easily checks that at least one of the variables $T_{r,\alpha}$ or $T_{s,\beta+1}$ is a generator of $\mathcal{I}_{r',\alpha'}$.

Conversely, let $P \in \cap \mathcal{I}_{r,\alpha}$, where $(r,\alpha) \in \mathcal{H}$, $r \le 2t$. We will show, namely, that for every $(k,\alpha) \in \mathcal{H}$ such that $(0,0) < (k,\alpha) \le (2t, \varepsilon_{2t})$ (lexicographic order), one has

$$P = M_{k,\alpha} + M'_{k,\alpha} \text{ with } M_{k,\alpha} \in \mathcal{I},\ M'_{k,\alpha} \in \mathcal{I}'_{k,\alpha'}$$

where $\mathcal{I}'_{k,\alpha} = (T_{s,\beta} \in \mathcal{I}_{k,\alpha} \mid (s,\beta) > (k,\alpha))$. This will clearly suffice to show that $P \in \mathcal{I}$ since we may apply it for $(k,\alpha) = (2t, \varepsilon_{2t})$.

The proof is by induction on (k,α), with the lexicographic order (which is obviously a total well-order). The initial step of the induction will be subsumed in the general step, so we deal with the latter.

Thus, suppose, inductively, that $P = M_{k,\alpha} + M'_{k,\alpha}$ with $(k,\alpha) \in \mathcal{H}, (0,0) < (k,\alpha) < (2t, \varepsilon_{2t})$. We have two cases:

• $\alpha \ne \varepsilon_k$. As $M'_{k,\alpha} \in \mathcal{I}'_{k,\alpha}$, we may write

$$M'_{k,\alpha} = A_{k,\alpha+2}T_{k,\alpha+2} + M_1,$$

where $M_1 \in \mathcal{I}'_{k,\alpha+1} \subset \mathcal{I}_{k,\alpha+1}$ (recall that $T_{k,\alpha+2} = T_{k+2,0}$ when $\varepsilon_k = \alpha + 1$).

But, since $P \in \mathcal{I}_{k,\alpha+1}$ and $M_{k,\alpha} \in \mathcal{I} \subset \mathcal{I}_{k,\alpha+1}$, we must have $A_{k,\alpha+2}T_{k,\alpha+2} \in \mathcal{I}_{k,\alpha+1}$. Consequently, $A_{k,\alpha+2} \in \mathcal{I}_{k,\alpha+1}$, say,

$$A_{k,\alpha+2} = \sum_{\substack{(s,\beta) \in \mathcal{H} \\ (s,\beta) \le (k,\alpha)}} A'_{s,\beta}T_{s,\beta} + A_1,\ A_1 \in \mathcal{I}'_{k,\alpha+1}.$$

It follows that

$$P = M_{k,\alpha} + \sum_{\substack{(s,\beta) \in \mathcal{H} \\ (s,\beta) \le (k,\alpha)}} A'_{s,\beta}\underbrace{T_{s,\beta}T_{k,\alpha+2}}_{\in \mathcal{I}} + Q,$$

with $Q = M_1 + A_1 T_{k,\alpha+2} \in \mathcal{I}'_{k,\alpha+1}$.

Therefore, P is of the form $M_{k,\alpha+1} + M'_{k,\alpha+1}$ in this case.

• $\alpha = \varepsilon_k$. Here, the immediate successor of (k, α) is $(k + 1, 0)$. The argument is entirely analogous as in the first case and will show that P has the form $M_{k+1,0} + M'_{k+1,0}$.

(ii) We claim that, for each, $(r, \alpha) \in \mathcal{H}$, $\dim k[\underline{T}]/\mathcal{I}_{r,\alpha} = 3$. Indeed, the generators of this prime miss but three variables among the $\underline{T}$. $\qquad\square$

We apply these results to the original ideal I of the given monomial curve in $\mathbb{P}^3$. For this, recall that the *analytic spread* of the ideal I is defined as $\ell(I) := \dim R[It]/\mathfrak{m}R[It]$, where $\mathfrak{m} = (x, y, z, w) \subset R = k[x, y, z, w]$.

Here is the main result of the section:

Theorem 2.8 *If I is not perfect (i.e., if the curve is not arithmetically Cohen-Macaulay) then $\ell(I) = 3$.*

Proof. Gathering the information, one has

$$
\begin{aligned}
\dim k[\underline{T}]/\mathcal{A} &= \dim k[\underline{T}]/\mathcal{I}, \quad \text{by Lemma 2.4} \\
&= 3, \quad \text{by Proposition 2.7.}
\end{aligned}
$$

By definition, $\mathcal{A}$ is contained in the presentation ideal of $R[It]/\mathfrak{m}R[It]$ as a k-algebra. Therefore, $\ell(I) \le 3$. On the other hand, it is well known that $\ell(I) = ht\, I (= 2)$ (if and) only if I is generated by an R-sequence. Since we are assuming that I is not perfect, it must be the case that $\ell(I) = 3$. $\qquad\square$

References

[1] H. Bresinsky and C. Huneke, Liaison of monomial curves in $\mathbb{P}^3$, J. Reine Angew. Math, **365** (1986), 33–66.

[2] H. Bresinsky and B. Renschuch, Basisbestimmung Veronesescher Projektionsideale mit allgemeiner Nullstelle, Math. Nachr., **96** (1980), 257-269.

[3] B. Buchberger, Gröbner bases-an algorithmic method in polynomial ideal theory, Chapter 6, *in* Multidimensional Systems Theory N.K. Bose (ed.), D.Reidel, 1985.

[4] S. Huckaba and C. Huneke, Powers of ideals having small analytic deviation, Amer. J. Math., *to appear.*

[5] S. Huckaba and C. Huneke, Rees algebras of ideals having small analytic deviation, *preprint.*

[6] M. Lejeune-Jalabert, Effectivité de calculs polynômiaux, Cours de D.E.A., Institut Fourier, Université de Grenoble, 1984–85.

[7] F. Mora, Classification of monomial curves, Rend. Acad. Naz. XL **101** (1983), 13–28.

[8] M. Morales, Syzygies of monomial curves and a linear diophantine problem of Frobenius, Max-Planck-Institut fur Mathematik, Bonn, *preprint*, 1987.

[9] M. Morales, Sysygies des plongements de Veronese quasi-homogènes (*provisory title*), *preprint*, 1990.

[10] M. Morales and A. Simis, Symbolic powers of monomial curves in $\mathbb{P}^3$ lying on a quadric surface, Comm. in Algebra **20(4)** (1992), 1109–1122.

[11] M. Morales and A. Simis, The second symbolic power of an arithmetically Cohen–Macaulay monomial curve in $\mathbb{P}^3$, Comm. in Algebra, *to appear*.

[12] M. Morales and A. Simis, Blow-up algebras of certain non-perfect determinantal loci, *in preparation*.

[13] F. Pauer and M. Pfeifhofer, The theory of Gröbner bases, l'Enseignement Mathématique, **34** (1988), 215–232.

[14] L. Robbiano, Gröbner basis: a foundation for Commutative Algebra, *preprint*, 1989.

Philippe Gimenez, Marcel Morales
Université de Grenoble I
Institut Fourier
38402 St. Martin d'Hères Cedex, France
pgimenez@frgren81.bitnet   morales@frgren81.bitnet

Aron Simis
Instituto de Matemática
Universidade Federal da Bahia
40210 Salvador, Bahia, Brazil
useraron@lncc.bitnet

Computational Complexity of Sparse Real Algebraic Function Interpolation

D. Grigoriev   M. Karpinski*   M. F. Singer†

Abstract.   We analyze the computational complexity of the problem of interpolating real algebraic functions given by a black box for their evaluations, extending the results of [GKS 90b, GKS 91b] on interpolation of sparse rational functions.

1   Introduction

We start the definition of a t-sparse real algebraic function.

Definition: 1. $Y(X_1, \ldots, X_n)$ is a t-sparse real algebraic (multivalued) function if its graph $\Gamma_Y \subset (\mathbb{R}_+)^{n+1}$ projects surjectively onto the positive orthant $(\mathbb{R}_+)^n$ and lies in the variety $\{f = 0\} \cap (\mathbb{R}_+)^{n+1}$ where f is a t-sparse fractional power polynomial

$$f = \sum_{i=1}^{t} \gamma^{(i)} X_1^{\alpha_1^{(i)}} \ldots X_n^{\alpha_n^{(i)}} Y^{\beta^{(i)}}$$

where $\alpha_j^{(i)}, \beta^{(i)} \in \mathbb{Q}$, $\gamma^{(i)} \in \mathbb{R}$ and the exponent vectors $(\alpha_1^{(i)}, \ldots, \alpha_n^{(i)}, \beta^{(i)})$ are pairwise distinct. By $\{f = 0\}$ we denote a set of points $\vec{x}$ satisfying $f(\vec{x}) = 0$. Moreover, let μ be a common denominator of all the rational numbers $\alpha_j^{(i)}, \beta^{(i)}$. Changing the coordinates $X_i \to X_i^{1/\mu}$, $Y \to Y^{1/\mu}$ (note that this is a diffeomorphism of $\mathbb{R}_+^{n+1}$) we get that $\tilde{f}(X_1, \ldots, X_n, Y) = f(X_1^\mu, \ldots, X_n^\mu, Y^\mu)$ is a polynomial in $X_1, \ldots, X_n, Y$. By this change of the coordinates we obtain a new algebraic function $\tilde{Y}$ and its graph $\Gamma_{\tilde{Y}}$. In addition we suppose that $\Gamma_{\tilde{Y}}$ is an irreducible (in the Zariski topology over $\mathbb{R}$, see [BCR 87]) component of the semialgebraic set $\{\tilde{f} = 0\} \cap \mathbb{R}_+^{n+1}$.

*Supported in part by Leibniz Center for Research in Computer Science, by the DFG, Grant KA 673/4-1 and by the SERC Grant GR-E 68297

†Supported in part by NSF Grant DMS-9024624

We call f a t-sparse representation of Y. If t is the least possible we call f a minimal t-sparse representation.

2. We are also given a black box that for each $(x_1, \ldots, x_n) \in (\mathbb{R}_+)^n$ gives the set of all values of Y at this point together with the partial derivatives up to the order t (if they exist; if not it gives the value ∞).

When we say that we are given a t-sparse real algebraic function we mean that we are given such a black box together with the integer t for a function as described in 1.

Unlike the case of rational functions [GKS 90b, GKS 91b] the values of Y at rational points can be irrational, thus we need a different (from the rational case) computational model. Moreover, together with the values of Y we need the values of its several partial derivatives. Also we need a zero-test for the arithmetic expressions of the values.

One computational model could be the following. An algorithm is given which for any rational point $\vec{x} \in \mathbb{Q}_+^n$ provides an algorithm which outputs a sequence $\{\eta_m \in \mathbb{Q}\}_{0 \leq m \in \mathbb{Z}}$ such that $\lim_{m \to \infty} \eta_m = Y(\vec{x})$ and the speed of convergency is uniform in some cube $(\vec{x} - \vec{\delta}, \vec{x} + \vec{\delta})$ (but the speed itself and $\vec{\delta}$ could be unknown). Then one can get similar algorithms converging (also locally uniformly) to the successive derivatives. For this model we need an assumption of the existence of a zero-test (namely, a test to determine if such a sequence converges to zero).

If we suppose the coefficients $\gamma^{(i)} \in \mathbb{Q}$ of f to be rational then the values in rational (even algebraic) points are algebraic and it is reasonable to represent each of the values of Y and its derivatives by its minimal polynomial and an interval in which the minimal polynomial has a unique root (see e.g. [GV88]), or by the means of Thom's lemma (see e.g. [HRS 90]), i.e. by the minimal polynomial and a succession of signs of derivatives of the minimal polynomial.

The third approach could be to consider the values in an abstract way (see e.g. [BSS 88]) and to treat them as the symbols for real numbers.

Anyway, independent of the way of representation, we assume that carrying out one arithmetic operation involving the outputs of black boxes has a unit cost, similarly to what is usually adopted in interpolation problems for black boxes (see e.g. [BT 88, GKS 90a, GKS 90b]).

We design an algorithm for finding the exponent vectors of all minimal (normalized) t_1-sparse representations of a t-sparse (so $t_1 \leq t$) real algebraic function Y (see the theorem at the end of the paper). It extends the interpolation algorithms for polynomials ([BT 88], [GKS 90a]) and for rational functions

([GKS 90b, GKS 91b]).

We indicate briefly the further contents of the paper:
In Section 2 we present a zero-test for t-sparse real algebraic functions. Namely, we prove that a set of points $\{1, \ldots, B\}^n$ plays a role of a zero-test set and give a bound on B. The proof invokes the bounds from [K 91] (Proposition 1) on the sum of Betti numbers of a real algebraic variety given by a sparse polynomial. In Section 3 we prove that any minimal t-sparse representation of an algebraic function has rational exponents. This implies (as is shown in Section 4), that there is a finite number of the minimal t-sparse representations.
In Section 4 we describe an algorithm which finds the exponent vectors of all the minimal t-sparse representations of a real algebraic function (interpolation algorithm). It uses a Wronskian formulation of linear dependence (see e.g. [K 73]) which appeared to be helpful also for sparse rational function interpolation ([GKS 90b, GKS 91a, GKS 91b]) and which allows to describe the family of exponent vectors as a solution of a system (over $I\!R$) of a polynomial equations. The complexity estimates of this algorithm are stated in the Theorem at the end of Section 4.

Acknowledgments. The authors thank N. Ivanov, M. Kontsevich and N. Vorobjov (jr.) for useful discussions.

2 Zero-test

Let g be a T-sparse fractional-power polynomial in the variables $X_1, \ldots, X_n, Y$ with the same denominator μ of the exponents of f (cf. definition 1). We describe a test to determine whether g vanishes on Γ_Y. Observe that this is equivalent to testing whether the dimension of $\{\tilde{g} = 0\} \cap \Gamma_{\tilde{Y}}$ is n since $\Gamma_{\tilde{Y}}$ is irreducible ($\tilde{g}$ is defined similar to $\tilde{f}$). Our zero-test relies on the results of Khovanskii. For our purposes we need the following

Proposition 1. (see Corollary 5, p. 92 and Theorem, p. 1 [K 91])
Let $h \in I\!R[X_1, \ldots, X_n]$ be a t-sparse polynomial such that $\{h = 0\} \subset I\!R^n$ is a nonsingular hypersurface. Then the sum of Betti numbers of $\{h = 0\}$ does not exceed $2^{\frac{t^2}{2}} n^{O(n)}$.

Note that in the above proposition, the i-th Betti number b_i ($\{h = 0\}$) is defined as the rank of i-th cohomology group H^i ($\{h = 0\}$, $I\!R$) with real coefficients, see e.g. [ES 52], [D 80], [BCR 87]). A similar bound is true if we change the hypothesis above to consider singular varieties that are compact.

Corollary 2. *Let $h \in I\!R[X_1, \ldots, X_n]$ be a t-sparse polynomial such that $\{h = 0\} \subset I\!R^n$ is compact. Then the sum of Betti numbers of $\{h = 0\}$ does not*

exceed $2^{(O(tn)^2)}$.

Proof.   We follow closely the arguments in Theorem 2 [M 64] or Proposition 11.5.4 [BCR 87]. Assume that $\{h = 0\}$ lies in a ball of radius R. Let $K(\epsilon, \delta) = \{f^2 + \epsilon^2(\sum_{i=1}^{n} x_i^2) \leq \delta^2\} \subset I\!\!R^n$ and let $\partial K(\epsilon, \delta) = \{f^2 + \epsilon^2(\sum_{i=1}^{n} x_i^2) = \delta^2\}$. For sufficiently small ϵ and almost all δ, $\partial K(\epsilon, \delta)$ is a nonsingular hypersurface. Apply proposition 1, we have that the sum of the Betti numbers of $\partial K(\epsilon, \delta)$ is at most $2^{(O(tn)^2)}$.

Let H^* be the sum of the cohomology groups. Alexander duality (see e.g. [D 80]) implies that rank $H^*(K(\epsilon_i, \delta_i)) = \frac{1}{2}$ rank $H^*(\partial K(\epsilon_i, \delta_i))$.

Let ϵ_i approach 0 monotonically and select δ_i so that δ_i/ϵ_i approaches R monotonically. We then have $K(\epsilon_i, \delta_i) \supset K(\epsilon_{i+1}, \delta_{i+1})$ and $\bigcap_i K(\epsilon_i, \delta_i) = K$. Therefore $H^*(K)$ is the direct limit (see e.g. [ES 52]) of the groups $H^* K(\epsilon_i, \delta_i)$ and so rank $H^*(K) = \lim(\text{rank } H^* K(\epsilon_i, \delta_i))$. This proves the corollary.   □

We now formulate the main result of this section.

Lemma 3.

If $\dim(\{\tilde{g} = 0\} \cap \Gamma_{\tilde{Y}}) \leq n - 1$ *(e.g. if* $g \not\equiv 0$ *on* Γ_Y*) then for at least one of the values* $x_1 = 1, 2, \ldots, B \leq 2^{(tT)^{O(n)}}$ *we have* $\dim(\{\tilde{g} = 0\} \cap \Gamma_{\tilde{Y}} \cap \{X_1 = x_1\}) \leq n - 2$.

Before proceeding to the proof of lemma 3 we describe a zero-test based on lemma 3. Continuing to apply lemma 3 one shows by induction on the dimension that there exists a point $(x_1, \ldots, x_n) \in \{1, 2, \ldots, B\}^n$ such that for each point $(x_1, \ldots, x_n, y) \in \Gamma_Y$ (recall that Y is defined everywhere on $I\!\!R_+^n$) $g(x_1, \ldots, x_n, y) \neq 0$ (thus the zero-test considers all these points $\{x_1, \ldots, x_n\} \in \{1, \ldots, B\}^n$). Notice that we supposed that $\Gamma_{\tilde{Y}}$ is irreducible, this was used only to reformulate the condition that g does not vanish on Γ_Y as $\dim(\{\tilde{g} = 0\} \cap \Gamma_{\tilde{Y}}) \leq n-1$ and just this inequality on the dimension is used as an inductive hypothesis. Observe also that at each step of the induction we obtain the same bound B for the number of values of the current coordinate X_i since at each step we deal with a substitution of some values $x_1, \ldots, x_{i-1}$ instead of $X_1, \ldots, X_{i-1}$ into the power-fractional polynomials f, g that does not increase their sparsity.

Now we proceed to the proof of lemma 3. We start with a definition. For each point $\vec{x}$ of $f(\vec{x}) = 0$ we define the multiplicity $m_f(\vec{x})$ of $\vec{x}$ on f as the minimal number k such that some partial derivative of f of order k does not vanish at $\vec{x}$. If we have a polynomial and write $f = \sum f_i$ where each f_i is homogeneous of degree i in $(\vec{X} - \vec{x})$ where $\vec{X} = (X_1, \ldots, X_n)$, then $m_f(\vec{x})$ is the smallest i such that $f_i \not\equiv 0$. Note if $f = g \cdot h$, then $m_f(\vec{x}) = m_g(\vec{x}) + m_h(\vec{x})$.

Lemma 4. (cf. [GKO 91])

If $f \not\equiv 0$ is t-sparse, then for each $\vec{x} \in (\mathbb{R}_+)^n$, $m_f(\vec{x}) \leq t - 1$.

Proof. Let $f = \sum_{i=1}^{t} c_i \vec{X}^{\vec{\alpha}_i}$ where $\vec{\alpha}_i = (\alpha_{1i}, \ldots, \alpha_{ni})$. Let $\vec{a} = (a_1, \ldots, a_n)$ be

a vector such that $\vec{a} \cdot \vec{\alpha}_i \neq \vec{a} \cdot \vec{\alpha}_j$ if $i \neq j$ and let $D = \sum_{i=1}^{t} a_i X_i \frac{\partial}{\partial X_i}$. It is enough

to show that if $\vec{x} \in (\mathbb{R}_+)^n$ and $f(\vec{x}) = D(f)(\vec{x}) = \ldots = D^{t-1}(f)(\vec{x}) = 0$ then
$f \equiv 0$. We have

$$
\begin{pmatrix}
1 & 1 & \cdots & 1 \\
\vec{a} \cdot \vec{\alpha}_1 & \vec{a} \cdot \vec{\alpha}_2 & \cdots & \vec{a} \cdot \vec{\alpha}_t \\
\vdots & & & \\
(\vec{a} \cdot \vec{\alpha}_1)^{t-1} & (\vec{a} \cdot \vec{\alpha}_2)^{t-1} & \cdots & (\vec{a} \cdot \vec{\alpha}_t)^{t-1}
\end{pmatrix}
\begin{pmatrix}
c_1 \vec{x}^{\vec{\alpha}_1} \\
c_2 \vec{x}^{\vec{\alpha}_2} \\
\vdots \\
c_t \vec{x}^{\vec{\alpha}_t}
\end{pmatrix}
=
\begin{pmatrix}
f(\vec{x}) \\
(Df)(\vec{x}) \\
\vdots \\
D^{t-1}(f)(\vec{x})
\end{pmatrix}
$$

Since the first matrix is a vandermonde matrix and $\vec{x} \in (\mathbb{R}_+)^n$, we have the
conclusion of lemma 4. $\qquad\square$

Note that in Lemma 4 it is enough to assume that no coordinate of $\vec{x}$ is zero.

Let $h \in \mathbb{R}[X_1, \ldots, X_n, Y]$ be a polynomial and let $V_1 \subset \mathbb{R}^{n+1}$ be an irreducible (over $\mathbb{R}$) component in the Zariski topology of the variety $\{h = 0\}$
such that $\dim V_1 (= \dim_{\mathbb{R}} V_1) = n$. Let $h = \prod h_i^{m_i}$ be a factorization of h where
$h_i \in \mathbb{R}[X_1, \ldots, X_n, Y]$ are irreducible over $\mathbb{R}$. Denote by $\bar{V}_1 \subset \mathbb{C}^{n+1}$ the closure
of V_1 in the Zariski topology, then $\dim_{\mathbb{C}} \bar{V}_1 = n$ and $\bar{V}_1$ is efined and irreducible
over $\mathbb{R}$. Then the generator $\bar{h} \in \mathbb{R}[X_1, \ldots, X_n, Y]$ such that $\bar{V}_1 = \{\bar{h} = 0\}_{\mathbb{C}^{n+1}}$
is irreducible and $\bar{h} \mid h$ since h vanishes on V_1 and thereby on $\bar{V}_1$. Let $\bar{h} = h_1$
for definiteness and we say that the polynomial h_1 corresponds to V_1, observe
that $V_1 = \{h_1 = 0\}$.

Let for some $x_1 > 0$, $\dim(\Gamma_{\tilde{y}} \cap \{\tilde{g} = 0\} \cap \{X_1 = x_1\}) = n - 1$. Let U be an
irreducible component of the variety $\Gamma_{\tilde{y}} \cap \{\tilde{g} = 0\} \cap \{X_1 = x_1\}$ of the dimension
$\dim(U) = n - 1$. Suppose that $V_1, \ldots, V_s$ are all the irreducible components
of the variety $\{\tilde{g} = 0\}$ such that $U \subset V_j$, then $s \geq 1$. Observe that for each
$1 \leq i \leq s$ either $\dim V_i = n$ or $\dim V_i = n - 1$. In the latter case $V_i = U$ since
a linear function $X_1 - x_1$ vanishes on a subvariety of the irreducible variety V_i
of the complete dimension $n - 1$. Thus either $\dim V_i = n$ for all $1 \leq i \leq s$ or
$s = 1$ and in this case $V_1 = U$. Suppose that $V_{s+1}, \ldots, V_{s_1}$ are all the irreducible
components of $\{\tilde{f} = 0\}$ such that $U \subset V_j$, then $s_1 - s \geq 1$. The same observation
concerns $V_{s+1}, \ldots, V_{s_1}$. Consider $\tilde{f}\tilde{g} = \prod h_i^{m_i}$ a factorization over $\mathbb{R}$. To each
V_j, $1 \leq j \leq s_1$ with the dimension $\dim V_j = n$ corresponds some h_{i_j} as above.
For almost all the points $y \in V_j$, $m_{h_{i_j}}(y) = 1$ (since almost all (in the sense of
Zariski topology) points of V_j and also of $\bar{V}_j$ are nonsingular, that is the gradient

of h_{i_j} does not vanish) therefore for almost all the points $y \in V_j$, $m_{\tilde{f}\tilde{g}}(y) = m_{i_j}$.

Define $M = \max\{m_{i_j} + 1\}$ where the maximum is taken over all the polynomials h_{i_j} which correspond to the irreducible components $V_{j_1}, \ldots, V_{j_q}$ among V_j, $1 \leq j \leq s_1$ with dimension n (in the case $q = 0$, when there are no such components we set $M = 1$). Consider the real algebraic variety $\tilde{U} = \tilde{U}_M \subset \{\tilde{f}\tilde{g} = 0\} \subset \mathbb{R}^{n+1}$ consisting of all the points y with the multiplicity $m_{\tilde{f}\tilde{g}}(y) \geq M$. Let us show that $\tilde{U} \supset U$. Namely, for every point $\tilde{x} \in U$, $m_{\tilde{f}\tilde{g}}(\tilde{x}) \geq m_{i_{j_1}} + \ldots + m_{i_{j_q}}$ and in the case when $q \geq 2$ obviously $m_{\tilde{f}\tilde{g}}(\tilde{x}) \geq M$. If $q = 1$ then the families $V_1, \ldots, V_s$ and $V_{s+1}, \ldots, V_{s_1}$ cannot consist both of the same single irreducible variety of dimension n, since otherwise this variety would be a subvariety of $\Gamma_{\tilde{Y}}$ (notice that here we do not make use of irreducibility of $\Gamma_{\tilde{Y}}$), but $\dim(\Gamma_{\tilde{Y}} \cap \{\tilde{g} = 0\}) \leq n - 1$ by the hypothesis of lemma 3. Thus in the case $q = 1$, one of two families $V_1, \ldots, V_s$ and $V_{s+1}, \ldots, V_{s_1}$ consists of a single irreducible variety of dimension n and another family consists of a single variety coinciding with U. Then $m_{\tilde{f}\tilde{g}}(\tilde{x}) = m_{\tilde{f}}(\tilde{x}) + m_{\tilde{g}}(\tilde{x}) \geq m_{j_1} + 1 = M$. In the case $q = 0$, $m_{\tilde{f}\tilde{g}}(\tilde{x}) \geq 1 = M$ is obvious, which shows $\tilde{U} \supset U$.

Therefore, for each V_j, $1 \leq j \leq s_1$ we have $\dim(V_j \cap \tilde{U}) = n - 1$. Observe that lemma 4 implies $\sum\limits_{1 \leq p \leq q} m_{i_{j_p}} \leq m_{\tilde{f}\tilde{g}}(\tilde{x}) \leq tT - 1$ since $\tilde{f}\tilde{g}$ is tT-sparse. Hence $\tilde{U}$ is defined by $tT\binom{tT-1+n}{n} \leq ((tT)^{n+1})$-sparse polynomial, since the relations defining $\tilde{U}$ involve the derivatives of $\tilde{f}\tilde{g}$ of orders less than tT.

Let $\tilde{U} = \bigcup\limits_{1 \leq l \leq r} \tilde{U}^{(l)}$ be a decomposition into irreducible (over $\mathbb{R}$) components. Each $\tilde{U}^{(l)}$ is a subvariety of one of the irreducible components of $\{\tilde{f} = 0\}$ or $\{\tilde{g} = 0\}$. If $\tilde{U}^{(l)}$ is contained in some component V of $\{\tilde{f} = 0\}$ or $\{\tilde{g} = 0\}$ which differs from $V_1, \ldots, V_{s_1}$ then $\dim(\tilde{U}^{(l)} \cap U) \leq \dim(V \cap U) \leq n - 2$. If $\tilde{U}^{(l)} \subset V_j$ for one of $1 \leq j \leq s_1$ then $\dim \tilde{U}^{(l)} \leq n - 1$ (see above) and either $\tilde{U}^{(l)} \supset U$ or $\dim(\tilde{U}^{(l)} \cap U) \leq n - 2$. If $\tilde{U}^{(l)} \supset U$ then $\tilde{U}^{(l)} = U$ since a linear function $X_1 - x_1$ vanishes on the subvariety U of the complete dimension $n - 1$ of the irreducible variety $\tilde{U}^{(l)}$ (cf. above). Observe that there exists $\tilde{U}^{(l)}$ such that $\tilde{U}^{(l)} \supset U$ (since $\tilde{U} \supset U$), therefore U is an irreducible component of $\tilde{U}$.

Now we can summarize what was proved above in the following.

Lemma 5.   *For each $x_1 > 0$ such that $\dim(\Gamma_{\tilde{Y}} \cap \{\tilde{g} = 0\} \cap \{X_1 = x_1\}) = n - 1$ and for each irreducible (over $\mathbb{R}$) component U with $\dim U = n - 1$ of the variety $\Gamma_{\tilde{Y}} \cap \{\tilde{g} = 0\} \cap \{X_1 = x_1\}$ there is an index $1 \leq i \leq tT$ such that U is an irreducible component of the variety $\tilde{U}_i$ consisting of the points $\tilde{x}$ with multiplicity $m_{\tilde{f}\tilde{g}}(\tilde{x}) \geq i$. The variety $\tilde{U}_i$ can be defined by an $(tT)^{O(n)}$-sparse polynomial.*

Thus, let $\tilde{U} = \tilde{U}_i = \bigcup_{1 \leq l \leq r} \tilde{U}^{(l)}$ be defined by a polynomial $h \in \mathbb{R}[X_1, \ldots, X_n, Y]$, let $h|_{X_1 = x_1} = \prod h_j^{m_j}$ be the decomposition of the polynomial $h|_{X_1 = x_1}$ into its irreducible (over $\mathbb{R}$) factors $h_j \in \mathbb{R}[X_2, \ldots, X_n, Y]$. As was proved earlier there is a factor of $h|_{X_1 = x_1}$ (let it be h_1 for definiteness) such that $U = \{h_1 = 0\} \cap \{X_1 = x_1\}$ since $\dim(U) = n - 1$ and U is an irreducible component of the variety $\tilde{U} \cap \{X_1 = x_1\} = \{h|_{X_1 = x_1}\} \cap \{X_1 = x_1\}$. Almost all the points of U are nonsingular (in the hyperplane $\{X_1 = x_1\}$ (in this context we sometimes say nonsingular omitting to mention a hyperplane)). By the implicit function theorem, h_1 takes both positive and negative values in a neighborhood in $\{X_1 = x_1\}$ of any nonsingular point.

Represent $\tilde{U} = \tilde{U}_i = \bigcup_{1 \leq l \leq r_1} \tilde{U}^{(l)} \cup \bigcup_{r_1 + 1 \leq l \leq r} \tilde{U}^{(l)}$ where $\tilde{U}^{(1)}, \ldots, \tilde{U}^{(r_1)}$ are all the irreducible components among $\tilde{U}^{(1)}, \ldots, \tilde{U}^{(r)}$ satisfying lemma 5 (so they include U), in particular each of them has the dimension $n - 1$ and lies in a hyperplane of the form $\{X_1 = x_1'\}$. Fix some $R > 0$ with the property that the closed ball B_R with the radius R contains at least one nonsingular point from any irreducible component $\tilde{U}^{(1)}, \ldots, \tilde{U}^{(r_1)}$ for all the varieties $\tilde{U}_i$, $1 \leq i \leq tT$ (cf. lemma 5).

Add a coordinate X_0 and consider the restriction of the polynomials $\tilde{f}\tilde{g}$ and h to the sphere S^{n+1} of the radius R in the space $\mathbb{R}^{n+2}$ with the coordinates $X_0, X_1, \ldots, X_n, Y$. Each of the varieties considered above, e.g. $\tilde{U} = \tilde{U}_i$ is transformed to a subvariety $\tilde{U}^{(S^{n+1})}$ of the sphere S^{n+1} given by the same polynomial h. It is clear how to describe $\tilde{U}^{(S^{n+1})}$ geometrically. Let π^+ be a homeomophism of the ball B_R onto the upper half of the sphere S^{n+1}, similar define π_-. Then $\tilde{U}^{(S^{n+1})} = \pi^+(B_R \cap \tilde{U}) \cup \pi_-(B_R \cap \tilde{U})$. Similarly one gets $\tilde{U}^{(l)(S^{n+1})}$.

Denote the sphere $S^n = S^{n+1} \cap \{X_1 = x_1\}$. Then $U^{(S^{n+1})} \subseteq S^n$ is $(n-1)$-dimensional variety and $U^{(S^{n+1})} = S^n \cap \{h_1 = 0\}$. As it was shown above h_1 takes both positive and negative values on S^n, hence the complement $S^n \setminus U^{(S^{n+1})}$ has at least two connected components, in other words the reduced homology group $\tilde{H}_0(S^n \setminus U^{(S^{n+1})})$ is nontrivial (in fact it is a free $\mathbb{R}$-module with the rank one less than the number of connected components). The Alexander duality principle (see [D 80]) implies $\tilde{H}_0(S^n \setminus U^{(S^{n+1})}) = H^{n-1}(U^{(S^{n+1})})$, in particular the latter group is nontrivial, thus $b_{n-1}(U^{(S^{n+1})}) \geq 1$.

Applying the Mayer-Vietoris formula (see [ES 52]) we obtain the inequality for Betti numbers

$$b_{n-1}(\tilde{U}^{(S^{n+1})}) \geq \sum_{1 \leq l \leq r_1} b_{n-1}(\tilde{U}^{(l)(S^{n+1})}) + b_{n-1}\left(\bigcup_{r_1 + 1 \leq l \leq r} (\tilde{U}^{(l)})^{(S^{n+1})} \right)$$

taking into account that the dimension of the variety

$$(\tilde{U}^{(l)(S^{n+1})} \cap (\bigcup_{1 \leq l_1 \leq r_1, \, l_1 \neq l} \tilde{U}^{(l)(S^{n+1})} \cup \bigcup_{r_1+1 \leq l \leq r} \tilde{U}^{(l)(S^{n+1})}))$$

for $1 \leq l \leq r_1$ does not exceed $n - 2$, and so $(n - 1)$-th cohomology group
of this variety is trivial. Let us sum these inequalities for all the varieties
$\tilde{U} = \tilde{U}_i$, $1 \leq i \leq tT$. Because of the proved above $b_{n-1}(\tilde{U}^{(S^{n+1})}) \geq r_1$. By the
corollary 2

$$(tT)2^{(tT)^{O(n)}} \geq \sum_{1 \leq i \leq tT} b_{n-1}(\tilde{U}_i^{(S^{n+1})})$$

and the right side of the latter inequality bounds from above (cf. lemma 5)
the number of hyperplanes of the form $\{X_1 = x_1\}$ such that $\dim(\Gamma_{\tilde{y}} \cap \{\tilde{g} = 0\} \cap \{X_1 = x_1\}) = n - 1$, this completes the proof of lemma 3.   □

3   Rationality of the exponents of a normalized minimal sparse representation

As in [GKS 90b, GKS 91b], we extend the notion of sparsity and say that
a real algebraic function Y (see the introduction) is t-quasisparse if $Q = 1 + \sum_{1 \leq i \leq t-1} c^{(i)} X_1^{a_1^{(i)}} \cdots X_n^{a_n^{(i)}} Y^{b^{(i)}} = 0$ for suitable reals $a_1^{(i)}, \ldots, a_n^{(i)}, b^{(i)}, c^{(i)} \in \mathbb{R}$
where the exponent vectors $(a_1^{(i)}, \ldots, a_n^{(i)}, b^{(i)})$ are pairwise distinct and distinct
from 0. Allowing real exponents, we call Q a normalized t-quasisparse representation. In fact, one could consider quasisparse representations of not only
algebraic functions, but we do not need it here.

We prove in this section that if Q is a minimal t-quasisparse representation,
then actually all $a_j^{(i)}, b^{(i)} \in \mathbb{Q}$. We start with the case $n = 1$.

Lemma 6.   If a real algebraic function $Y : \mathbb{R}_+ \rightarrow \mathbb{R}_+$ is minimal t-
quasisparse and satisfies $1 + \sum_{1 \leq i \leq t-1} c^{(i)} X^{a^{(i)}} Y^{b^{(i)}} = 0$ then $a^{(i)}, b^{(i)} \in \mathbb{Q}$ unless
$t = 2$ (the latter means that Y equals to a monomial in X).

Proof.   We can consider continuation of Y on $\mathbb{C}$ (and get $\Gamma_Y \subset \mathbb{C}^2$) and
also get an algebraic function (satisfying the same polynomial relation). We
can also analytically continue the relation Q. As usually in the neighborhood
of a point of Γ_Y where $X = 0$ or $Y = 0$ (so the function $X^{a^{(i)}}$ or $Y^{b^{(i)}}$ have
singularities), one should understand the relation Q to hold in a neighborhood
with a branch cut deleted (i.e. having a curve starting from the singular point
deleted).

Since the Newton polygon process and Puiseux series can be generalized to take into account fractional-power polynomials, we let $Y = cX^a + \sum \gamma_j X^{j/\nu}$ be the Puiseux series of an algebraic function $Y(X)$ in a neighborhood of $X = 0$. Let the leading term be cX^a and ν be a common denominator of the (rational) exponents (including a). If we let $Y_1 = Y/cX^a$, then Y_1 satisfies the relation

$$1 + \sum_{1 \leq i \leq t-1} c^{(i)} c^{b^{(i)}} X^{a^{(i)} + b^{(i)} a} Y_1^{b^{(i)}} = 0$$

and Y_1 is also minimally t-quasisparse.

Setting $\tilde{X} = X^{1/\nu}$, then Y_1 is analytical in a neighborhood of 0 as a function of $\tilde{X}$ and $Y_1(0) = 1$, therefore Y_1^b is also analytical in a neighborhood of 0 for each $b \in \mathbb{R}$. Hence the equality

$$1 + \sum_{1 \leq i \leq t-1} c^{(i)} c^{b^{(i)}} \tilde{X}^{\nu(a^{(i)} + b^{(i)} a)} \cdot Y_1^{b^{(i)}} = 0$$

can be reduced to an equality

$$1 + \sum_{\nu(a^{(i)} + b^{(i)} a) \in \mathbb{Z}} c^{(i)} c^{b^{(i)}} \tilde{X}^{\nu(a^{(i)} + b^{(i)} a)} Y_1^{b^{(i)}} = 0$$

where the summation ranges over all $\nu(a^{(i)} + b^{(i)} a) \in \mathbb{Z}$. Thus, because of minimal t-quasisparsity of Y_1 we get that $a^{(i)} + b^{(i)} a \in \mathbb{Q}$ for all $1 \leq i \leq t-1$. Since $Y_1 \not\equiv \text{const}$ (otherwise Y is a monomial in X which is equivalent to $t = 2$) one can change the roles of X, Y_1 and consider X as an algebraic function of Y_1. Let $c_1 Y_1^b$ be the first term of the Puiseux series expansion of X in the neighborhood of $Y_1 = 0$, denote $X_1 = X/c_1 Y_1^b$, then $b \in \mathbb{Q}$ and $X_1(0) = 1$. We get

$$1 + \sum_{1 \leq i \leq t-1} c^{(i)} c^{b^{(i)}} c_1^{a^{(i)} + b^{(i)} a} X_1^{a^{(i)} + b^{(i)} a} Y_1^{b^{(i)} + b(a^{(i)} + b^{(i)} a)} = 0 .$$

As above one proves that $b^{(i)} + b(a^{(i)} + b^{(i)} a) \in \mathbb{Q}$, hence $b^{(i)} \in \mathbb{Q}$, finally one concludes that $a^{(i)} \in \mathbb{Q}$, that proves lemma 6.   □

Observe that the statement of the lemma holds also for an algebraic function Y over a field $k(X)$ where $k \subset \mathbb{C}$. Now we treat algebraic functions in many variables.

Corollary 7. *Let $k \subset \mathbb{C}$ be a field and Y be minimal t-quasisparse and algebraic over $k(X_1, \ldots, X_n)$. Assume Y is not a monomial. If $Q(X_1, \ldots, X_n, Y) = 0$, then all the exponents $a_j^{(i)}, b^{(i)} \in \mathbb{Q}$.*

Proof. We argue by induction on n.
For $n = 1$, this follows from lemma 6 and the observation after it. Assume for some i, $Y = X_j^\alpha \tilde{Y}$ where $\tilde{Y}$ is algebraic over $k(X_1, \ldots, X_{j-1}, X_{j+1}, \ldots, X_n)$. This implies α is rational. We then have

$$1 + \sum_{1 \leq i \leq t-1} c^{(i)} X_1^{a_1^{(i)}} \cdots X_j^{a_j^{(i)} + b^{(i)} \alpha} \cdots X_n^{a_n^{(i)}} \tilde{Y}^{b^{(i)}} = 0$$

Since $\tilde{Y}$ does not depend on X_j and Y is minimally t-quasisparse, we have that $a_j^{(i)} + b^{(i)}\alpha = 0$.

By induction each $b^{(i)} \in \mathbb{Q}$, so $a_j^{(i)} \in \mathbb{Q}$. The induction hypothesis implies that all other exponents are rational as well.

Now assume that for all j, $Y \neq X_j^\alpha \tilde{Y}$ for any $\tilde{Y}$ algebraic over $k(X_1, \ldots, X_{j-1}, X_{j+1}, \ldots, X_n)$. Apply lemma 6 to Y considered as an algebraic function in X_j over $k(X_1, \ldots, X_{j-1}, X_{j+1}, \ldots, X_n)$ (without loss of generality we can suppose that k is finitely generated over $\mathbb{Q}$, so one can consider the field $k(X_1, \ldots, X_{j-1}, X_{j+1}, \ldots, X_n)$ as a subfield of $\mathbb{C}$). This implies that $a_j^{(i)}, b^{(i)} \in \mathbb{Q}$. The Corollary is therefore proved.   □

4 Finding the exponents of minimal t-sparse representations

Assume (as in the introduction) that Y is minimally t-sparse and let $f(X_1, \ldots, X_n, Y) = 1 + \sum_{1 \le i \le t-1} \gamma^{(i)} X_1^{\alpha_1^{(i)}} \cdots X_n^{\alpha_n^{(i)}} Y^{\beta^{(i)}} = 0$ be a normalized t-sparse representation of Y. Introduce variables $a_1^{(i)}, \ldots, a_n^{(i)}, b^{(i)}$, $1 \le i \le t-1$ that take their values in $\mathbb{R}$ and define operators $D_l = X_l \frac{d}{dX_l}$, $1 \le l \le n$. For any choice of the operators $\mathcal{D}_1, \ldots, \mathcal{D}_{t-1}$ such that $\mathcal{D}_j = D_1^{j_1} \cdots D_n^{j_n}$ where $1 \le \text{ord}(\mathcal{D}_j) = j_1 + \ldots + j_n \le t-1$, denote the generalized Wronskian

$$W_{\mathcal{D}_1, \ldots, \mathcal{D}_{t-1}} = \frac{\det(\mathcal{D}_j(X_1^{a_1^{(i)}} \cdots X_n^{a_n^{(i)}} Y^{b^{(i)}}))_{1 \le i, j \le t-1}}{X_1^{a_1^{(1)} + \ldots + a_1^{(t-1)}} \cdots X_n^{a_n^{(1)} + \ldots + a_n^{(t-1)}} \cdot Y^{b^{(1)} + \ldots + b^{(t-1)} - (t-1)^2}}$$

$$\in \mathbb{Z}[a_1^{(1)}, \ldots, a_n^{(1)}, b^{(1)}, \ldots, a_1^{(t-1)}, \ldots, a_n^{(t-1)}, b^{(t-1)}, \{DY\}_{0 \le \text{ord}(D) \le t-1}]$$

Observe that

$$\deg_{a_1^{(1)}, \ldots, b^{(t-1)}}(W_{\mathcal{D}_1, \ldots, \mathcal{D}_{t-1}}) \le t-1$$

$$\deg_Y(W_{\mathcal{D}_1, \ldots, \mathcal{D}_{t-1}}) \le (t-1)^2$$

$$\deg_{\{DY\}_{1 \le \text{ord}(D) \le t-1}}(W_{\mathcal{D}_1, \ldots, \mathcal{D}_{t-1}}) \le t-1$$

From [K 73], p. 83, it follows that $1 + \sum_{1 \le i \le t-1} c_i X_1^{a_1^{(i)}} \cdots X_n^{a_n^{(i)}} Y^{b^{(i)}} = 0$ (so the exponents $a_1^{(i)}, \ldots, a_n^{(i)}, b^{(i)}$ provide a normalized t-sparse representation) for suitable $c_i \in \mathbb{R}$ iff $W_{\mathcal{D}_1, \ldots, \mathcal{D}_{t-1}} = 0$ for all choices of $\mathcal{D}_1, \ldots, \mathcal{D}_{t-1}$ where

$1 \leq \mathrm{ord}(\mathcal{D}_j) \leq t-1$, $1 \leq j \leq t-1$. Denote

$$W = \sum_{1 \leq \mathrm{ord}(\mathcal{D}_j) \leq t-1,\, 1 \leq j \leq t-1} W^2_{\mathcal{D}_1, \dots, \mathcal{D}_{t-1}}.$$

Consider a minimal K such that a fractional-power polynomial $\frac{\partial^K f}{\partial Y^K}$ does not vanish identically on Γ_Y. Such K exists and moreover $K \leq t-1$. Indeed, rewrite $f = \sum_{1 \leq s \leq t} Y^{\eta^{(s)}} f_s$, where f_s are fractional-power polynomials in $X_1, \dots, X_n$, if $\frac{\partial f}{\partial Y}, \dots, \frac{\partial^{t-1} f}{\partial Y^{t-1}}$ vanish on Γ_Y then by lemma 4 every f_s also vanishes on Γ_Y which is impossible since Y is defined on $\mathbb{R}^n_+$. Then

$$0 = \frac{d}{dX_l} \frac{\partial^{K-1} f}{\partial Y^{K-1}} = \frac{\partial^K f}{\partial X_l \partial Y^{K-1}} + \left(\frac{\partial^K f}{\partial Y^K} \right) \frac{dY}{dX_l}.$$

Continuing applying the operators D_l we get by induction on $r = \mathrm{ord}(\mathcal{D}_j)$ that $\mathcal{D}_j Y$ can be expressed in the form $h / (\frac{\partial^K f}{\partial Y^K})^{2r-1}$ where h can be considered as a polynomial in $t+1$ monomials

$$\mathcal{M} = \{ Y, X_1^{\alpha_1^{(1)}} \cdots X_n^{\alpha_n^{(1)}} \cdot Y^{\beta^{(1)}-t-r}, \dots, X_1^{\alpha_1^{(t)}} \cdots X_n^{\alpha_n^{(t)}} \cdot Y^{\beta^{(t)}-t-r} \}$$

of the degree $t + O(r)$.

Substituting these expressions in W we obtain an expression $\hat{W}$ of the form $\hat{h} / (\frac{\partial^K f}{\partial Y^K})^{2t^2}$ where $\hat{h}$ belongs to

$$\mathbb{Z}[a_1^{(1)}, \dots, b^{(t-1)}][Y, X_1^{\alpha_1^{(1)}}, \dots, X_n^{\alpha_n^{(1)}}, Y^{\beta^{(1)}-2t}, \dots, X_1^{\alpha_1^{(t)}}, \dots, X_n^{\alpha_n^{(t)}}, Y^{\beta^{(t)}-2t}]$$

of degree $O(t^2)$ in the monomials from $\mathcal{M}$ with $r = t-1$.

Apply lemma 3 taking as $g = \hat{W} \cdot (\frac{d^K f}{dY^K})^{2t^2+1} = \hat{h}(\frac{d^K f}{dY^K})$. Then one can bound the sparsity T of g as follows: $T \leq t^{O(t)}$. Lemma 3 implies that there is a point $(x_1, \dots, x_n) \in \{1, \dots, B_1\}^n$. where $B_1 \leq 2^{t^{O(nt)}}$ such that $g(x_1, \dots, x_n, y) \neq 0$ for any value y of the function Y in the point $(x_1, \dots, x_n)$, provided that g does not vanish identically on Γ_Y. Since $(\frac{d^K f}{dY^K})(x_1, \dots, x_n) \neq 0$ all the derivatives $(\mathcal{D}_j y)(x_1, \dots, x_n)$ are defined, thus $W(x_1, \dots, x_n)$ is defined and $W(x_1, \dots, x_n) \neq 0$. Thus, we obtain the following

Lemma 8. $a_1^{(1)}, \dots, a_n^{(1)}, b^{(1)}, \dots, a_1^{(t-1)}, \dots, a_n^{(t-1)}, b^{(t-1)}$ *are the expo-nents of some normalized t-sparse representation of Y if and only if the vectors $(a_1^{(i)}, \dots, a_n^{(i)}, b^{(i)})$ are pairwise distinct and distinct from the zero vector (we call this the nontriviality condition on $a_1^{(1)}, \dots, b^{(1)}$) and the following system holds:*

$$W(x) = 0, \quad x \in \mathcal{J} \tag{1}$$

where $\mathcal{J}$ is the set of points $x \in \{1, \ldots, B_1\}^n$ where $B_1 \leq 2^{t^{O(nt)}}$ for which $(\mathcal{D}Y)(x)$ are defined for all the operators $\mathcal{D}$ of the orders at most $t-1$.

Remark that $W(x) \in I\!R[a_1^{(1)}, \ldots, b^{(t-1)}]$ and we get this polynomial of degree at most $O(t)$ in $O(nt)$ variables by plugging for $(\mathcal{D}Y)(x)$ the black-box values, provided that they are defined.

Corollary 7 implies that all the solutions $a_1^{(1)}, \ldots, b^{(t-1)}$ of a system of polynomial inequalities (1) (under the nontriviality condition) are rationals, therefore (1) has only a finite number of solutions. We say that $\deg f \leq æ$ (and so $\deg Y \leq æ$, see the introduction) if the absolute values of numerators and denominators of the rational numbers $\alpha_j^{(i)}, \beta^{(i)}$ do not exceed $æ$. The algorithm solves the system (1) (with the nontriviality conditions) using [GV88] in $B_1^n t^{o(nt)}(\log \deg Y)^{O(1)} \leq 2^{t^{O(nt)}}(\log \deg Y)^{O(1)}$ arithmetic operations with the depth $t^{O(nt)}(\log \deg Y)^{O(1)}$ ([HRS 90]). Denote also by M the maximum of absolute values of the outputs of the black-box during the computation, then the bounds from [GV88] imply that $\deg Y \leq M^{2^{t^{O(nt)}}}$. Observe also that [GV88] entails that (1) (with nontriviality condition) has at most $t^{O(nt)}$ solutions, thus the normalized t-sparse representations of Y.

If it is only known that Y is t-sparse, then the algorithm tests successively $t_1 = 1, 2, \ldots \leq t$ for minimal t_1-sparsity.

Summarizing we formulate the main result of the paper:

Theorem.   *a)*   *For t-sparse real algebraic function Y one can find $t_1 \leq t$ and the exponent vectors of all its normalized minimal t_1-sparse ($t_1 \leq t$) representations with $2^{t^{O(nt)}}(\log \deg Y)^{O(1)}$ arithmetic operations and with the depth $t^{O(nt)}(\log \deg Y)^{O(1)}$. The number of all minimal normalized sparse representations does not exceed $t^{O(nt)}$.*
b)   *One can also bound $\deg Y \leq M^{2^{t^{O(nt)}}}$ where M is the maximum of absolute values of the outputs of the black-box during the computation.*

References

[BCR 87]   Bochnak, J., Coste, M., and Roy, M. F., *Géométrie algébrique réelle*, Springer-Verlag, 1987.

[BT 88]   Ben-Or, M., Tiwari, P., *A deterministic algorithm for sparse multivariate polynomial interpolation*, Proc. STOC ACM, 1988, pp. 301-309.

[BSS 88]   Blum, L., Shub, M., and Smale, S., *On a theory of computation over the real numbers; NP completeness, recursive functions and universal machines*, Proc. IEEE FOCS, 1988, pp. 387-397.

[D 80]   Dold, A., *Lectures on algebraic topology*, Springer-Verlag, 1980.

[ES 52]   Eilenberg, S., and Steenrod, N., *Foundation of algebraic topology*, Princeton, 1952.

[GKO 91]   Grigoriev, D., Karpinski, M., and Odlyzko, A., *Nondivisibility of sparse polynomials is in NP under the Extended Riemann Hypothesis*, Preprint Max-Planck-Institute für Mathematik N6, 1991; to appear in J. AAECC **3** (1992).

[GKS 90a]   Grigoriev. D., Karpinski, M., and Singer, M., *Fast parallel algorithms for sparse multivariate polynomial interpolation over finite fields*, SIAM J. Comput., 1990, v. 19, N6, pp.1059-1063.

[GKS 90b]   Grigoriev. D., Karpinski, M., and Singer, M., *Interpolation of sparse rational functions without knowing bounds on exponents*, Proc. IEEE FOCS, 1990, pp. 840-847.

[GKS 91a]   Grigoriev. D., Karpinski, M., and Singer, M., *The interpolation problem for k-sparse sums of eigenfunctions of operators*, Adv. Appl. Math., 1991, v. 12, pp.76-81.

[GKS 91b]   Grigoriev, D., Karpinski, M., and Singer, M., *Computational complexity of sparse rational interpolation*, submitted to SIAM J. Comput.

[GV88]   Grigoriev, D., Vorobjov, N. (Jr.), *Solving systems of polynomial inequalities in subexponential time*, J. Symb. Comput., 1988, v. 5, pp. 37-64.

[HRS 90]   Heintz, J., Roy, M. F., Solerno, P., *Sur la complexité du principe de Tarski-Seidenberg*, Bull. Soc. Math. France, 1990, 118, pp. 101-126.

[K 73]   Kolchin, E., *Differential algebra and algebraic groups*, Academic Press, 1973.

[K 91]   Khovanskii, A., *Fewmonomials*, Transl. Math. Monogr., AMS **88**, 1991.

[M 64]   Milnor, J., *On the Betti numbers of real varieties*, Proc. AMS
               15, 1964, pp. 275-280.

Dima Grigoriev
Steklov Mathematical Institute
Fontanka 27, St. Petersburg, 191011 Russia
and
Dept. of Computer Science
University of Bonn
5300 Bonn 1, Germany
grigorev@cs.bonn.edu

Marek Karpinski
Dept. of Computer Science
University of Bonn
5300 Bonn 1, Germany
and
International Computer Science Institute
Berkeley, California, USA
marek@cs.bonn.edu

Michael Singer
Dept. of Mathematics
North Carolina State University
Raleigh, NC 27695-8205, USA
singer@math.ncsu.edu

Shade, Shadow and Shape

J.-P. Henry   M. Merle

1 Introduction

We shall motivate the use of geometric cues in intelligent vision, then explain how mathematics can be used to define and unify treatment of these cues.[1] [2]

1. a very classical cue : the apparent contour,

2. a less classical cue : the shade line separating full light from shadow,

3. a far less used cue : the cast shadow line.

They appear as a triple of 3 curves, defined through 2 different projections, in the image. We shall recall briefly what has been done in the now classical case where only the first cue is used. This includes many mathematical results, some ancient, that we shall only sketch and some very recent works applied to vision and inspired by the pioneering ideas of Jan Koenderink. In the main part of the text we shall develop a mathematical treatment of the local situations arising from the properties of the 3 curves precedently defined. We prove that, up to change of coordinates, there is only *a finite number* of stable *"Sol y Sombra Patterns"* that can arise in generic situations :

1. a **Sol y Sombra Pattern** will be defined as a suitable *equivalence class of local singularity of the triple of curves ;*

2. a **situation** will be defined as a set-up {equation of surface, direction of light, direction of eye or camera}

3. **generic situations** will be defined, first by impressing a *topology* on a suitable space of applications, hence on the space of situations, then by deciding that around generic situations there must be a neighbourhood of situations where the induced Sol y Sombra Patterns must stay equivalent.

---

[1]We are glad to thank Robert Fournier (INRIA, Sophia-Antipolis) for the use of ZICVIS, a versatile, user-friendly, software for drawing and viewing curves and lighted surfaces, a goody for geometers.

[2]Supported by the E.E.C. ESPRIT project BRA 3001 "INSIGHT"

4. finally, we shall say that a Sol y Sombra Pattern is *stable by small perturbations* (small motions of the observer or of the object or of the light), if, for near by situations, the corresponding Sol y Sombra Patterns are in the same equivalence class.

We shall also begin the classification of "Sol y Sombra Patterns" (singularity of the triple of curves) that are not stable but can be seen when moving the "situation" (by moving the object or the observer or the light spot) along a line : These are the *singularities of codimension 1*.

2   A mathematical insight in vision?

2.1   Motivations

What is intelligent vision ? What do we really use to understand scenes ? What are the pertinent **features** to extract from images to detect objects or to distinguish between them ?
And second question but not less important, how do we organize these cues to classify objects ?

2.1.1   Why Classify ?

Recognition means cognition and matching, therefore memory and a process, an *algorithm*, to match an image from the camera (the retina) with an image in the memory. Any (even the less sophisticated or the most specialized) Visual System needs some database to perform recognition. This database as well as the process of visual matching needs organisation and hierarchy, and therefore classification.

2.1.2   How to classify :

Some usual cues   Of course it seems that edges are important, and polyhedral objects provide images with lots of edges, so they should be and have been basic experiment support (see theoretical analyses in [KD], [K]). Textures and especially noticeable changes of textures are also detected even in the absence of occlusions (for a discussion of the relation between these two clues, see [T-R 90]). In motion perception or in stereo vision edges or discontinuity of texture are generally used, first they are detected then matched.

Discontinuity as an origin of cues   The key word appears to be discontinuity, discontinuity of texture, or discontinuity of orientation (two faces of a polyhedral object) or discontinuity of depth (an object on a background) or discontinuities in lighting.

Discontinuity out of smoothness : What about smooth objects with constant texture (or slowly varying texture) ? What are the discontinuities that they generate ?

Mathematicians can bring unifying and efficient concepts to answer this question.

There are several heuristic reasons to this :

First of all discontinuities are no longer physical incongruities that most mathematicians try to avoid, they are mathematical objects of interest especially under the name of **singularities**. Ideas of René Thom (see [T2], [T3] for original works, and [Cn], [Bn] or [Gi] for pedagogical introductions) have shown that classification of elementary singularities (Catastroph Theory) could be used to understand many situations (see [Hi]).

The second reason is related to the other key-word *classification*. Mathematicians have developped numerous tools for this task. Metrics and topology have been extently used to give precise meaning to the usual words of "neighbourhood" or proximity and "distance". *Isomorphism* is used for classifying in *equivalence classes*. To what extent objects in the same equivalence class are identical depends on the "rigidity" of the isomorphism that is used. For example an analytic isomorphism is very strong and strict, a topological one is much looser, two objects can be topologically isomorphic but not analytically. We shall talk of equivalence (analytic, differentiable, topological) when 2 objects are isomorphic (analyticaly, differentiably, topologicaly).

Another natural idea that is related to the idea of "proximity of forms" or of " small perturbations" has been formalized using *deformation*. This is another way of measuring proximity of forms. Unstead of looking at 2 views as 2 different points in some sort of big space of all views, and measuring their distance, we can think of them as 2 points *with a path* from one to the other. It becomes easier to measure their variation along the path, using infinitesimal calculus.

2.2 Towards a mathematical formulation

We shall now proceed to define the cues that we intend to use in the rest of this paper as mathematical objects. We shall begin with a first very classical cue in vision studies.

2.2.1 The apparent or occluding contour

We have seen that discontinuities were good cues. We can be a little more precise and use very simple mathematical terms. What is a view ? A view is a 2D representation of a 3D object. This can be seen as a projection either from a near point (conical projection) or an infinite point. The center of projection is the eye or the camera.

The apparent contour and the edges are lines of points where this projection is singular : either the tangent plane is seen as a line by the projection point or there is no (or two or even three secant tangent planes, that is too many), this is a singularity for the projection : at all those point there is no "isomorphism" between the tangent plane and the image plane (the television screen for example) even at a very low level, because they do not have the same dimension.

On a smooth object, which we shall modelize as a smooth surface, other discontinuities can be seen : curves separating lighted parts from shaded ones. These can be seen again as curves of points which are singular for another projection, projection from the sun or from a light spot (at least when there is one point-like light source). The shade line is made of points where the sun rays are tangent to the surface, i.e. included in the tangent plane.

Now we have several curves on smooth surfaces, and we can now proceed to classify them and relate their "forms" to the forms of the projected surfaces.

Our aim is to introduce a description and a coarse classification of all stable Sol y Sombra Patterns of objects which are lighted by a one point source at infinite distance. In fact by Sol y Sombra Pattern we will mean the apparent contour **and** the shadow edges of a smooth surface.

2.2.2   The shade and cast shadow curves

We look at a smooth object, i.e. a smooth generic surface, defined by polynomial equations in 3-space, in Monge form we can define the surface M as the image of a 2-plane by

$$(x, y) \longrightarrow (x, y, f(x, y))$$

and take a first orthogonal projection λ of this surface onto a plane ; the critical locus for this projection is a curve on the surface, the "shade curve" S, which in most situations separates shaded parts from lighted parts on the surface (Critical points for λ are points where light rays are tangent to the surface). Now if we take the set of all points where this light ray cuts again for the first time the smooth surface M we obtain a second curve, which we will call the "cast shadow curve", (or sometimes for short, the "shadow curve") O, which, together with the "shade curve", bounds the shaded part on the surface M.

2.2.3   The Sol y Sombra Pattern

Take another projection π of the surface orthogonally onto a plane, which will be the image by camera, and look at the critical locus of this projection C and at the projections of this curve Γ, the *outline* (apparent contour of the

surface which is the set of critical values of π since the surface is smooth),
and of the *shade and shadow curves*.

By this process we get three plane curves :

1. the image Σ of the shade curve, which we will call the shade contour,

2. the image Ω of the shadow curve, which we will call the shadow contour,

3. the apparent contour Γ of the surface M (intersection of visual rays tangent to M with the image plane).

This apparent contour (which is sometimes called outline or profile of the surface) is also the image by π of what is called by computer-vision authors the "rim" or the contour generator.

The singularities of apparent contours of smooth surfaces have been studied by several authors (Whitney [W], Mather [Ma 73], [Ma 76], Gaffney and Ruas [G], Bruce, Giblin [B-G1], Arnold and his school ...) and classified (for a survey see [Bn])

In this paper we go two steps forward, by adding the shade and cast shadow curves to the apparent contour and we call this artefact the Sol y Sombra Pattern.

2.2.4 Classifying the singularities of the contour, shade and shadow curves

We intend to classify the types of their common local singularities, i.e what we have called "Sol y Sombra Patterns" in the introduction. This study needs three steps :

1. Relate the singularity of this triple of curves to a convenient map, and construct a space of maps that modelize the "Sol y Sombra Patterns".

2. Define one (or several) equivalence relations on this space in a compatible way with analytic equivalence ("isomorphism") of local singularities.

3. Compute and describe the local singularities that remain isomorphic by small changes of the map.

We shall determine these types and show that their number is finite.

3   Mathematical recalls and notations

Our task is a little more complex than, but related to, the usual task of classifying the singularities of the sole apparent contour curve. We recall, in this section, the usual treatment of this problem, and its solution : there are only a finite number of stable (resp. codimension 1) singularity types for the apparent contour of generic surfaces.

Don't forget that in our application to vision, a View is a *projection* of a surface to a plane (a portion of a plane : the TV set on which we monitor the view of the camera). Precisely we will study local features and we will say that two views are equivalent if and only if there exists (local) diffeomorphisms of the surface and its image which identify the two projections. In particular the two apparent contours will be (locally) diffeomorphic as embedded curves in the plane.

We are interested in finding cases when the projection is stable by small perturbations (the view looks alike, when moving a little the camera or the object). We postulate, after Thom (Catastrophe theory , [T2]) that any feature that is not stable either is not seen by the camera, or is not detected because of noise and blurring. This postulation should be discussed in relation with scaling problems.

When one wants to study smooth objects (surfaces), it is possible to *classify* all projections of generic surfaces, or to classify generic projections of all surfaces (notice that the two cases overlap for generic projections of generic surfaces [W]).

In this paper we shall restrict ourselves to the first case. Some differences may appear in the classification according to the choice of the type of equivalence and topology on the set of maps to classify.

In the following subsections, we shall give very few historical milestones, but we shall not recall the mathematical results referring the reader either to the original papers (quoted in the bibliography) or to the very good introduction by Demazure[D2].

3.1   Some historical ramblings

Previous and related work in **transversality, stability and preparation theorems,** can be coarsely classified in three groups.

3.1.1   Fondamental theorems

The seminal task was Hassler Whitney's ([W]). The main mathematical work of defining concepts, relating geometric intuitions to precise statements, finding the good set-up for separating necessary hypothesis from uncertain grounds, and finally proving very hard theorems, is mainly due to René Thom and John Mather ([Ma 69], [Ma 70], [Ma 73], [Ma 76]), with collateral work of J.-C. Tougeron and alii.

3.1.2 Specialised results

The papers of Arnold and his school ([A1], [A6], [Pl1], [Pl2], see also [Bn] for a survey) or C.T.C. Wall, Bruce and alii ([Wa], [B-G1],[G]), *on this subject* is rather on the **R&D** side, using Mather's and Thom's theorems in specified contexts, adding lemmas and computations in specialised set-ups.

3.1.3 Applied results

There is now a third firesquad using again Thom's and Mather's results and generalisations of the specialised lemmas of the second group, and aiming to situations approximating more and more the real world ([Ri1], [Ri2], [Ta1], [Ta2], [B-G3]). The present work can be included in that group.

3.2 Classification of the apparent contour singularities

3.2.1 Notations for the classical case

A smooth surface M, imbedded in an euclidean space E_3 of dimension 3, being given, we want to describe the singularities of its apparent contour for a given projection on E_2.

In a neighborhood of a point of M we can choose coordinates such that locally M is defined as the image of the (x, y) plane by the morphism :

$$(x, y) \longrightarrow (x, y, z = f(x, y))$$

the Oz axis is outside the tangent plane, and the direction of projection is given by the $x'x$ axis :

$$(x, y, z) \longrightarrow (y, z)$$

We shall take f to be polynomial in x, y.

The apparent contour for the Ox direction is made of points where the tangent plane to M contains the $x'x$ direction, and is given on M by the equation:

$$\partial f(x, y)/\partial x = 0$$

and we want its image Γ by projection on the (y, z) plan.

The first thing to specify when speaking of classification is when do we decide to put two such curves of E_2 in the same class?

One of the stronger condition for equivalence is the existence of a local embedded analytic isomorphism between the two curves.

$$
\begin{array}{ccc}
R^2 & \overset{\sim}{\longrightarrow} & R^2 \\
\uparrow & & \uparrow \\
\Gamma_1 & \longrightarrow & \Gamma_2
\end{array}
$$

In that case there exists an analytic change of coordinates such that the germ of Γ_1 goes on the germ of Γ_2.

Of course, it will be nice to be able to decide if such an isomorphism exists or even more, if the two germs Γ_1 and Γ_2 are topologically equivalent as *embedded* subspaces in R^2.

We only deal here with a sufficient condition, namely, there exists a commutative diagram:

$$
\begin{array}{ccccc}
(x,y) & R^2 & \xrightarrow{R} & R^2 & (X,Y) \\
& \downarrow G & \downarrow G & \downarrow F & \downarrow F \\
G(x,y) & R^2 & \xleftarrow{L} & R^2 & (Y,Z)
\end{array}
$$

where L and R are analytic isomorphisms, R is origin preserving, and such that

$$G = L \circ F \circ R$$

The two morphisms G and F have their sets of critical values, Γ_1 and Γ_2 isomorphic. The apparent contour of M

$$M \;=\; \{(x,y,z) \mid z = f(x,y)\}$$

is the locus of the critical values of

$$
\begin{array}{cc}
R^2 & (X,Y) \\
\downarrow & \downarrow \\
R^2 & (Y, f(X,Y))
\end{array}
$$

Using morphisms of type

$$(x,y) \qquad\qquad \xrightarrow{R} \quad (X = x + \epsilon P(x,y), Y = y + \epsilon Q(x,y))$$
$$\downarrow$$
$$(Y + \epsilon\phi(Y,Z), Z + \epsilon\psi(Y,Z)) \;\xleftarrow{L}\qquad\qquad (Y, Z = f(X,Y))$$

we see (using Taylor expansion) that the morphism

$$(X,Y) \longrightarrow (Y, f(X,Y))$$

is infinitesimally equivalent (i.e. when $\epsilon^2 = 0$), to the morphism :

$$
\begin{pmatrix} x \\ y \end{pmatrix} \to \begin{pmatrix} y + \epsilon\phi(y,f) + \epsilon Q(x,y) \\ f(x,y) + \epsilon\psi(y,f) + \epsilon P(x,y)\partial f/\partial x + \epsilon Q(x,y)\partial f/\partial y \end{pmatrix}
$$

As we want to compare (classify) morphisms which are of the form

$$
\begin{pmatrix} x \\ y \end{pmatrix} \longrightarrow \begin{pmatrix} y \\ f(x,y) + \epsilon h(x,y) \end{pmatrix}
$$

so as to work on the sole deformation of $f(x,y)$, we can obtain such an equivalent representation of the surface M by taking $Q(x,y) = -\phi(y,f)$. We can conclude that the two surfaces :

$$z = f(x,y)$$

and

$$z = f(x,y) + \epsilon\psi(y,f) + \epsilon P(x,y)\partial f(x,y)/\partial x + \epsilon\phi(y,f)\partial f(x,y)/\partial y$$

will have the locally diffeomorphic apparent contours on the (y,z) plan.

3.2.2 Stable contours

As a special case of the preceding remark we get that if we define

$$\mathcal{K} \; := \; R\{y, f\} + R\{x, y\}\partial f(x, y)/\partial x + R\{y, f\}\partial f(x, y)/\partial y$$

the equality

$$\mathcal{K} \; = \; R\{x, y\}$$

implies that the analytic type of the apparent contour is infinitesimally stable, and by Mather's result is stable : any small deformation will give an equivalent germ with a locally diffeomorphic apparent contour.

Stable singularities of the apparent contour are unavoidable (inescapable), i.e. they are seen from every viewpoint (after may be a small perturbation). They are known since works by H. Whitney (see [W]) to be

the **etale** map

$$(x, y) \longrightarrow (x, y, z = x)$$

the **fold** map

$$(x, y) \longrightarrow (x, y, z = x^2)$$

the **cusp** map

$$(x, y) \longrightarrow (x, y, z = xy - x^3)$$

Remark In the case of stable singularities, the problem mentioned in 3.2.1 disappears: the three stable singularities are also stable ones with respect to the $\mathcal{A}_{top}$-equivalence. This is no longer true for singularities of greater codimension (there exists analytic moduli, such as *cross ratio*, which are not topological moduli).

3.2.3 Codimension 1

If we allow the observer or the camera (or the object) to move along a curve, we will get more singularity types, all those with orbit of codimension 1 in their deformation space. These are obtained if $\mathcal{K}$ is of codimension 1 in $R\{x, y\}$.

We get 3 other situations :

the **lips**

$$z = x^3 + xy^2$$

the **beaks**

$$z = x^3 - xy^2$$

and the **camel**

$$z = x^4 + xy \quad .$$

If the camera can be moved in all directions, i.e. in a projective space of dimension 2, it is necessary to look at singularities with orbits of codimension 2 in their deformation space. If the camera can be moved from infinity to a short distance of the object then we get one parameter more, namely the distance to the object and classification has to be carried on to codimension 3 (Platonova, Sherback [Pl2]).

3.2.4   Generalizations

Several generalizations have already been worked on, for example following ideas of Koenderink see [K]), J.H. Rieger ([Ri1], [Ri2]) has studied these projections up to greater codimension (typically between 3 and 6), this beeing necessary as he wants to look at one parameter family of generic surfaces, because he wants to introduce a scale parameter to modelize modifications of surfaces biased by refinements on their representations.

J.W. Bruce and P.J. Giblin (see [B-G2]) have investigated the singularities of generic surfaces with boundary. One of their students, F. Tari (see [Ta1] [Ta2]) has considered, in his Ph.D. thesis (Liverpool, May 1990), pairs of smooth surfaces and also triples of tranversal surfaces and has related this with [B-G2] and classified these situations at least up to codimension 2.

J.H. Rieger ([Ri2]) has also studied singularities of projections of piecewise-smooth surfaces. O. A. Platonova has worked on classifications of projections of smooth surfaces via several definitions of this type of classification. Platonova has found with Scherback eleven types of elementary singularities (see [Pl2]) of projections of generic surfaces by pencils of line, which is less than can appear in a generic 3-parameter family of projections[3] (classified by Goryunov).

4   Classifying views of lighted smooth surfaces

4.1   The mathematical set-up

4.1.1   Notations and hypothesis

Like in the classical case we can assume that our object is a smooth surface M locally embedded in R^3 and that, in a neighbourhood of the origin, it is the image of R^2 by the morphism:

$$
\begin{array}{ccc}
R^2 & \xrightarrow{\;f\;} & R^3 \\
(x,y) & \longrightarrow & (x,y,z = f(x,y))
\end{array}
$$

---

[3]Projections from all points of the euclidean 3-space are related and do not form a generic 3-parameter family of projections

We assume that the camera is at the infinity in the direction of the x-axis (projection on the (y,z)-plane) and that the sun direction is the y-axis, (with the sun on the $y > 0$ side), the only restriction we have thus made is that the camera is not in line with the sun.

In our coarse way to take images, what we call the image of the lighted surface is the union of three plane curves:

1. the image Σ of the shade curve, which we call the shade contour; this is the image of points on M where the sun ray is tangent to M

2. the image Ω of the shadow curve, which we call the shadow contour; it is the image of the points on M where a tangential sun ray cuts the surface M out of the shade curve

3. the apparent contour Γ of the surface M; it is the image of points on M where the visual ray is tangent to M.

We want to classify the local singularity types of the *union* of 3 analytic curves:

1. The curve of critical values of the x-axis projection π on M i.e. the critical value curve of the morphism $\bar{f}$,

$$
\begin{array}{ccc}
R^2 & \xrightarrow{\bar{f}} & R^2 \\
(x,y) & \longrightarrow & (y, f(x,y))
\end{array}
$$

will be called $\Gamma(\bar{f})$ or Γ if no confusion is possible.

2. The shade contour Σ, which is the image by π of the visible part of the shade curve S. The shade curve S has been defined as the critical value curve S of the projection λ of y-axis.

   The equations in R^3 of S will be:

$$
\begin{array}{rcl}
z & = & f(x,y) \\
0 & = & \partial f(x,y)/\partial y
\end{array}
$$

3. The shadow contour Ω, image by π of the shadow curve O.

   The equations in R^3 of $O \cup S$ will be given by a projection of an analytic set of R^4:

$$
\left\{ (x,y,y',z) ;\;
\begin{array}{rcl}
z & = & f(x,y') \\
z & = & f(x,y) \\
0 & = & \partial f(x,y')/\partial y
\end{array}
\right\}
$$

We can also define O as the inverse image of the locus of critical values of the projection λ of kernel the y-axis.

4.1.2   Classification principles

We would like to use, as equivalence relation, analytic equivalence, i.e. existence of an embedded analytic isomorphism between the union of 3 curves, sending apparent contour on apparent contour, shade curve on shade curve and shadow curve on shadow curve. Of course, analytic (or even topological) equivalence of the union of the three contours would be a better criterion, but harder to deal with. Moreover, stability with respect to analytic equivalence is a sufficient condition for topological stability (see 3.2.2).

$$
\begin{array}{ccc}
R^2 & \xrightarrow{\sim} & R^2 \\
\uparrow & & \uparrow \\
\Gamma_1 \cup \Sigma_1 \cup \Omega_1 & \xrightarrow{\sim} & \Gamma_2 \cup \Sigma_2 \cup \Omega_2
\end{array}
$$

The Right-Left equivalence on the morphism $\bar{f}$ defined by

$$\bar{f} \simeq L \circ \bar{f} \circ R$$

where R and L are analytic isomorphisms of R^2, will of course induce an embedded isomorphism of the critical curve of $\bar{f}$ with the critical curve of $L \circ \bar{f} \circ R$.

As $\Gamma(\bar{f})$ can also be defined as the locus of critical values of the morphism $\bar{f}$, we get a sufficient condition for the equivalence of $\Gamma(\bar{f})$ with $\Gamma(L \circ \bar{f} \circ R)$, that is of the two apparent contours. What can we say about the other two curves ?

4.1.3   A criterion for invariance of shade and shadow

We need extra conditions to insure that the shade and shadow curves are invariant by a combined change of coordinates on the source space and on the target space. A first necessary condition is that they are such by an *infinitesimal* change of coordinates.

First of all, we remark that the ideal S (the **"shadow ideal"**) defining the union of the shade and shadow curves is given by :

$$
\begin{array}{ll}
S & \subset \ R\{x,y,y',z\} \\
S & = \ (z - f(x,y'), \ z - f(x,y), \ \partial f(x,y')/\partial y)R\{x,y,y',z\}
\end{array}
$$

Remark If the sun is in the $y > 0$ direction, then the shadow curve is better defined by adding the inequality $y' \geq y$ which says that the point (x,y,z) is in the shadow of the point (x,y',z). In the following, we will study both parts of the shadow curve, as we can switch from one to the other by reversing the propagation of the light.

We want to compare, for a change of coordinates on the source and on the target, the transform of the *shadow ideal* with the shadow ideal of the transform.

To simplify notations we will write P' instead of $P(x, y')$ for an element of the algebra $R\{x, y\}$ and write ϕ' instead of $\phi(y, f(x, y'))$, for an element of the subalgebra $R\{y, f\} \subset R\{x, y\}$.

For ε small, that is computing $(mod\ \varepsilon^2)$, we have to compare the following ideals in $R\{x, y, y', z\}$:

$$\mathcal{I}_1 = \left\{ \begin{array}{l} \dfrac{\partial f}{\partial y}(x + \varepsilon P', y' + \varepsilon \phi') \\[2mm] \dfrac{1}{y' - y}(f(x + \varepsilon P', y' + \varepsilon \phi') - f(x + \varepsilon P, y + \varepsilon \phi)) \end{array} \right.$$

and the shadow ideal of the embedding associated to $g(x, y)$, where

$$g(x, y) = f((x + \varepsilon P(x, y), y + \varepsilon \phi(y, f)) + \varepsilon \psi(y, f)$$

which is

$$\mathcal{I}_2 = \left\{ \begin{array}{l} \dfrac{\partial g}{\partial y}(x, y') \\[2mm] \Delta_y g \end{array} \right.$$

where Δ_y is the partial difference operator on y:

$$\Delta_y(g(x, y)) := \frac{g(x, y) - g(x, y')}{y - y'} = \frac{g - g'}{y - y'}.$$

$$\mathcal{I}_2 = \left\{ \begin{array}{l} \partial f/\partial x(x + \varepsilon P', y' + \varepsilon \phi')\varepsilon \partial P/\partial y + \varepsilon \partial \psi/\partial y(y', f') + \varepsilon \partial \psi/\partial z \partial f/\partial y \\ \qquad\qquad + (\partial f/\partial y(x + \varepsilon P', y' + \varepsilon \phi'))(1 - \varepsilon \partial \phi/\partial y) \\ \Delta_y(f(x + \varepsilon P, y + \varepsilon \phi) + \varepsilon \psi(y, f)) \end{array} \right.$$

Comparing these two ideals give two conditions insuring invariance of shade and shadow.

$$\frac{\partial f}{\partial x}\frac{\partial P}{\partial y} + \frac{\partial \psi}{\partial y} \quad \in \quad \frac{\partial f}{\partial y} R\{x, y\}$$

$$\Delta_y \psi \quad \in \quad (\frac{\partial f}{\partial y}(x, y'), \Delta_y f)\, R\{x, y, y'\}$$

Let's observe that, using the Taylor formula, the second condition implies

$$\frac{\partial \psi}{\partial y} \in \frac{\partial f}{\partial y} R\{x, y\}.$$

Hence we get a new formulation of the invariance of the shadow ideal by an infinitesimal change of coordinates. Gluing together with 4.1.2 we get the following

Proposition 4.1.4 *An infinitesimally small deformation of the function f given by $g = f + \varepsilon h$ induces an infinitesimally small trivial deformation of*

the shade, shadow and apparent contour curves if and only if there exists
functions $P \in R\{x, y\}$, $\phi \in R\{y, f\}$ and $\psi \in R\{y, f\}$ such that

$$(o) \qquad h = P\frac{\partial f}{\partial x} + \phi\frac{\partial f}{\partial y} + \psi$$

$$(i) \quad \frac{\partial f}{\partial x}\frac{\partial P}{\partial y} \in \frac{\partial f}{\partial y} R\{x, y\}$$

$$(ii) \quad \Delta_y \psi \in (\frac{\partial f}{\partial y}(x, y'), \Delta_y f) R\{x, y, y'\}$$

Using this result we want to give (at least sufficient) conditions for a function f to give a stable image, that is conditions which insure that *every*
infinitesimal perturbation of f induces an infinitesimally small trivial deformation of the shade, shadow and apparent contour curves

Lemma 4.1.5 *The two conditions (i) and (ii) are naturally satisfied if*

1. *$P(x, y)$ depends of y only through powers of $f(x, y)$, that is belongs to
   the subalgebra $R\{x, f(x, y)\}$ of $R\{x, y\}$*

2. *$\psi \in R\{f(x, y)\}$.*

Proof : *If*

$$P(x, y) = \mathcal{P}(x, f(x, y))$$

we get

$$\frac{\partial P}{\partial y} = \frac{\partial \mathcal{P}}{\partial z}\frac{\partial f}{\partial y}$$

and if ψ only depends of $z = f(x, y)$:

$$\frac{\partial \psi}{\partial y} = \frac{\partial \psi}{\partial z}\frac{\partial f}{\partial y}$$

*condition (i) is easily verified, so is (ii) because it is satisfied by all powers
of f.* ∎

For the study of stable shade, shadows and contours we will need the following consequence:

Proposition 4.1.6 *If the module*

$$\mathcal{K}' = R\{x, f\}\frac{\partial f}{\partial x} + R\{y, f\}\frac{\partial f}{\partial y} + R\{f\}$$

*is equal to the whole ring $R\{x, y\}$ all infinitesimal perturbations of the map-
ping $(x, y) \longrightarrow (y, f(x, y)$ will not change the shade, shadow and apparent
contours. Hence the singularity associated to f is infinitesimally stable.*

4.2   Stable patterns

From now on, we will talk of **Sol y Sombra Patterns** as a short-term for the analytic embedded type of the union of the three curves, apparent contour, shade and shadow.

We will define **stable Sol y Sombra Patterns** to be Sol y Sombra Patterns such that there exists an equivalent Sol y Sombra Pattern lying in any small deformation of it.

4.2.1   Computation method

A necessary condition for a surface M to provide a stable Sol y Sombra Pattern, is that its associated apparent contour must be stable. So we have first to consider

the **etale** map

$$(x, y) \longrightarrow (x, y)$$

the **fold** map

$$(x, y) \longrightarrow (x, y^2)$$

the **cusp** map

$$(x, y) \longrightarrow (x, xy - y^3)$$

compute for each type the shade and shadow. If it is stable we have got a stable Sol y Sombra Pattern, if it is not, we look at small perturbations of this surface until we have enumerated all stable Sol y Sombra Patterns.

We will first look for stable Sol y Sombra Patterns with contour isomorphic to the contour of the first type (the "etale"), i.e. empty (not going through the point).

4.2.2   Far from the rim

Let us first consider the embedding of M, given by

$$(x, y, z = x)$$

in that case 0 is not on the apparent contour of M for the projection π which is generic, but this surface is tangent to the sun rays, which is a very instable situation (gliding rays). We look for perturbations of the morphism $(x, y) \longrightarrow (y, x)$ which are stable Sol y Sombra Patterns. We get those perturbations by adding to x terms of increasing degree.

We obtain three different stable "Sol y Sombra Patterns"

"Sunny side" (or full shade)   If the initial form of f is linear and not divisible by x, the pattern is stable and is equivalent to the mapping

$$(x, y, z = x + y)$$

the point is not on the rim,
the point is in full light (or in full shadow).

"Shade line"   If the initial form of $f - x$ is a quadratic form which doesn't vanish along $x = 0$, then the pattern is stable and equivalent to the mapping

$$(x, y, z = x + y^2)$$

the point is not on the rim,
the point is on the shade line, which is a fold for the shading, the shadow line is not going through this point.

Shade crease or "shaded box pleat"   If the initial form of $f - x$ is a quadratic form vanishing on $x = 0$, we first notice that the mapping $(x, y) \longrightarrow (y, x + xy)$ is not a stable one (4.3.2). Hence assume that the initial form of $f - x - xy$ is a cubic form not divisible by x. The module

$$K' = R\{x, f\}\frac{\partial f}{\partial x} + R\{y, f\}\frac{\partial f}{\partial y} + R\{f\}$$

is equal to the whole ring $R\{x, y\}$ (just notice first that $\partial f/\partial y R\{y, f\}$ is the ideal $\partial f/\partial y R\{x, y\}$ and choose a basis of the quotient $R\{x, y\}/\partial f/\partial y$ generated by powers of y)
Then by 4.1.6 we get a stable pattern, equivalent to the mapping

$$(x, y) \longrightarrow (y, x + xy + y^3).$$

The rim

$$\frac{\partial f}{\partial x} = 1 + y = 0$$

does not contain the point, the shade is given by

$$S = \left\{(x, y, z) ; \begin{array}{ccc} z & = & x + xy + y^3 \\ \frac{\partial f}{\partial y} & = & x + 3y^2 = 0 \end{array}\right\}$$

which gives

$$\Sigma = \left\{(y, z) ; z = -3y^2 - 2y^3\right\}$$

the shadow curve Ω is the projection of the part of

$$\left\{(x, y, y', z) ; \begin{array}{ccc} z & = & x + xy + y^3 \\ z & = & x + xy' + y'^3 \\ \frac{\partial f}{\partial y}(x, y') & = & x + 3y'^2 = 0 \end{array}\right\}$$

not contained in S, on the (y, z) plan.

$$\Omega = \left\{ (y, z) \ ; \ z = -\frac{3}{4}y^2 + \frac{1}{4}y^3 \right\}$$

We obtain two half parabolas with the same tangent but different curvatures, (there is no symmetry with respect to the normal line).

Remark If the initial form of $f - x - xy$ is a cubic form divisible by x, or if it is in the ideal m^4, then f is not a stable pattern.

4.2.3   On the apparent contour

The simplest equation for the fold of the apparent contour is :

$$(x, y, z = x^2)$$

But this surface is not stable for shade.
To have a fold on the apparent contour, the derivative $\partial f / \partial x$ must vanish at the origin.

The apparent contour fold If the initial form of f is linear and not divisible by x (in order to have $(\partial f / \partial x)(0, 0) = 0$, the pattern is stable and is equivalent to the mapping

$$(x, y) \longrightarrow (y, x^2 + y)$$

gives a stable Sol y Sombra Pattern (the shade curve is empty).

"the crescent moon" If the initial form of f is a quadratic form non divisible by x, then we get an unstable pattern (cf 4.3.2). If not, we get a stable one, equivalent to the mapping

$$(x, y) \longrightarrow (y, x^2 \pm xy)$$

The apparent contour is a smooth line, with a shadow curve which is half a parabola, with the same tangent line, and different curvature, beginning at the summit.

The cusp fold of the apparent contour Another stable local Sol y Sombra Pattern is of course a cusp fold of the apparent contour which is far from the shade edges, with a local equation which is given by :

$$f(x, y) = y + xy + x^3$$

4.3   The codimension 1 singularities

4.3.1   Far from the rim, shade-shadow singularities

To get an unstable pattern in the "far from the rim" case (cf 4.2.2) (that is 0 not on the apparent contour), we must take an f such that

1. $f(x, y) - x$ is a cubic form

2. $f(x, y) - x - xy$ is a cubic form divisible by x.

Shade folds confluence or the "Diabolo singularity"   If $f(x, y) - x$ is a cubic form non divisible by x, we get a codimension 1 pattern, equivalent to

$$(x, y, z = x + x^2 y - y^3)$$

the rim

$$\frac{\partial f}{\partial x} = 1 + 2xy = 0$$

does not contain the point, the shading rim is given by

$$S = \left\{ (x, y, z) ; \begin{array}{rcl} z & = & x + x^2 y - y^3 \\ \frac{\partial f}{\partial y} & = & x^2 - 3y^2 = 0 \end{array} \right\}$$

which gives two branches for the shade contour

$$\Sigma_1 = \left\{ (y, z) ; z = +\sqrt{3}\, y + 2y^3 \right\}$$

$$\Sigma_2 = \left\{ (y, z) ; z = -\sqrt{3}\, y - 2y^3 \right\}$$

the shadow curve Ω is the projection of the part of

$$\left\{ (x, y, y', z) ; \begin{array}{rcl} z & = & x + x^2 y - y^3 \\ \frac{\partial f}{\partial y}(x, y') & = & x^2 - 3y'^2 = 0 \end{array} \right\}$$

not contained in S, on the (y, z) plan.

$$\Omega_1 = \left\{ (y, z) ; z = -\frac{\sqrt{3}}{2} y - \frac{1}{4} y^3 \right\}$$

$$\Omega_2 = \left\{ (y, z) ; z = +\frac{\sqrt{3}}{2} y - \frac{1}{4} y^3 \right\}$$

In the equivalence module y is missing and therefore this situation can be deformed by a one parameter family

$$(x, y, z = x - u^2 y + x^2 y - y^3)$$

on a scene with 2 "shade crease" points $(x = u, z = 0)$ and $(x = -u, z = 0)$ of type

$$(X, y, z = X + Xy + y^3)$$

The grazed camel If $f(x, y) - x - xy$ is a cubic form divisible by x, the function f can be written down

$$x + xy + xm^2 + m^4$$

A typical form is the "shade" analog of the camel singularity of the apparent contour:

$$(x, y, z = x + xy + y^4)$$

The shading rim is given by

$$S = \left\{ (x, y, z) ; \begin{array}{rcl} x & = & -4y^3 \\ z & = & -4y^3 - 3y^4 \end{array} \right\}$$

4.3.2 On the rim, rim-shade singularities

The half-moon singularity In codimension 1 by deformation of the "fold of the apparent contour" case given by

$$(x, y, z = x^2)$$

(in that case 0 is on the apparent contour of M) we can get an unstable (codimension 1) situation :

$$(x, y, z = x^2 - y^2)$$

the rim

$$\frac{\partial f}{\partial x} = 2x = 0$$

contains the point 0, and is the smooth curve $z = -y^2$. The shade curve is given by :

$$S = \left\{ (x, y, z) ; \begin{array}{rcl} z & = & x^2 \\ \frac{\partial f}{\partial y} & = & -2y = 0. \end{array} \right\}$$

this is the z axis another smooth curve orthogonal to the apparent contour. Computation of the tangent space of the deformation shows that xy is the only missing term. The module

$$K' = R\{x, f\}\frac{\partial f}{\partial x} + R\{y, f\}\frac{\partial f}{\partial y} + R\{f\}$$

is equal to the sum of the two algebras $R\{x, y^2\}$ and $R\{x^2, y\}$.

5   Application to vision

We can now look forward using our new tools for intelligent Vision.

We only intend to add some preprocessing or eventually post processing in a number of situations where they could help, and that we shall now attempt to enumerate.

5.1   Shape from grey levels and Sol y Sombra Patterns

The most natural motivation for determinating the Sol y Sombra Patterns is to use them in relation with usual procedures of *Shape from Shading*. The apparent contour is generally used as lines of determinacy of the normals, which gives limit conditions for integration of vector fields on the plane image of the surface. The curves of the contour are not only solutions of ordinary differential equations related to the computation of the *image irradiance equation* ([Ho2] ch. 11, and [Ho1]), i.e. base characteristics, but even more they are *characteristic strips*, which means that the surface orientation is known along this curve.

The shade curves provide information of the same type as we know that along points of the shade curve the sun rays are tangent to the surface, which means that the normal to the surface is orthogonal to the sun direction. The tangent to the curve gives another information on the normal to the surface at this point. Therefore they can also be used as initial curve, with a major advantage on occluding contours, because the slope of the surface is not infinite at these points. But the cast shadow curve does not provide the same sort of information, and dicriminating them, is of the upmost necessity. This question introduces our last consideration, labelling.

5.2   Labelling Sol y Sombra Patterns

We shall not develop this point. It will be the subject of a following paper. Let us only say we shall need classification of singularities of the triple of curves, not only local singularities, as in this article, but also semi-local ones : They are due to projections of 2 or 3 different points of the surface or of different surfaces on the *same point* of the image. After this again mathematical task, it is possible to discriminate between permissible patterns and impossible ones. This allows to label the three curves in a very similar way to what has been initiated by Clowes and Huffman for edges of polyhedral objects, (world of children's blocks), extended by Waltz ([Wz]) to polyhedral objects with shadows, and by Malik ([Mk]) to opaque objects with piecewise smooth surfaces.

References

[AR]   ABRAHAM (Ralph) and Joel ROBBIN, Transversal Map-
         pings and Flows, Benjamin, New-York, 1967.

[A1]   ARNOLD (Vladimir), Lectures on Bifurcations in Versal Fam-
         ilies, Russian Math. Surveys, 27, 1972.

[A2]   ARNOLD (Vladimir), Critical Points of Smooth Functions,
         Proc. Intern. Congr. Math. Vancouver 1974, pp. 19-39.

[A6]   ARNOLD (Vladimir), Catastrophe Theory, Springer Verlag,
         1984.

[A7]   ARNOLD (Vladimir I.), *Wave front evolution and equivariant
         Morse lemma*, Comm. Pure Appl. Math., **29**, (1976), 557-582.

[A8]   ARNOLD (Vladimir I.), Singularities of systems of rays, Rus-
         sian Math. Surveys, 38, 1983.

[AGV]   V. ARNOLD, A. VARCHENKO, S. GOUSSEIN-ZADE *Sin-
         gularités des applications différentiables*, traduit du russe, Edi-
         tions Mir, Moscou.

[Bn]   BENNEQUIN (Daniel), *Caustique mystique* (d'après Arnold
         et al.), Séminaire Bourbaki 84-85, n° 634.

[B-G1]   J.W. BRUCE and P.J. GIBLIN, "Outlines and their duals",
         Proc. London Math. Soc. **50** (1985), 552-570.

[B-G2]   J.W. BRUCE and P.J. GIBLIN, *"Curves and Singularities"*,
         Cambridge University Press, (1984).

[B-G3]   J.W. BRUCE and P.J. GIBLIN, "Projections of Surfaces with
         boundary", *Proc. London Math. Soc.*, (3) **60** (1990), 392-416.

[Cn]   CHAPERON (Marc), Introduction à la théorie des catastro-
         phes, Journées X-UPS, vol.2 (1978-79), Centre de Mathéma-
         tiques, Ecole Polytechnique, 1984.

[Cr]   CHENCINER (Alain), "Travaux de Thom et Mather sur la
         stabilité topologique", Séminaire Bourbaki, 25 ème année,
         1972-1973, exposé 424,

[D1]   DEMAZURE (Michel), "Classification des germes à point
         critique isolé et à nombre de modules 0 ou 1 (d'après
         V.I. Arnold)", Séminaire Bourbaki 1973-1974, exposé 443,
         Springer Lecture Notes in Math. **431** (1975), 124-142.

[D2]   M. DEMAZURE, "Géométrie, Catastrophes et bifurcations",
            Ecole Polytechnique, Cours, Édition 1987, published under the
            title :"Catastrophes et bifurcations", Éditions Ellipses, Paris.

[Du83]   DUFOUR (Jean-Paul), "Famille de courbes planes différen-
            tiables", Topology, vol. 22, 4, 449-474, 1983, Printed in Great
            Britain, Pergamon Press Ltd.

[Du89]   DUFOUR (Jean-Paul), "Modules pour les famille de courbes
            planes", Annales de l'Institut Fourier, 39, 1, (1989), 225-238.

[G]   GAFFNEY (Terence), "The structure of $TA(f)$, classification
            and application to differential geometry, Orlik, P (ed.), Singu-
            larities Proc. Symp. Pure Math. 40 (part 1) (1983), 409-428.

[Gi]   GILMORE (Robert), Catastrophe Theory for Scientists and
            Engineers, Wiley, New-York, 1981.

[GG]   GOLUBITSKY (Martin) and Victor GUILLEMIN, Stable
            Mappings and Their Singularities, Graduate Texts in Mathe-
            matics, vol. 14, Springer-Verlag, 1974, second corrected print-
            ing 1980.

[GR]   GORJUNOV, in "Singularitiy theory and its applications"
            Advances in Soviet Math. Vol 1 (Arnold ed.) AMS 1990.

[HM]   HENRY (Jean-Pierre), and Michel MERLE, "Fronces et dou-
            bles plis", in Michel Merle Thèse d'État, Paris, Février 1990.

[HT]   HENRY (Jean-Pierre), and Bernard TEISSIER, "Suffisance
            des familles de jets et Équisingularité", in Jean-Pierre HENRY
            Thèse d'État, Paris, Septembre 1979, et Séminaire F. Norguet
            (1974-75), Springer Lecture Notes in Math. 482 (1976), 351-
            357.

[Hi]   HILTON (Peter John), editor, "Structural Stability, the The-
            ory of Catastrophes, and Applications in the Sciences, Pro-
            ceedings of the conference held at Battelle Seattle Researh
            Center 1975, Springer Lecture Notes in Math. 525 (1976).

[Ho1]   HORN (Berthold Klaus Paul), Height and Gradient from
            Shading, International Journal of Computer Vision, 5 :1,
            (1990), 37-75, Kluwer Academic Publishers.

[Ho2]   HORN (Berthold Klaus Paul), Robot Vision, MIT Press, MA ;
            and McGrawHill : New-York (1986).

[Ho3]   HORN (Berthold K. P.), Obtaining shape from shading infor-
            mation., in P. H. Winston (ed.) The Psychology of Computer
            Vision, McGrawHill : New-York (1975), 115-155.

[Ir]   IRWIN (M.), Smooth Dynamical Systems, Academic Press, 1980.

[K]   J.J. KOENDERINK, "What does the occluding contour tell us about solid shape," *Perception*, **13**, (1984) 321-330.

[KD]   J.J. KOENDERINK and A.J. Van DOORN, "The shape of smooth objects and the way contours end." *Perception*, **11**, (1984) 129-137.

[KD]   J.J. KOENDERINK and A.J. Van DOORN, "The singularities of the visual mapping," *Bio. Cybernet.* **24** Springer-Verlag (1976) 51-59.

[Mk]   MALIK (Jitendra), Interpreting Line Drawings of Curved Objects, *International Journal of Computer Vision*, **1**, (1987), 73-103, Kluwer Academic Publishers, Boston.

[Ma]   MATHER (John), Stratifications and Mappings, *Dynamical Systems*, Academic Press, Inc., New-York and London, (1973).

[Ma 69] MATHER (John), Stability of C^∞ mappings III, *Publ. I.H.E.S.*, **35**, (1969), 127-156.

[Ma 70] MATHER (John N.), Stability of C^∞ mappings IV, *Publ. I.H.E.S.*, **37**, (1970), 223-248.

[Ma 73] MATHER (John N.), *Generic projections*, Annals of Math., **98** no 2, September, (1973) 226-245.

[Ma 76] MATHER (John N.), How to Stratify Mappings and Jet Spaces, L.N.M. n°535, pp 128-176, Springer Verlag, 1976.

[Pl1]   PLATONOVA (O. A.), Singularities of the mutual disposition of a surface and a line, *Russian Math. Surveys*, **36** :1, (1981), 248-249.

[Pl2]   PLATONOVA (O. A.), Singularities of projections of smooth surfaces, *Russian Math. Surveys*, **39** :1, (1984), 177-178.

[Ri1]   J.H. RIEGER, "Families of maps from the plane to the plane", *J. London Math. Soc.*, (2) **36** (1987), 351-369.

[Ri2]   J.H. RIEGER, "On the classification of views of piecewise smooth objects", *Image and vision computing*, vol. **5** no 2 (1987), 91-97.

[RK]   ROSENFELD (A.) and A.C. KAK, *Digital Picture Processing*, Vols 1 & 2, second edition, Academic Press, New-York 1982.

[Ta1]   TARI (Fari), Projections of piecewise smooth surfaces, to appear in *J. London Math. Soc.*

[Ta2]   TARI (Fari), *Some applications of Singularity Theory to the Geometry of Curves and Surfaces*, Ph. D. Thesis, Liverpool, 1990.

[T1]   THOM (René), *Sur la théorie des enveloppes*, Journal de Mathématiques Pures et Appliquées, XLI-2 (1962).

[T2]   THOM (René), Stabilité structurelle et morphogénèse: essai d'une théorie générale des modèles, Benjamin, Reading (Mass), 1972 ou Addison-Wesley, Reading, Mass. 1973.

[T3]   THOM (René), Modèles mathématiques de la morphogénèse, Ch. Bourgois, Paris, 1981.

[T1]   TOUGERON (Jean-Claude), "Stabilité des applications différentiables", d'après Mather, Séminaire Bourbaki, 1967-1968, exposé 336, *Addison-Wesley Benjamin*.

[T-R 90] TODD (James T.) and Francene D. REICHEL "Visual Perception of smoothly curved surfaces from double-projected contour patterns.", Journal of experimental Psychology : Human Perception and Performance.,Vol. 16, No 3, 1990, 665-674.

[Tu]   TUENO (Jean-Paul), Thèse de 3 ème Cycle, Université de Montpellier, Novembre 1989.

[Wa]   C.T.C. WALL, " Geometric Properties of Generic Differential Manifolds", Vol III, *Springer lecture Notes in Math.*, **597** (1977), 707-774.

[Wz]   WALTZ, (David) " Understanding line drawings of scenes with shadows", in P. H. Winston (ed.) *The Psychology of Computer Vision*, McGrawHill : New-York (1975), 19-91.

[W]   WHITNEY (Hassler), "On singularities of mappings of Euclidean spaces. I. Mappings of the plane into the plane", *Ann. of Math.* **62** (1955), 374-410.

Jean-Pierre HENRY   Michel MERLE
Centre de Mathématiques   Laboratoire de Mathématiques
U. R. A. au C.N.R.S n° 169   U. R. A. au C.N.R.S n° 168
École Polytechnique   Université de Nice-Sophia Antipolis
91128 PALAISEAU Cedex   06108 NICE Cedex 2

Arrangements of singularities and proper partitions of Dynkin diagrams

P. Jaworski

1 Introduction.

We associate to every isolated singularity of an analytic function and to every tame polynomial a bilinear form, namely the intersection form $\langle .,. \rangle$ in the vanishing homology group H.

It is well-known that any decomposition of an isolated singularity induces a partition of the corresponding Dynkin diagram. Analogously for the tame polynomial F there is a partition of the "global" Dynkin diagram into "local" Dynkin diagrams which correspond to the critical points of F. The converse of this fact is valid only under additional conditions ([7, 5, 6]).

Using the method of Dynkin diagrams one reduces the complicated problems from the elimination theory (seeking for a polynomial with given critical values and critical points) to much simpler questions concerning combinatorics and graphs. This can be further simplified by finding the invariants of the equivalence relation between Dynkin diagrams.

The aim of this paper is to study such invariants. It shows that the most useful connection between them is the trace of the monodromy operator. As an application of our method we describe in the last section all possible arrangements of critical points of the polynomials of the type $T_{3,3,2}$.

2 The basic facts about Dynkin diagrams.

We recall the basic facts about Dynkin diagrams of singularities and tame polynomials (i.e. such polynomials for which the set where their gradient is small is bounded - see [2]). Let the germ

$$f : (C^3, 0) \to C$$

have an isolated singularity at the origin. We consider the two-dimensional homology group of a nonsingular level (in some neighbourhood of the origin), the so called vanishing homology group. Analogously; let $F : C^3 \to C$ be a tame polynomial; then we consider the two-dimensional homology group of any non-singular level. This group is generated by vanishing cycles ([2]). We describe the intersection form $\langle\ ,\ \rangle$ with the help of a Dynkin diagram i.e. the graph with ordered vertices and weighted edges. The vertices correspond to the base elements of the homology group and the weights of edges are equal to their intersection numbers. If the intersection number is "0" (resp. negative) then we omit the line (use the dotted line resp.). We use the double lines for the intersection numbers equal to ± 2.

We shall consider *only* the Dynkin diagrams constructed for *distinguished bases* (see [1] s.2.2 and s.2.8). Since the selfintersection of any distinguished base element is -2, thus the Dynkin diagram completely determines the intersection form. On fig.1 we show the Dynkin diagrams of simple singularities (the ordering of vertices is arbitrary - see s.3.6).

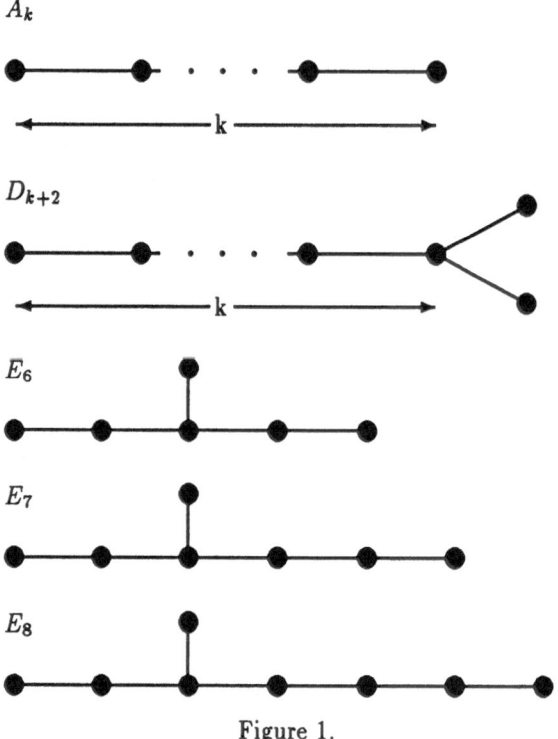

Figure 1.

Two Dynkin diagrams are equivalent (have the same type) if they are associated to two bilinear forms such that one is obtained from the second by a successive use of canonical base changes α_i and β_i , and by changes of signs of base elements (compare [1] s.2.6). We recall that α_i , i=1,..,μ-1, transforms the

base e_j , $j = 1, .., \mu$, to the base e'_j, where

$$e'_i = e_{i+1} + \langle e_i, e_{i+1} \rangle e_i,$$

$$e'_{i+1} = e_i,$$

$$e'_j = e_j \text{ for } j \neq i, i+1,$$

where $\langle \, , \, \rangle$ denotes the intersection form.
The transformation β_{i+1} , $i = 1, \ldots, \mu - 1$, is an inverse of α_i i.e.

$$e'_i = e_{i+1},$$

$$e'_{i+1} = e_i + \langle e_i, e_{i+1} \rangle e_{i+1},$$

$$e'_j = e_j \text{ for } j \neq i, i+1.$$

We remark that the above transformations define the action of the Artin brai
group on the space of bilinear forms and on the set of Dynkin diagrams (see [1
s.2.6).
Any deformation of a singularity S induces a partition of some Dynkin diagra
E of S:

$$E \longrightarrow E_1 / \ldots / E_k,$$

where the connected components of each subdiagram E_i are Dynkin diagrams
singularities lying at i-th critical level. (The weights of edges joining vertices
the same sub diagram are left unaltered, the other edges are removed.) Mo
over the induced partition is proper i.e. the vertices of E which belong to t
same subdiagram are enumerated by successive numbers (see [7, 5]). We add
the similar construction for tame polynomials.
In such a way we reduce the question of nonexistence of a given arrangem
of singularities for polynomials from a certain family, having the same glo
Dynkin diagram, to the question of nonexistence of a given proper partition
this diagram. Usually there is a huge number of equivalent diagrams. But
need not investigate all of them. In the next section we develop the meth
which allow us to reduce the investigation to only some basic ones. The cru
point is to study the invariants of the equivalence relation between Dynkin
agrams.

3   Equivalences of Dynkin diagrams.

3.1   Notation.

If D is a Dynkin diagram with μ vertices then by $L(D)$ we denote the lat
$L = Z(e_1, .., e_\mu)$ with a bilinear product $\langle \, , \, \rangle$ such that

1. $\langle e_i, e_i \rangle = -2$,
2. for $i \neq j$ $\langle e_i, e_j \rangle$ equals the weight of the edge joining i-th and j-th vertices.

By a distinguished base of the lattice L we mean every base which may be obtained from the $\{e_1, \ldots, e_\mu\}$ by canonical transformations α_i, β_j and changes of signs of base elements.

3.2 Subdiagrams.

By a Dynkin subdiagram of a given Dynkin diagam D we mean the Dynkin diagram D' containing some of the vertices of D, say $i_1 < i_2 < \ldots < i_k$ and the edges which join them; moreover the i_j-th vertex of D is j-th of D' . We denote $D' = D(i_1, \ldots, i_k)$. By a proper subdiagram of a given Dynkin diagram D we mean the Dynkin subdiagram containing the subsequent vertices of D, for example $i+1, i+2, .., i+k$.

Lemma 1 *Any Dynkin subdiagram of a given Dynkin diagram D is a proper subdiagram of a Dynkin diagram equivalent to D.*

Proof:

Let D' be a Dynkin subdiagram of D containing k vertices, say $i_1 < i_2 < \ldots < i_k$. We apply to D subsequently the transformations:

$$\beta_{i_1}, \beta_{i_1-1}, \ldots, \beta_2; \ldots \ldots; \beta_{i_k}, \beta_{i_k-1}, \ldots, \beta_k.$$

The proper subdiagram containing first k vertices of the transformed Dynkin diagram is equal to D' .

3.3   The main invariants.

The invariants of the equivalence of Dynkin diagrams may be divided into two groups; those which depend on the ordering of vertices, and those which do not. The most important invariants which do not depend on the ordering are the following:

1. $\mu(D)$ - the number of vertices of a Dynkin diagram, i.e. the dimension of the lattice $L(D)$.
2. $\#(D)$ - the number of connected components of the Dynkin diagram.
3. μ_+, μ_-, μ_0 - the dimensions of the positive, negative and isotropic subspaces of L over real numbers (i.e. of $L \otimes R$).
We call a diagram elliptic if $\mu_+ = \mu_0 = 0$ and k-parabolic if $\mu_+ = 0$ and $\mu_0 = k$.

The composition of reflections at hyperplanes orthogonal to the base vectors e_i :

$$H(e) = h_\mu(h_{\mu-1}(\ldots h_1(e))), \quad h_i(e) = e + \langle e, e_i \rangle e_i,$$

is an example of an invariant which depends on the ordering of vertices. We shall call this isomorphism of L the "monodromy operator" (compare [1] §2.5).

Lemma 2 *The monodromy operator is an invariant.*

Proof:

1. The change of sign of a base element e_i does not change H.
Indeed, let $e_i' = -e_i$ then $h_i'(e) = e + \langle e, -e_i \rangle (-e_i) = e + \langle e, e_i \rangle e_i = h_i(e)$.
Thus $h_i' = h_i$ and $H' = H$.

2. The base transformation α_i does not alter H.
It is enough to show that $h_{i+1}' \circ h_i' = h_{i+1} \circ h_i$.
We have $e_i' = e_{i+1} + \langle e_{i+1}, e_i \rangle e_i$, $\quad e_{i+1}' = e_i$, hence $\langle e_i', e_{i+1}' \rangle = -\langle e_i, e_{i+1} \rangle$.
Thus

$$h_{i+1}' \circ h_i'(e) = h_{i+1}'(e + \langle e, e_i' \rangle e_i') =$$
$$= e + \langle e, e_i' \rangle e_i' + \langle e, e_{i+1}' \rangle e_{i+1}' + \langle e, e_i' \rangle \langle e_i', e_{i+1}' \rangle e_{i+1}' =$$
$$= e + \langle e, e_{i+1} + \langle e_{i+1}, e_i \rangle e_i \rangle (e_{i+1} + \langle e_{i+1}, e_i \rangle e_i) +$$
$$+ \langle e, e_i \rangle e_i - \langle e, e_{i+1} + \langle e_{i+1}, e_i \rangle e_i \rangle \langle e_{i+1}, e_i \rangle e_i =$$
$$= e + \langle e, e_i \rangle e_i + \langle e, e_{i+1} \rangle e_{i+1} + \langle e, e_i \rangle \langle e_i, e_{i+1} \rangle e_{i+1} = h_{i+1} \circ h_i(e).$$

Since the equivalence is generated by α_i, $\beta_{i+1} = (\alpha_i)^{-1}$ and sign changes hence the monodromy operator is an invariant.

We derive from the monodromy operator H the numerical invariants, the symmetric functions of eigenvalues of H. The most useful of them is the trace of the monodromy operator H (i.e. the sum of its eigenvalues). We put $tr\, D = tr H$.
Let H_i denote the composition of reflections $h_1, h_2, \ldots, h_i$ (i.e. $H_\mu = H$) and D' the subdiagram of D containing all vertices with the exception of the last one.

Lemma 3 *1. $tr\, D = \mu - \sum_{i=1}^{\mu} \langle H_i e_i, e_i \rangle$, where $\mu = \mu(D)$,*
2. $tr\, D - tr\, D' = 1 + \langle H_{\mu-1} e_\mu, e_\mu \rangle$.

Proof:
By $(e)_i$ we denote the i-th coordinate of a vector e.

$$tr\, D = tr\, H = \sum_{i=1}^{\mu} (H e_i)_i, \quad H e_i = H_i e_i + \sum_{j=i+1}^{\mu} c_j e_j,$$

hence

$$(He_i)_i = (H_ie_i)_i = (H_{i-1}e_i)_i + \langle H_{i-1}e_i, e_i \rangle = 1 + \langle H_{i-1}e_i, e_i \rangle = 1 - \langle H_ie_i, e_i \rangle.$$

Therefore

$$tr\ D = \mu - \sum_{i=1}^{\mu} \langle H_ie_i, e_i \rangle,$$

$$tr\ D - tr\ D' = 1 - \langle H_\mu e_\mu, e_\mu \rangle = 1 + \langle H_{\mu-1}e_\mu, e_\mu \rangle.$$

3.4   Linearity of the trace.

The crucial property of the trace is its linearity. The trace is additive in according to the union of Dynkin diagrams.
We use the notation compatible with the notation of partitions: $(D_1, D_2, \ldots, D_k)$ means a disjoint union of Dynkin diagrams ordered lexicographicaly i.e if $i < j$ then the vertices of D_j follow the vertices of D_i .
If D_1, D_2 , are proper subdiagrams of a given Dynkin diagram D such that the vertices of D_2 follow the vertices of D_1 then $D_1 \cup D_2$ means the Dynkin subdiagram of D containing all vertices of D_1 and D_2 .

Lemma 4 *1.* $tr(D_1, D_2) = tr\ D_1 + tr\ D_2$,
2. $tr\ D_0 \cup (D_1, D_2) = tr\ D_0 \cup D_1 + tr\ D_0 \cup D_2 - tr\ D_0$,
3. $tr\ (D_1, D_2) \cup D_0 = tr\ D_1 \cup D_0 + tr\ D_2 \cup D_0 - tr\ D_0$.

Proof:
    We show how to prove the second point, the proofs of others are quite similar. We shall use the decomposition:

$$L(D_0 \cup (D_1, D_2)) = L(D_0) \oplus L(D_1) \oplus L(D_2).$$

Let $(e_1, \ldots, e_k)$ ($(e_{k+1}, \ldots, e_m)$, $(e_{m+1}, \ldots, e_n)$ respectively) be the base of $L(D_0)$ ($L(D_1)$, $L(D_2)$ resp.). We have

$$tr\ D_0 = k - \sum_{i=1}^{k} \langle H_ie_i, e_i \rangle,\ \ tr\ D_0 \cup D_1 = m - \sum_{i=1}^{m} \langle H_ie_i, e_i \rangle.$$

$$tr\ D_0 \cup D_2 = n + k - m - \sum_{i=1}^{k} \langle H_ie_i, e_i \rangle - \sum_{i=m+1}^{n} \langle h_i \circ \ldots \circ h_{m+1} \circ H_k e_i, e_i \rangle.$$

$L(D_1)$ and $L(D_2)$ are orthogonal hence for $i = m+1, \ldots, n$

$$\langle H_ie_i, e_i \rangle = \langle h_i \circ \ldots \circ h_{m+1} \circ H_k e_i, e_i \rangle.$$

Therefore

$$tr \, D_0 \cup (D_1, D_2) = n - \sum_{i=1}^{k} \langle H_i e_i, e_i \rangle - \sum_{i=k+1}^{m} \langle H_i e_i, e_i \rangle - \sum_{i=m+1}^{n} \langle H_i e_i, e_i \rangle =$$

$$= (m - \sum_{i=1}^{k} \langle H_i e_i, e_i \rangle - \sum_{i=k+1}^{m} \langle H_i e_i, e_i \rangle) +$$

$$+ (k + n - m - \sum_{i=1}^{k} \langle H_i e_i, e_i \rangle - \sum_{i=m+1}^{n} \langle H_i e_i, e_i \rangle) - (k - \sum_{i=1}^{k} \langle H_i e_i, e_i \rangle) =$$

$$= tr \, D_0 \cup D_1 + tr \, D_0 \cup D_2 - tr D_0.$$

3.5   Trace of an elliptic or 1-parabolic diagram.

Lemma 5 *1. If D is an elliptic or 1-parabolic Dynkin diagram then*

$$tr \, D + \#D \geq 0.$$

2. If $D' = D(1, \ldots, \mu - 1)$ $(\mu = \mu(D))$ is elliptic then

$$tr \, D + \#D \geq tr \, D' + \#D'.$$

Proof:

We start with the second point of the lemma.

Since the trace is linear thus we may assume that D is connected. Let $D_1, \ldots, D_k$ be connected components of D' and $L_1, \ldots, L_k$ the associated sublattices of $L' = L(D')$. We remark that L_i's are pairwise orthogonal. So

$$H_{\mu-1} e_\mu = e_\mu + f_1 + \ldots + f_k,$$

where f_i is an element of L_i . D is connected hence all f_i are nonzero, i.e. $\langle f_i, f_i \rangle \leq -2$. Now

$$-2 = \langle H_{\mu-1} e_\mu, H_{\mu-1} e_\mu \rangle = -2 + 2 \sum \langle f_i, e_\mu \rangle + \sum \langle f_i, f_i \rangle.$$

Hence

$$2 \sum \langle f_i, e_\mu \rangle = - \sum_{i=1}^{k} \langle f_i, f_i \rangle \geq 2k = 2\#D'.$$

Therefore

$$tr \, D = tr \, D' + (1 + \langle H_{\mu-1} e_\mu, e_\mu \rangle) =$$

$$= tr \, D' + (1 - 2 + \sum \langle f_i, e_\mu \rangle) \geq tr \, D' - 1 + \#D'.$$

Next one may prove the first point by induction on number of vertices of D.

From the above lemma we obtain the following useful fact:

Lemma 6 *Let $D = (D_1, D_2) \cup (D_3, \ldots, D_{k+2})$ be an elliptic or 1-parabolic connected Dynkin diagram. If $tr\ D = tr\ D_1 = tr\ D_2 = -1$ then only one subdiagram $(D_1, D_2) \cup D_i$ $(i = 3, \ldots, k+2)$ is connected.*

Proof:

From lemma 4 we obtain by induction:

$$tr\ D = \sum_{i=1}^{k} tr(D_1, D_2) \cup D_{i+2} - (k-1)tr(D_1, D_2).$$

Thus from lemma 5

$$-2k + 1 = \sum_{i=1}^{k} tr(D_1, D_2) \cup D_{i+2} \geq -\sum_{i=1}^{k} \#(D_1, D_2) \cup D_{i+2}.$$

Hence $2k - 1 \leq \sum_{i=1}^{k} \#(D_1, D_2) \cup D_{i+2}$. But on the other hand D is connected so the subdiagrams $(D_1, D_2) \cup D_{i+2}$ have at most two components and at least one of them is connected. Thus

$$\sum_{i=1}^{k} \#(D_1, D_2) \cup D_{i+2} \leq 2k - 1;$$

therefore there must be an equality, i.e. just one subdiagram is connected.

3.6   Appendix: the permutation of vertices.

In general the change of the ordering of vertices changes the type of the Dynkin diagram. But there are some exceptions.

Lemma 7 *If the two subsequent vertices, say i and $i+1$, are not connected by an edge (i.e. $\langle e_i, e_{i+1} \rangle = 0$) then we may permute them.*

Proof:

We apply the transformation α_i or β_{i+1} .

Lemma 8 *The cyclic change of ordering of vertices, $1 \to 2, 2 \to 3, \ldots, \mu \to 1$, does not change the type of a diagram.*

Proof:

We apply the sequence of canonical transformations β_μ, $\beta_{\mu-1}, \ldots, \beta_1$ and change the sign of the first element. We obtain the new base e'_i, $i = 1, \ldots, \mu$;

$$e'_1 = -e_\mu, \quad e'_j = e_{j-1} + \langle e_{j-1}, e_\mu \rangle e_\mu \quad \text{for } 1 < j,$$

The new base may be obtained from the old one by the hyperplane reflection h_μ ,and renumbering. h_μ preserves the scalar product $\langle \, , \, \rangle$ hence the weights of edges are unaltered.

Next we recall the following result of O.V.Lyashko. A Dynkin diagram is called tree-like if the underlying graph has no cycles and all edges have weight one.

Lemma 9 ([7]) *If D is a tree-like Dynkin diagram then a permutation of its vertices preserves its type.*

Corollary 1 *The trace of a tree-like Dynkin diagram equals to -1.*

4   Preparation Theorem.

Let D' be a proper subdiagram of a connected Dynkin diagram D spanned by first μ -1 vertices. Let L and L' be the lattices associated to D and D' ; $L = L' \oplus Ze_\mu$.

Theorem 1 (The preparation theorem) *Let $\mu(D) = \mu$, $tr\ D = -1$.*
T_μ . If D' is elliptic and $tr\ D' = -\#D'$ then there exists a distinguished base of L' , $e_1, \ldots e_{\mu-1}$, such that the Dynkin diagram of $\langle \, , \, \rangle$ in the extended base $e_1, \ldots e_{\mu-1}$, e_μ is tree-like.
V_μ . Let $\mu(D) = \mu$, $tr\ D = -1$. If D is connected and elliptic and e is an element of the lattice L such that $\langle e, e \rangle = -2$, then there exists a distinguished base of L $e_1, \ldots e_\mu$ such that:
i/ $e_\mu = e$,
ii/ the Dynkin diagram of $\langle \, , \, \rangle$ in the base $e_1, \ldots e_\mu$ is tree-like.

Proof.

$V_1, \ldots, V_{\mu-1} \Rightarrow T_\mu$.
Let $D_1, \ldots D_k$ be connected components of D' and let $L_1, \ldots, L_k$ be the corresponding sublattices of L', we remark thet they are pairwise orthogonal. We assume that the vertices of each D_i are ordered subsequently.
We write the image of e_μ under the monodromy operator of L' in the following form:

$$H_{\mu-1}(e_\mu) = e_\mu + \sum_{i=1}^{k} f_i \text{ where } f_i \in L_i \ (f_i \neq 0).$$

$$\text{So} \quad \langle e_\mu, H_{\mu-1}(e_\mu) \rangle = -2 + \sum_{i=1}^{k} \langle e_\mu, f_i \rangle.$$

$$\text{Since} \quad 1 + \langle e_\mu, H_{\mu-1}(e_\mu) \rangle = tr\, D - tr\, D' = -1 + k$$

$$\text{hence} \quad \sum_{i=1}^{k} \langle e_\mu, f_i \rangle = k.$$

On the other hand (f_i's are pairwise orthogonal!)

$$-2 = \langle H_{\mu-1}(e_\mu), H_{\mu-1}(e_\mu) \rangle = -2 + \sum_{i=1}^{k} 2\langle e_\mu, f_i \rangle + \sum_{i=1}^{k} \langle f_i, f_i \rangle.$$

$$\text{Thus} \quad \sum_{i=1}^{k} \langle f_i, f_i \rangle = -2k.$$

D is elliptic hence for each i the product $\langle f_i, f_i \rangle$ is negative and even. Therefore for each i $\langle f_i, f_i \rangle = -2$. Hence for each i we have

$$-2 = \langle e_\mu + f_i, e_\mu + f_i \rangle = -2 + 2\langle e_\mu, f_i \rangle - 2.$$

Thus $\langle e_\mu, f_i \rangle = 1$.

Next we apply V_{μ_i}. Let $e_{i,j}$, $j = 1, .., \mu_i$, be a distinguished base of L_i such that the last element equals to f_i and the associated Dynkin diagram of $(\ ,\ )$ restricted to L_i is tree-like. The cycles $\{e_{1,1}, \ldots, e_{1,\mu_1}, e_{2,1}, \ldots, e_{k,\mu_k}, e_\mu\}$ form a distinguished base of L in which $(\ ,\ )$ has a tree-like Dynkin diagram. Indeed; the monodromy operator H is an invariant, hence

$$H(e_\mu) = -(k+1)e_\mu + \sum_{i=1}^{k} e_{i,\mu_i}.$$

$$\text{So} \quad \langle e_\mu, e_{i,j} \rangle = \begin{cases} 1 & \text{if } j = \mu_i, \\ 0 & \text{otherwise.} \end{cases}$$

$T_\mu \Rightarrow V_\mu$.
We apply such a cyclic change of ordering that the subdiagram D' spanned by first $\mu - 1$ vertices is connected. D' is elliptic thus

$$-1 \le tr\, D' = tr\, D - (1 + \langle H_{\mu-1}e_\mu, e_\mu \rangle) = -2 - \langle H_{\mu-1}e_\mu, e_\mu \rangle.$$

D is elliptic hence

$$-1 \le \langle H_{\mu-1}e_\mu, e_\mu \rangle \le 1.$$

Thus $tr\, D' = 1$. Hence from T_μ we obtain that D is equivalent to a tree-like diagram. D is elliptic hence it may only be one of the following: A_μ, D_μ or E_6,

E_7, E_8. Therefore e is a vanishing cycle and there exists a distinguished base $e'_1, \ldots, e'_\mu$ such that $e = e'_\mu$ (see [3] or [1] s.3.6). Next we aplly once more T_μ and obtain that the above base may be chosen in such a way that the associated Dynkin diagram is tree-like.

Corollary 2 *If* $\operatorname{tr} D = -1$ *and* $D' = A_k$, $k < 7$, *then* D *is elliptic.*

5 Applications. The critical points of $T_{3,3,2}$ polynomials.

We consider the family of polynomials:

$$F_{a,b,c}(x,y,z) = xyz + x^3 + y^3 + z^2 + a_2x^2 + a_1x + b_2y^2 + b_1y + c_1z.$$

We shall call this family $T_{3,3,2}$. For any choice of parameters $F_{a,b,c}$ is a tame polynomial (in the sense of Broughton - [2]) hence the global Dynkin diagram does not depend on parameters. We show it at fig.2. Diagram "a" is obtained using the method of A'Campo and Gusein-Zade (see [1] §4 Th.1) as in [4]. In the diagram "a" vertices • follow □ and □ follow ⊙. The second diagram - "b" is equivalent to it. Indeed: We choose an ordering of vertices of "a" which fulfils the above rule and moreover the vertice □ in the lower row has number 5. We apply the cyclic change of the ordering $1 \to 7$, $7 \to 6, \ldots$ and the sequence of transformations α_6, α_5, α_4. The diagram "b" is tree like hence every ordering of vertices is admissible. We shall denote the equivalence class of the above diagrams by $T_{3,3,2}$. All of them are 1-parabolic. Indeed: let e_i, $i = 1, \ldots, 7$ be the base of the lattice corresponding to the diagram "a" with the same ordering of vertices as before, then the vector

$$e = 2e_1 + e_2 + e_3 + e_4 + e_5$$

is isotropic. Moreover the subdiagram $D(2, \ldots, 7)$ is a disjoint union of A_5 and A_1 diagrams hence it is elliptic.

a 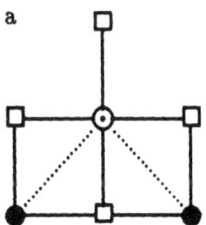 b

Figure 2.

Proposition 1 *There is a one-to-one correspondence between the proper partitions of $T_{3,3,2}$ diagrams and arrangements of critical points of polynomials from the family $T_{3,3,2}$.*

For a proof see [6]. Compare lemmas 3, 8 [5] and s.3.3 [4].

Proposition 2 *Let F be a polynomial belonging to the $T_{3,3,2}$ family. Then:*
1. F has only simple critical points.
2. F has at least two different critical values.
3. On one critical level the polynomial F may have only the following critical points:
E_6, A_5A_1 , $3A_2$ *or their decompositions:*
D_5, A_5, A_4A_1, A_32A_1, $2A_2A_1$; D_4, A_4, A_3A_1, $2A_2$, A_22A_1, $4A_1$;
A_3, A_2A_1, $3A_1$; A_2, $2A_1$; A_1.
Moreover every such "case" occurs for some choice of parameters.

Proof:
    First two points are proved in [4] s.3.3.
Point 3. The diagram "b" is the only tree-like diagram of $T_{3,3,2}$ type. The diagrams of critical levels are elliptic hence they must be contained in it (see theorem 1).

    Next we shall describe all possible arrangements of types of critical points for polynomials from the $T_{3,3,2}$ family.

Proposition 3 *All arrangements $X_1/\ldots/X_k$, $X_i = (X_{i,1},\ldots,X_{i,j_i})$ such that:*
i. each X_i is listed in Proposition 2;
ii. $\sum \mu(X_i) = 7$;
exist with the following exceptions: $2A_2/A_2A_1, 2A_2/3A_1, 4A_1/A_2A_1$.

Proof:
    1. The existence part.
For every arrangement which fulfils i. and ii. different then $2A_2/A_2A_1, 2A_2/3A_1$, $4A_1/A_2A_1$ one can find a corresponding proper partition of the diagram 2a or 2b (if necessary we apply a cyclic change of vertices, etc.).

    2.The nonexistence part.
From Lemma 6 we have that in $(A_2A_1)\cup 4A_1$ (resp. $2A_2\cup 3A_1$) only one vertice may join both subdiagrams A_2 and A_1 (resp. both A_2) hence the only possible cases are those which are shown at fig.3 (resp.4). We denote by $\square$ (resp. by $\bigcirc$) the connected components of the first (resp. of the second) subdiagram.

Analogously for $2A_2 \cup A_2A_1$ the only possible cases are shown at fig.5. Next from proposition 1 point 3 we obtain that neither of them may exist. Indeed the complement of the vertex (or vertices) marked by x is at 3.a - A_23A_1 , at 3.b D_4A_1, at 3.c D_4A_2 at 3.d D_4A_2 or A_4A_2 ; at 4.a and b A_3A_2 ; at 5.a $2A_3$, at 5.b A_4A_2 or D_4A_2 no one of them is a subdiagram of $T_{3,3,2}$ type diagram.

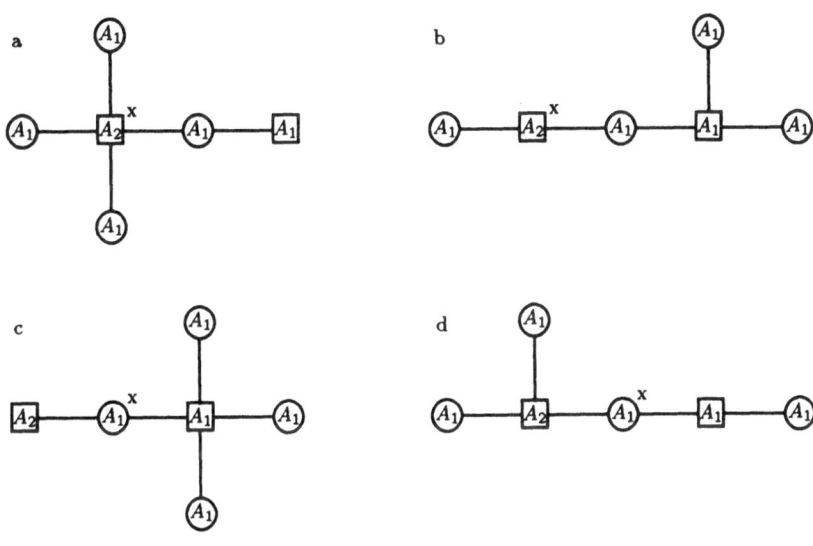

Figure 3.

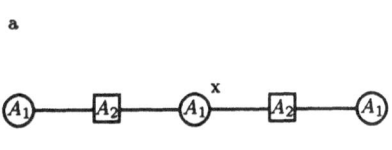

Figure 4.

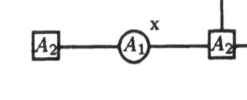

Figure 5.

References

[1] V.I.Arnold, S.M.Gusein-Zade, A.N.Varchenko, Singularities of Differentiable Maps v.2, Birkhauser 1988.

[2] S.A.Broughton, On the topology of polynomial hypersurfaces. In: Singularities. Proc. of Symposia in pure math. 40/1 (1983) 167-178.

[3] S.M.Gusein-Zade, Distinguished bases of simple singularities. Funct.Analiz 14:4 (1980) 73-74 (in Russian).(Func.Anal.Appl. 14:4 (1980) 307-308 English tr.)

[4] P.Jaworski, Distribution of critical values of miniversal deformations of parabolic singularities. Invent. math. 86 (1986) 19-33.

[5] P.Jaworski, Decomposition of parabolic singularities. Bull.Sc.math. 112 (1988) 143-176.

[6] P.Jaworski, Decompositions of hyperbolic singularities, to appear.

[7] O.V.Lyashko, Decompositions of simple singularities of functions. Funct.Analiz 10:2 (1976) 49-56 (in Russian).

PIOTR JAWORSKI
UNIVERSITY OF WARSAW
INSTITUTE OF MATHEMATICS
UL.BANACHA 2
02-097 WARSZAWA
POLAND

Versal deformations of powers of volume forms

V. P. Kostov   S. K. Lando

Abstract. Deformations of powers of volume forms of the kind

$$F(x, \lambda)(dx)^\alpha, x \in C^n, F(x,0) = f(x), dx = dx_1 \wedge \ldots \wedge dx_n, \alpha \in C$$

are investigated. If f has an isolated singularity of multiplicity μ at the origin, then the form $f(x)(dx)^\alpha$ has a μ-parameter versal deformation for almost every value of α. Exceptional values of α form a discrete set of negative rational numbers. Given f, the versal deformation can be obtained algorithmically. For non-exceptional values of α it is the same as the versal deformation of the germ of a function f.

Introduction

The subjects of this paper are the powers of volume forms $f(x)(dx)^\alpha$ and their deformations. Here $x = (x_1, ..., x_n) \in C^n, dx = dx_1 \wedge \ldots \wedge dx_n$ is a standard volume form on C^n, α – a complex parameter, f – a germ of a holomorphic function having an isolated singularity of multiplicity μ at the origin. Objects of this kind are familiar to mathematicians. For $\alpha = 1$ they are usual volume forms. For $\alpha = -1$ they are vector fields if $n = 1$ and they are plane Poisson structures if $n = 2$ [Arn2]. For $\alpha = 1/2$ they are semiforms [GSt] and so on.

The case $\alpha = 0$ corresponds to the usual germs of holomorphic functions that we consider to be well-known. In papers [Var2, Lan1] the complete solution of the problem of reducing powers of differential forms to a normal form was given. The versality theorem for a one-dimensional argument was proved by A.B. Givental' [Gi1] for $\alpha = 1/n, n \in N$ and for arbitrary $\alpha \neq 0$ by V.P.Kostov ([Kos1] – in the holomorphic case, [Kos2] – in the smooth case).

In the present paper the versality theorem for deformations of powers of volume forms is proved. The paper is based on the second author's thesis [Lan2] and on the new solution of equation (0.1) given by the first author. This solution provides an algorithm for finding the versal deformation in the resonant case. Solving the homological equation

$$\beta dF \wedge \omega + F d\omega + \sum_{i=1}^{\mu} \theta_i \partial F / \partial \lambda_i = F dx \qquad (0.1)$$

(for the resonant case see **2.2.**) is the crucial moment of the proof (solving with respect to unknown $(n-1)$-form ω and functions θ_i depending on the parameters; here $\beta = 1/\alpha$); Lemma 3.2.1. from the proof is interesting in itself. The theorem is reduced to this equation by the standard homotopy method. But standard methods fail in proving the solvability of homological equations. The reason is that the image of the differential operator

$$\beta dF \wedge + Fd \qquad (0.2)$$

in the left hand side of equation (0.1) proves not to be a module over the ring of analytic functions. The fact does not allow us to apply the Weierstrass preparation theorem (see e.g.[AVG]). That is why special techniques are needed for solving the equation. The differential operator (0.2) is in fact a differential and is called by A.B.Givental' the *twisted differential*. A.B.Givental' [Gi2] investigated corresponding twisted complexes, twisted cohomologies and introduced twisted Picard-Lefshetz formulas. Our solving of the homological equation is, in fact, calculation of highest cohomologies of a twisted complex.

Acknowledgements. V.I.Arnol'd attracted our attention to the problem and guided our work. We are grateful to S.V.Chmutov, A.Galligo, A.B.Givental' and, especially, to A.N.Varchenko for fruitful discussions.

1 Definitions and formulation of the results.

1.1 Action of the group of biholomorphisms.

Definition. A *deformation* of a germ $f(x)$ of a holomorphic function is a germ of a holomorphic function $F : C^n \times C^k \to C$, for which $F(.,0) \equiv f$. A deformation of a form $f(x)(dx)^\alpha$ is a form $F(x,\lambda)(dx)^\alpha, F(.,0) \equiv f$.

Consider a group consisting of germs of families of biholomorphisms g : $x \mapsto g(x,\lambda)$, such that

1) family g depends holomorphically on parameters $\lambda \in C^k$,

2) $g(x,0) \equiv x$.

The group acts on the set of deformations of a fixed form $f(x)(dx)^\alpha$ with a fixed number of parameters according to the rule

$$g : F(x,\lambda)(dx)^\alpha \to F(g(x,\lambda),\lambda)(dg(x,\lambda))^\alpha,$$

The germ of the form $(dg(x,\lambda))^\alpha$ is uniquely determined by the condition $(dg(x,0))^\alpha \equiv (dx)^\alpha$. Instead of powers $f(dx)^\alpha$ of differential forms we will further investigate equivalent objects: forms of the kind $f^\beta dx$, where $\beta = 1/\alpha$, $\beta \in C \backslash \{0\}$.

Definition. A deformation $F^\beta dx$ of a germ $f^\beta dx$ is called *versal* if any other deformation is equivalent to one induced from it; this means that for any deformation $F' : C^n \times C^{k'} \to C$, $F'(.,0) = f$ there exist a germ of a holomorphic function of the parameters $\theta : (C^{k'},0) \to (C^k,0)$ and a germ of a family of biholomorphisms $g : C^n \times C^{k'} \to C^n$, such that

1)$g(x,0) \equiv x$;

2)$F^\beta(x,\lambda')dx = f^\beta(g(x,\lambda'),\theta(\lambda'))J(g)dx$. Here $J(g)$ denotes the Jacobian of the biholomorphism g.

1.2   I.N.Bernstein's polynomial and resonance numbers.

I.N.Bernstein [Ber1, Ber2, BG] proved the following statement, that was conjectured by I.M.Gel'fand.

Theorem. *For any polynomial $f : C^n \to C$ there exist a polynomial differential operator $D(x, \beta, \partial/\partial x)$ and a one-variable polynomial B not identically equal to zero such that*

$$D(x, \beta, \partial/\partial x) \circ f^\beta \equiv B(\beta)f^{\beta-1}.$$

Bjork [Bjo] extended the theorem to the case of holomorphic germs f with isolated singularities at the origin. It is easy to see that all such polynomials B form an ideal in the ring of one-variable polynomials. Let $\tilde{b}$ be the generator of the ideal with leading coefficient equal to 1. Then $\tilde{b}(0) = 0$, so $\tilde{b}(\beta) = \beta b(\beta)$ for some polynomial $\tilde{b}$.

Definition. The polynomial b is called *the reduced Bernstein polynomial* of the germ f. A number $\beta \in C$ is called *a resonance number* for the germ f if there exists $k \in N$ such that $\beta + k$ is a root of the reduced Bernstein polynomial.

Theorem [Mal]. *The roots of the reduced Bernstein polynomial are rational numbers, smaller than 1.*

As a consequence, resonance numbers are real, rational, negative and form a finite number of arithmetic progressions with difference (-1).

1.3   Versality Theorems.

Theorem 1.3.1. *If β is non-resonant, then the form $f^\beta dx$ has a μ-parameter versal deformation $F^\beta dx$ where $F = F_0 + \sum \lambda_j \varphi_j$. Here F_0 is a normal form of the germ of a function f with respect to the biholomorphisms defined in 1.1. (for $\lambda = 0$). The monomials φ_j form a basis of the local algebra $\mathcal{A} = C\{x\}/(\partial F_0/\partial x_1, ..., \partial F_0/\partial x_n)$. Their choice is described below.*

Theorem 1.3.2. *If β is resonant, then the versal deformation is of the form*

$$F^\beta(1 + \sum_j \tilde{\eta}_j \varphi_{k_j} F^{n_j})dx$$

the sum being finite, F being as above, φ_{k_j} being among φ_j, n_j being non-negative integers, $\tilde{\eta}_j = \eta_j^0 + \eta_j$, where η_j^0 are constants and η_j are parameters. For $\lambda = \eta = 0$ the constants η_j^0 give the classification of the volume forms $f^\beta dx$, i.e. for different η^0 the corresponding volume forms $F_0^\beta(1 + \sum \eta_j \varphi_{k_j} F_0^{n_j})dx$ are not analytically equivalent. Given the polynomial F_0, the numbers k_j, n_j can be found algorithmically. The algorithm is described in 3.6.. It can also be used for finding the roots of the reduced Bernstein polynomial of F_0.

Corollary 1.3.3. *Every analytic family* $\tilde{f}^\beta(x,\varepsilon)dx$, $\tilde{f}(x,0) \equiv f(x)$ *can be transformed by an analytic diffeomorphism* $x \mapsto g(x,\varepsilon)$ *to one induced from the versal deformation from Theorem 1.3.2. (i.e. obtained by a substitution* $\lambda_j = \lambda_j(\varepsilon) \in \mathcal{O}(\varepsilon)$).

Corollary 1.3.4. *Set* $\varphi_j = x^{m_j} \equiv x_1^{m_{j_1}} \ldots x_n^{m_{j_n}}$. *Let the germ* f *be quasi-homogeneous,* $\alpha_1, ..., \alpha_n$ *being the weights of* $x_1, ..., x_n$, $\alpha_j \in Q$. *Then in the sum* $\sum \nu_j x^{m_j} F^{n_j}$ *from the versal deformation of the volume form the summation is taken over exactly these* (m_j, n_j) *which satisfy the condition*

$$(m_j, \alpha) + \sum \alpha_k + \nu_j + \beta = 0, (m_j, \alpha) \equiv m_{j_1}\alpha_1 + \ldots + m_{j_n}\alpha_n.$$

The corollary follows from [Var2].

Define the monomials φ_j (the reader is assumed to be acquainted with the theory of Gröbner bases, see [Bu1],[Bu2]). Let $\mathcal{B}$ be a Gröbner basis of the ideal $\mathcal{J} \equiv (\partial F_0/\partial x_1, \ldots, \partial F_0/\partial x_n)$ with respect to the degree lexicographic ordering $1 < x_1 < \ldots < x_n < x_1^2 < x_1 x_2 < \ldots < x_2^2 < x_2 x_3 < \ldots < x_n^2 < x_1^3 < \ldots$ and let I be the ideal generated by the leading monomials of the elements of $\mathcal{B}$. Then all the monomials of $C\{x_1, \ldots, x_n\}/I$ are the monomials $\varphi_j(1 = \varphi_1 < \varphi_2 < \ldots < \varphi_\mu)$. The number μ is the multiplicity of the singularity of F_0 at 0.

Remark. The Gröbner basis $\mathcal{B}$ consists of polynomials with leading coefficients equal to 1. The stretchings $x_j \mapsto \delta_j x_j$ for suitably chosen $\delta_j > 0$ allow us to assume that for every $g_k \in \mathcal{B}$ the sum of the modules of all their not leading coefficients is less than $\delta > 0$, where $(1 + \delta)^\mu - 1 < 1/2$. This condition is used in 3.1..

2   Inference to the homological equation.

2.1   Reduction Lemma.

Lemma 2.1.1. *Let* $F(., \lambda)$ *be a versal deformation of the germ* f. *Deformation* $F^\beta dx$ *of the germ of a form* $f^\beta dx$ *is versal iff any deformation of the kind*

$$F^\beta(x, \lambda)E(x, \lambda')dx$$

(where $E : C^n \times C^{k'} \to C, E(., 0) \equiv 1$) *may be induced from it up to equivalence.*

Proof: Let $F'(x, \lambda')$ be another deformation of germ f, $\lambda' \in C^{k'}$. We have to find a holomorphic mapping $\theta : \lambda' \to \theta(\lambda')$ and a family of local biholomorphisms $g : x \mapsto g(x, \lambda'), g(x, 0) \equiv x$, transforming the form $F^\beta dx$ into the form $F'^\beta dx$. As F is versal, there exist a transformation $\theta_1 : \lambda' \mapsto \theta_1(\lambda')$ and a family of local biholomorphisms $g_1 : x \mapsto g_1(x, \lambda')$, transforming deformation F into F':

$$F(g_1(x, \lambda'), \theta_1(x, \lambda')) \equiv F'(x, \lambda').$$

This transformation transforms the form $F^\beta dx$ into the form $F^\beta J(g_1)dx$. Choose a holomorphic function $E : C^n \times C^{k'} \to C$ satisfying two conditions:

1) $E(x, 0) \equiv 1$

2) $E(g_1(x, \lambda'), \lambda') \equiv 1/J(g_1)(x, \lambda')$

Such a function always exists: it is sufficient to set

$$E(x, \lambda') = 1/J(g_1)(g_1^{-1}(x, \lambda'), \lambda').$$

Then deformation $F'^\beta dx$ of the form $f^\beta dx$ is equivalent to one induced from deformation $F^\beta(x, \lambda)E(x, \lambda')dx$ depending on the parameters λ and λ'. The corresponding transformation of parameters is $\lambda = \theta_1(\lambda'); \lambda' = \lambda'$. The converse is obvious. The lemma is proved.

Of course, it is sufficient to consider only the case of one additional parameter $\lambda' \in C^1$.

2.2 The homological equation.

Consider the non-resonant case first. We will prove the versality theorem by the standard homotopy method [AVG]. Construct a germ of a vector field f at the origin $(0,0,0)$ of the space $C^n \times C^\mu \times C$, such that

$$1) \quad \vec{v} = \partial/\partial\lambda' + \sum_{i=1}^{\mu} \theta_i(\lambda, \lambda')\partial/\partial\lambda_i +$$

$$\sum_{i=1}^{n} H_i(x, \lambda, \lambda')\partial/\partial x_j = \partial/\partial\lambda' + \sum \theta_i \partial/\partial\lambda_i + V$$

$$2) \quad L_v(F^\beta E dx) = 0.$$

Here $V = \sum H_i \partial/\partial x_i$, and L_v denotes the differentiation along the vector field $\vec{v}$. To construct such a vector field it is necessary and sufficient to prove the solvability of the homological equation

$$0 = (F^\beta \partial E/\partial\lambda')dx + \beta \sum_{i=1}^{\mu} \theta_i \partial F/\partial\lambda_i F^{\beta-1} E dx + L_v(F^\beta E dx) \qquad (2.1)$$

Theorem. *A deformation $F^\beta E dx$ is equivalent to one induced from deformation $F^\beta dx$ iff equation (2.1) is solvable with respect to the unknown vector field V and functions $\theta_1, \ldots, \theta_\mu$.*

A proof of the theorem is obtained by means of the standard homotopy method [AVG]. The flow of the λ-part of the vector field $\vec{v}$ determines a parameter transformation $\theta(\lambda, \lambda')$ and the corresponding coordinate transformation g is determined by the flow of the vector field V, x-part of $\vec{v}$.

Denote the n-form $-(\partial E/\partial\lambda')dx$ by $d\psi$, where ψ is some $(n-1)$-form, and put $d\phi_i = (\partial F/\partial\lambda_i)E dx$ $(i = 1, ..., \mu)$. According to E. Cartan's formula

$L_v(F^\beta E dx) = F^{\beta-1}(\beta dF \wedge \omega + F d\omega)$, where $\omega = v \lfloor (Edx)$. Dividing equation
(2.1) by $F^{\beta-1}$, we rewrite it in the new notation:

$$\beta dF \wedge \omega + F dx + (\sum_{j=1}^{\mu} \theta_j \varphi_j) dx = \Psi dx \quad , \quad \varphi_j = d\Phi_j \quad , \quad \Psi dx = F d\psi \qquad (2.2)$$

In the resonant case we obtain (after a simplification based on the implicit
function theorem) the following homological equation:

$$\beta dF \wedge \omega + F d\omega + (\sum_{j=1}^{\mu} \theta_j \varphi_j) dx + \sum \xi_j \varphi_{k_j} F^{\eta_j} = \Psi dx \qquad (2.3)$$

$(n_j, k_j$ are defined in 1.3.). When we solve the homological equation, we con-
sider its right hand-side to be any form of the kind Ψdx, $\Psi \in \mathcal{O}(x, \lambda, \lambda')$ (or
$\Psi \in \mathcal{O}(x, \lambda, \xi, \lambda')$ in the resonant case), not necessarily of the kind $F d\psi$.

3   Solution of the homological equation.

In 3.1. we develop some properties of the series of types A and B defined there;
this is necessary for the estimations of the convergence in 3.2. and 3.4. In
3.2. we show that it is sufficient to consider the homological equation with
right hand-side as a series of F only. In 3.3. we consider equation (2.2.) for
$\lambda = 0$ and prepare the algorithm for finding the numbers k_j, n_j from Theorem
1.3.2.; the algorithm itself is in 3.6. In 3.4. we prove the versality theorems for
$Re\beta >> 0$ and in 3.5. for β arbitrary.

3.1   Series of types A and B.

Denote by $||P||$ the sum of the modules of the coefficients (which are assumed
constant) of the polynomial P. A series $\sum_{m \in (N \cup \{0\})^\mu} \lambda^m P_m(x)$ is called a
series of type A $(\lambda^m \equiv \lambda_1^{m_1} \ldots \lambda_\mu^{m_\mu})$ if P_m are polynomials of x of degrees
$\leq a_1(m_1 + \ldots + m_\mu) + a_2$, $a_1, a_2 \geq 0$ and $||P_m|| \leq cd^{(m_1 + \ldots + m_\mu)}, c > 0, d > 0$.
A series $\sum_{j=1}^{\mu} \sum_{m \in (N \cup \{0\})^\mu} \lambda^m Q_{m_j}(\partial F/\partial x_1, \ldots, \partial F/\partial x_n) \varphi_j$ is called a series
of type B if Q_{m_j} are polynomials of $\partial F/\partial x_i, i = 1, \ldots, n$ of degrees $\leq a_1(m_1 +$
$\ldots + m_\mu) + a_2$, $a_1, a_2 > 0$ and $||Q_{m_j}|| \leq cd^{(m_1 + \ldots + m_\mu)}, c > 0, d > 0$. We let the
reader prove

Proposition 3.1.1. *Consider for a fixed λ a series of type B as a series
of $\partial F/\partial x_j - \sum_{j=1}^{\mu} \sum_{m \in (N \cup 0)^\mu} c_{jm}(F_x)^m \varphi_j$. For $|\lambda|$ sufficiently small we have
$|c_{jm}| \leq C_0 \rho_0^{|m|}$, $|m| = m_1 + \ldots + m_n$, where $C_0 > 0, \rho_0 > 0$; when $\lambda \to 0$, then
one can choose ρ_0 such that $\rho_0 \to 0, C_0$ being fixed. If in addition $Q_{0j} \equiv 0$ for
$j = 1, ..., \mu$, then one can choose (for $\lambda \to 0$) ρ_0, C_0 such that $\rho_0 \to 0, C_0 \to 0$.*

Proposition 3.1.2. *For every monomial* $\varphi = x_1^{m_1} \ldots x_n^{m_n}$ *there exist n series of type A - $\omega_1, \ldots, \omega_n$ and μ functions $\theta_1(\lambda), \ldots, \theta_\mu(\lambda) \in \mathcal{O}(\lambda)$ such that*

$$\varphi = \partial F / \partial x_1 \omega_1 + \ldots + \partial F / \partial x_n \omega_n + \theta_1 \varphi_1 + \ldots + \theta_\mu \varphi_\mu$$

For the proof we need the following two propositions:

Proposition 3.1.3. *Every monomial* $\varphi = x_1^{m_1} \ldots x_n^{m_n}$ *(we set $|m| = m_1 + \ldots + m_n$) is equal to a finite sum $\sum_{j=1}^\mu \sum_{|m'| \leq m_0} c_{jm'} (F_{0x})^{m'} \varphi_j$; $m_0 \leq a|m|$, $\sum_{jm'} |c_{jm'}| \leq bd^{|m'|}$ the constants $a > 0, b > 0, d > 0$ depend only on F_0. The sum is unique.*

Proof: The existence and convergence of such sums follows from Nakayama's lemma. Their finiteness follows from the fact that for $|\partial F_0 / \partial x_1|, \ldots,$ $|\partial F_0 / \partial x_{n-1}|$ sufficiently small $\partial F_0 / \partial x_n$ can assume any value. Hence, the series is convergent for all values of $\partial F_0 / \partial x_n$ if the ones of $\partial F_0 / \partial x_j, j \neq n$ are fixed. At infinity the sum of the series grows no faster than a power – hence, the series is a polynomial in $\partial F_0 / \partial x_n$ (and in the same way – in $\partial F_0 / \partial x_1, \ldots, \partial F_0 / \partial x_{n-1}$). Finiteness implies uniqueness. The existence of a, b, d follows from the following: when we increase the power of φ by 1, i.e. $\varphi \mapsto x_k \varphi, 1 \leq k \leq n$, then we multiply the sum of φ by x_k and for $x_k \varphi_j$ which are not among the monomials φ_j substitute their sums which are a finite fixed set. Hence, m_0 increases by no more than a fixed constant and the sum $\sum_{jm'} |c_{jm'}|$ – no more than a constant times.

Proposition 3.1.4. *Every monomial* $\varphi = x_1^{m_1} \ldots x_n^{m_n}$ *can be presented as* $\varphi = (\partial F_0 / \partial x_1) \omega_{01} + \ldots + (\partial F_0 / \partial x_n) \omega_{0n} + \theta_{01} \varphi_1 + \ldots + \theta_{0\mu} \varphi_\mu$ *where ω_j are polynomials of $x; \theta_j \in C$. Under the assumptions of the remark in 1.3., there exists a constant $c_1 > 0$ such that for all φ we have $\|\omega_{01}\| + \ldots + \|\omega_{0n}\| + |\theta_{01}| + \ldots + |\theta_{0\mu}| \leq c_1$. There exist constants $c_2, c_3, c_4 > 0$ such that $\deg \omega_{0j} \leq c_2 |m| + c_3$ and $\deg \omega_{0j} \leq |m| + c_4$ for $|m|$ sufficiently large.*

The first statement follows from Proposition 3.1.3. The estimations follow from the following: let for all monomials φ of degree $|m| \leq m_0$ the polynomials ω_j and the numbers θ_j be found. For $\deg \varphi = m_0 + 1$ we can write $\varphi = x_k \varphi', \deg \varphi' \leq m_0$, hence

$$\varphi = (\partial F_0 / \partial x_1) x_k \omega_{01} + \ldots + (\partial F_0 / \partial x_n) x_k \omega_{0n} + \theta_{01} x_k \varphi_1 + \ldots + \theta_{01} x_k \varphi_\mu$$

(ω_{0j} and θ_{0j} correspond to φ_1). For $x_k \varphi_j$ which are not among the monomials φ_j put $x_k \varphi_j = \varphi_{kj} g_{\nu(k,j)} + \theta_{kj1} \varphi_1 + \ldots + \theta_{kj\mu} \varphi_\mu$ where $\varphi_{kj} = x_s^q$ for some $1 \leq s \leq n, q \leq \mu$ and $g_{\nu(k,j)}$ is an element of the Gröbner basis of $\mathcal{J}$. Hence, $\|\omega_{0j}\|$ can increase only because we must add to $x_k \omega_{0j}$ polynomials $\theta_{kjs} p_{kjs}$, where $\varphi_{kj} g_{\nu(k,j)} = p_{kj1} \partial F_0 / \partial x_1 + \ldots + p_{kjn} \partial F_0 / \partial x_n$. The set of polynomials p_{kjs} is finite. We have $\|g_{\nu(k,j)} - lm(g_{k,j})\| \leq \delta$, see the remark in 1.3.; 'lm' means 'leading monomial'. Find in the same way the polynomials ω_{0j} and numbers θ_{0j} for $x_{k_1}x_{k_2}\varphi'$, $x_{k_1}x_{k_2}x_{k_3}\varphi'$, ..., $x_{k_1}x_{k_2}\ldots x_{k_\mu}\varphi' = \varphi''$, $1 \leq k_j \leq n$. Let for φ' inequality $|\theta_{01}|+\ldots+|\theta_{0\mu}| \leq h$ hold. Then for φ'' we have $|\theta_{01}|+\ldots+|\theta_{0\mu}| \leq h(1+\delta)^\mu - h$. Really, after μ times repeating the procedure described above every time $|\theta_{01}|+\ldots+|\theta_{0\mu}|$ increases no more than $(1+\delta)^\mu$ times. We subtract h because after μ times performing it every of the initial terms $\theta_{0j}\varphi_j$ (corresponding to φ') has been multiplied by $x_{k_1}\ldots x_{k_\mu}$ and, hence, has ceased to be among the monomials φ_j. As $(1+\delta)^\mu - 1 < 1/2$, $\|\omega_{01}\| + \ldots + \|\omega_{0\mu}\| + |\theta_1| + \ldots + |\theta_\mu|$ increases as the sum of a geometric progression with ratio $< 1/2$ and, hence, is limited. The procedure described implies also the existence of c_2, c_3, c_4 – for $|m|$ sufficiently large $\deg\omega_{0j}$ increase exactly by 1 under $\varphi' \mapsto x_k\varphi'$.

Proof of Proposition 3.1.2.: Set $F = F_0 + F_1$. Then F_1 and $\partial F_1/\partial x_j$ are linear forms of λ and polynomials of x. Set

$$p_0 \equiv \varphi = (\partial F_0/\partial x_1)\omega_{01} + \ldots + (\partial F_0/\partial x_n)\omega_{0n} + \theta_{01}\varphi_1 + \ldots + \theta_{0\mu}\theta_\mu$$

Present $p_1 \equiv -((\partial F_1/\partial x_1)\omega_{01} + \ldots + (\partial F_1/\partial x_n)\omega_{0n})$ as $p_1 = (\partial F_0/\partial x_1)\omega_{11} + \ldots + (\partial F_0/\partial x_n)\omega_{0n} + \theta_{11}\varphi_1 + \ldots + \theta_{1\mu}\varphi_\mu$. Set $p_2 = -((\partial F_1/\partial x_1)\omega_{11} + \ldots + (\partial F_1/\partial x_n)\omega_{1n})$, then present p_2 in a similar form as p_0, p_1 etc. One verifies readily that $\varphi = \partial F/\partial x_1(\omega_{01} + \omega_{11} + \omega_{21} + \ldots) + \ldots + \partial F/\partial x_n(\omega_{0n} + \omega_{1n} + \omega_{2n} + \ldots) + (\theta_{01} + \theta_{11} + \theta_{21} + \ldots)\varphi_1 + \ldots(\theta_{0\mu} + \theta_{1\mu} + \theta_{2\mu} + \ldots)\varphi_\mu$, that $\omega_{0j}+\omega_{1j}+\omega_{2j}+\ldots$ satisfy all the conditions of the definition of a series of type A and that $\theta_{0j}+\theta_{1j}+\theta_{2j}+\ldots$ are convergent series of λ (θ_{1j} are not constants, but linear forms, θ_{2j} are bilinear forms in λ etc.). We leave the details for the reader.

Proposition 3.1.5. *Every monomial $\varphi = x_1^{m_1} \ldots x_n^{m_n}$ is equal to a series of type B.*

Proof: 1^0 It follows from Nakayama's lemma that every monomial φ can be presented as

$$\varphi = \sum_{j=1}^\mu \sum_{i\in(N\cup 0)^n} c_{ij}(\lambda)(F_x)^i\varphi_j \quad , \quad c_{ij} \in \mathcal{O}(\lambda)$$

where the series is convergent for (λ, F_x) sufficiently small. Hence, the series can be written as $\sum_j \sum_i \sum_m c_{ijm}\lambda^m(F_x)^i\varphi_j$ with $|c_{ijm}| \leq C'(\rho')^{|m|+|i|}$, $C' > 0, \rho' > 0$. We prove that $c_{ijm} = 0$ for $m > a'|i| + b'$ for some $a' > 0, b' > 0$ which implies that $\sum_i |c_{ijm}|$ increase proportionally to $(\rho')^{|m|}$ (if ρ' is assumed to be > 1, which is no loss generality). To this end we use Proposition 3.1.3. – present φ as

$$\varphi = \sum_{j=1}^\mu \sum_{|i|\leq m_0} c_{ij}(F_{0x})^i\varphi_j = S[\varphi]$$

Replace in $S[\varphi]$ F_{0x} by F_x (this gives $\tilde{S}[\varphi]$) and write

$$S[\varphi] = \tilde{S}[\varphi] + R(x, \lambda)$$

where $R = \sum_{|m| \le 2m_0} \lambda^m S_m$, where S_m is a sum of type $S[\varphi]$, $deg_{F_{0x}} S_m \le m_0$ and R_m is a polynomial of x depending only on the set $\{\varphi_1, \ldots, \varphi_\mu\}$. Present R_m as a series $S[R_m]$. Hence, $deg_{F_{0x}} S[R_m] \le aa''|m|$, a is from Proposition 3.1.3..

Replace in $S[R_m] F_{0x}$ by F_x – this gives $\tilde{S}[R_m], deg_{F_x} \tilde{S}[R_m] \le m_0 + aa''$. We have (see the definition of a series of type B) $\sum_j Q_{0j} \varphi_j = \tilde{S}[\varphi], \sum_j Q_j \varphi_j = \tilde{S}[R_m]$ for $|m| = 1$. Futher we proceed in the same way. At the p-th step we have $\sum_{m,j} \lambda^m Q'_{mj} \varphi_j$, where $Q'_{mj} \equiv Q_{mj}(F_x)$ for $|m| \le p$ and for $|m| > p$, Q'_{mj} are polynomials of $\partial F_0/\partial x_1, \ldots, \partial F_0/\partial x_n$ and $deg Q'_{mj} \le m_0 + aa''|m|$, with respect to F_x for $|m| \le p$ and with respect to F_{0x} for $|m| > p$. At each step we substitute F_x for F_{0x} in Q'_{mj} for $|m| = p+1$ and present the correcting term as a sum $\sum \lambda^m Q'_{mj} \varphi_j$ of the same kind, with the same a''; this sum is finite with respect to m. The proposition is proved.

3.2 The simplification.

Lemma 3.2.1. *Any holomorphic form ψdx, see (2.2.) and (2.3.), can be presented as a sum*

$$\psi dx = \Phi_1 dx + \Phi_2 dx \quad , \qquad \Phi_1 = \sum_{i=0}^{\infty} \sum_{j=1}^{\mu} c_{ij} F^i \varphi_j$$

where $c_{ij} \in \mathcal{O}(\lambda, \lambda')$ or $c_{ij} \in \mathcal{O}(\lambda, \xi, \lambda')$, the series being convergent in the neighbourhood of O in $C^n \times C^\mu \times C$ (in $C^n \times C^\mu \times C^k \times C$), and

$$\Phi_2 = \beta dF \wedge \omega * + F d\omega *$$

for some form $\omega *$ (holomorphic in (x, λ, λ') or $(x, \lambda, \xi, \lambda')$). Call Φ_1 a series of type C.

Corollary 3.2.2. *If equation (2.2.) (or (2.3.)) has a holomorphic solution for any right hand-side of the form $\Phi_1 dx$, see the lemma, then it has a holomorphic solution for any right hand-side ψdx.*

Really, if (H, θ) (or (H, θ, ξ)) is a solution to equation (2.2.) (respectively, to equation (2.3.)) with right hand-side $\Phi_1 dx$, then $(H + \omega*, \theta)$ (or $(H + \omega*, \theta, \xi)$) is a solution to this equation with right hand-side ψdx.

Proof of Lemma 3.2.1.: 1^0. Using Nakayama's lemma, we present the right hand-side ψdx as

$$\psi dx = (\sum_{\gamma \in (N \cup 0)^n} \sum_{j=1}^{\mu} d_{\gamma j}(F_x)^{\gamma} \varphi_j) dx \qquad (*)$$

where $\gamma = (\gamma_1, \ldots, \gamma_n)$, $\gamma_k \in N \cup 0$, $F_x^{\gamma} = (\partial F/\partial x_1)^{\gamma_1} \ldots (\partial F/\partial x_n)^{\gamma_n}$, $d_{\gamma j} \in \mathcal{O}(\lambda, \lambda')$ (or $d_{\gamma j} \in \mathcal{O}(\lambda, \xi, \lambda')$) and series $(*)$ is convergent in some closed polydisc centered at the origin of $OF_x \lambda \lambda'$ (or $OF_x \lambda \xi \lambda'$). Hence, we can suppose that there exist constants $C > 0, \rho > 0$ such that $|d_{\gamma j}| \leq C(\rho)^{|\gamma|}, |\gamma| = \gamma_1 + \ldots + \gamma_n$, for all (λ, λ') (or (λ, ξ, λ')) from the polydisc. Without loss of generality we put $C = 1$.

2^0. For $\beta \neq -k, k \in N$ we find the forms $\Phi_1 dx, \Phi_2 dx$ iteratively; the case $\beta = -k$ is considered in 10^0. Let ψdx be defined by $(*)$. Set $\mathcal{L}[\psi dx] = (\sum_{j=1}^{\mu} d_{0j} \varphi_j) dx = \Omega^0$, i.e. Ω^0 is 'the constant term' of $\psi dx \equiv \psi^0 dx$ with respect to 'the variables' $\partial F/\partial x_j$. Set

$$\psi dx \equiv \psi^0 dx = \Phi_2^0 dx + \psi^1 dx + \Omega^0$$

$$\Phi_2^0 dx = \beta dF \wedge \omega^0 + F d\omega^0 \qquad , \qquad \psi^1 dx = -F d\omega^0$$

These formulas don't define the $(n-1)$-form ω^0 uniquely. Denote by $\gamma \pm e_k$ the multiindex $(\gamma_1, \ldots, \gamma_{k-1}, \gamma_k \pm 1, \gamma_{k+1}, \ldots, \gamma_n)$. Set

$$\beta dF \wedge \omega^0 = (\partial F/\partial x_1)(\sum_{\gamma_1 > 0} \sum_{j=1}^{\mu} d_{\gamma j}(F_x)^{\gamma - e_1} \varphi_j) dx +$$

$$(\partial F/\partial x_2)(\sum_{\gamma_1 = 0, \gamma_2 > 0} \sum_{j=1}^{\mu} d_{\gamma j}(F_x)^{\gamma - e_2} \varphi_j) dx +$$

$$+ \ldots + (\partial F/\partial x_n)(\sum_{\gamma_1 = \ldots = \gamma_{n-1} = 0, \gamma_n > 0} \sum_{j=1}^{\mu} d_{\gamma j}(F_x)^{\gamma - e_n} \varphi_j) dx$$

3^0. Set $\Omega_1 = \mathcal{L}[-d\omega^0]$. The form $d\omega^0$ is well-defined as a series of type $(*)$. Really, the derivations $\partial/\partial x_j$ act as linear operators on the space $\{\varphi_1, \ldots, \varphi_\mu\}$. The products $(\partial^2 F/\partial x_i \partial x_s)\varphi_u = T_{isu}$ are sums of type $(*)$ with $d_{\gamma j} \in \mathcal{O}(\lambda)$ and at the same time series of type B, see Proposition 3.1.5.. Denote their set (which is finite) by S.

4^0. Set

$$-d\omega^0 = \Phi_2^1 dx + \psi^2 dx + \Omega^1 \quad , \quad \Phi_2^1 dx = (\beta + 1) dF \wedge \omega^1 + F d\omega^1 \quad , \quad \psi^2 dx = -F d\omega^1$$

Note that $F\Phi_2^1 dx = \beta dF \wedge (F\omega^1) + Fd(F\omega^1)$. Similarly set

$$-d\omega^k = \Phi_2^{k+1}dx + \psi^{k+2}fx + \Omega^{k+1} \quad , \quad \Omega^{k+1} = \mathcal{L}[-d\omega^k],$$

$$\Phi_2^{k+1}dx = (\beta + k + 1)dF \wedge \omega^{k+1} + Fd\omega^{k+1} \quad , \quad \psi^{k+2}dx = -Fd\omega^{k+1}, k = 0, 1, \ldots$$

The forms $\omega^1, \omega^2, \ldots$ are defined by $\psi^1 dx, \psi^2 dx, \ldots$ in the same way as ω^0 is defined by $\psi^0 dx$, see 2^0. We have $F^k \Phi_2^k dx = \beta dF \wedge (F^k \omega^k) + Fd(F^k \omega^k)$ and the series $\Phi_2 dx = (\Phi_2^0 + F\Phi_2^1 + F^2\Phi_2^2 + \ldots)dx$ and $\Phi_1 dx = \Omega^0 + F\Omega^1 + F^2\Omega^2 + \ldots$ stabilize as series of $(F, \partial F/\partial x_j)$ and F respectively. Formally, we have $\psi dx = \Phi_1 dx + \Phi_2 dx$ and $\Phi_2 dx = \beta dF \wedge \omega + Fd\omega, \omega = \omega^0 + F\omega^1 + (F)^2\omega^2 + \ldots$. It only remains to prove the convergence of the series $\Phi_1 dx, \Phi_2 dx$. We prove the convergence of ω as a series of $(F, \partial F/\partial x_j)$ which implies the one of $\Phi_1 dx$ and $\Phi_2 dx$.

5^0. *The proof of the convergence is based on the fact that although the derivations in the definitions of $\psi^{k+1}dx$ by ω^k spoil the convergence, their effect is compensated by the growth of the numbers $|\beta + k|$ involved in the definitions of* ω^k. Present ω^0 as a series of type $(*)$ denoted by $(*)_{\omega^0}$; we write $d_{\gamma j}^0$ instead of $d_{\gamma j}$. Then

$$max_\lambda |d_{\gamma j}^0| \le \rho^{|\gamma|+1}/|\beta| = \rho\rho^{|\gamma|}/|\beta|$$

Denote the coefficients $d_{\gamma j}$ of the series $(*)$ of $d\omega_0$ by $d_{\gamma j}'$. We have $d_{\gamma j}' = g_{\gamma j}' + h_{\gamma j}'$, $g_{\gamma j}'$ are obtained when in $(*)_{\omega^0}$ we differentiate the factors φ_j with respect to $x_1, \ldots, x_n$, $h_{\gamma j}'$ are obtained when in $(*)_{\omega^0}$ we differentiate the factors $(F_x)^\gamma$ and substitute $T_{s\nu j}$ for $(\partial^2 F/\partial x_s \partial x_\nu)\varphi_j$, see 3^0. We have $\partial\varphi_j/\partial x_\nu = c_{j\nu}\varphi_{m(j,\nu)}$ or 0, where $c_{j\nu}$ are constants, $1 \le m(j, \nu) \le \mu$. Hence, there exists a constant $c_0 > 0$ such that

$$max_\lambda |g_{\gamma j}'| \le c_0\rho\rho^{|\gamma|}/|\beta|$$

6^0. Estimate the coefficients $h_{\gamma j}'$. We have

$$\partial(F_x)^\gamma/\partial x_\nu = \sum_{s=1}^n \gamma_s (F_x)^{\gamma-e_s}(\partial^2 F/\partial x_s x_\nu)$$

After this derivation one must substitute for $(\partial^2 F/\partial x_s \partial x_\nu)\varphi_j$ their sums $T_{s\nu j}$ from S, see 3^0. Hence, there exist holomorphic functions $\delta_j(\gamma', \lambda) = \sum_{s=1}^n (\gamma_s' \delta_{js}(\lambda)) + \delta_{j0}(\lambda)$ depending only on the set S such that

$$h_{\gamma j}' = \sum_{|\gamma'| \le |\gamma|+1} \delta_j(\gamma', \lambda)d_{\gamma' j}^0$$

We can assume that all the sums $\sum_{j=1}^{\mu}\sum_{\gamma\in(N\cup0)^n} max_\lambda|d_{\gamma j}(\lambda)|$ where $d_{\gamma j}$ are the coefficients of the series $(*)$ for the elements of S are finite, see Proposition 3.1.4. (if not, we can consider equation (2.2.) or (2.3.) on a polydisc with a smaller radius with respect to λ). Hence, there exists a constant $c > 0$ such that

$$|h'|_{\gamma j} \leq \sum_{j=1}^{\mu} \sum_{|\gamma'|\leq|\gamma|+1} max_\lambda|\delta_j(\gamma',\lambda)|d_{\gamma'j}^0$$

$$\leq c(|\gamma|+1)max_\lambda max_j max_{|\gamma'|\leq|\gamma|+1}|d_{\gamma'j}^0| \leq c\rho^2(|\gamma|+1)\rho^{|\gamma|}/|\beta|$$

(we assume that $\rho > 1$ which is no loss of generality). Finally,

$$max_\lambda|d'_{\gamma j}| \leq max_\lambda|g'_{\gamma j}| + max_\lambda|h'_{\gamma j}| \leq (\rho^{|\gamma|}/|\beta|)(c_0\rho + c\rho^2(|\gamma|+1))$$

$$\leq \tilde{c}(|\gamma|+1)\rho^{|\gamma|}/|\beta|$$

for $\tilde{c} = c_0\rho + c\rho^2$. Hence,

$$max_\lambda|d_{\gamma j}^1| \leq max_\lambda max_{\gamma'':|\gamma''|=|\gamma|+1}|d'_{\gamma''j}|/|\beta+1| \leq \tilde{c}\rho(|\gamma|+2)\rho^{|\gamma|}/|\beta||\beta+1|$$

7^0. Suppose that the coefficients of the series $(*)$ of $\omega^k - d_{\gamma j}^k$ — satisfy the inequalities

$$max_\lambda|d_{\gamma j}^k| \leq (\tilde{c})^k(|\gamma|+k+1)\ldots(|\gamma|+2k)\rho^{|\gamma|}/|\beta||\beta+1|\ldots|\beta+k| \qquad (**)$$

(for $k = 1$ they are true). Denote the coefficients of the series $(*)$ of $d\omega^k$ by $d_{\gamma j}^{(k+1)}$ and set $d_{\gamma j}^{(k+1)} = g_{\gamma j}^{(k+1)}+h_{\gamma j}^{(k+1)}$, the definition of $g_{\gamma j}^{(k+1)}, h_{\gamma j}^{(k+1)}$ being similar to the one of $g'_{\gamma j}, h'_{\gamma j}$, see 5^0. As in 5^0., we obtain the inequality

$$max_\lambda|g_{\gamma j}^{(k+1)}| \leq c_0(\tilde{c})^k\rho(|\gamma|+k+1)\ldots(|\gamma|+2k)\rho^{|\gamma|}/(|\beta||\beta+1|\ldots|\beta+k|)$$

8^0. Estimate $h_{\gamma j}^{(k+1)}$. Similarly to 6^0 we have

$$\sum_{|\gamma'|\leq|\gamma|+1} \delta_j(\gamma',\lambda)d_{\gamma'j}^k = h_{\gamma j}^{(k+1)}$$

where δ_j are the same as in 6^0 (they depend only on S). Hence,

$$max_\lambda|h_{\gamma j}^{(k+1)}| \leq c(|\gamma|+1)max_\lambda max_j max_{|\gamma'|\leq|\gamma|+1}|d_{\gamma'j}^k| \leq$$

$$\leq c\rho^2(\tilde{c})^k(|\gamma|+k+1)\ldots(|\gamma|+2k+1)\rho^{|\gamma|}/(|\beta|\ldots|\beta+k|)$$

(because $|\gamma|+1 \leq |\gamma|+k+1$). Finally,

$$max_\lambda |d_{\gamma j}^{(k+1)}| \le max(|g_{\gamma j}^{(k+1)}| + |h_{\gamma j}^{(k+1)}|) \le$$

$$\le (\tilde{c})^k (c_0\rho + c\rho^2)(|\gamma| + k + 1)\ldots(|\gamma| + 2k + 1)\rho^{|\gamma|}/(|\beta|\ldots|\beta+k|) =$$

$$(\tilde{c})^{k+1}(|\gamma| + k + 1)\ldots|\gamma| + 2k + 1)\rho^{|\gamma|}/(|\beta|\ldots|\beta+k|)$$

and (as in 5^0)

$$max_\lambda |d_{\gamma j}^{k+1}| \le max_\lambda max_{\gamma'':|\gamma''|=|\gamma|+1} |d_{\gamma''j}^{(k+1)}|/|\beta + k + 1| \le$$

$$\le (\tilde{c})^{k+1}\rho(|\gamma| + k + 2)\ldots(|\gamma| + 2k + 2)\rho^{|\gamma|}/(|\beta|\ldots|\beta + k + 1|) =$$

$$= (\tilde{c})^{k+1}\rho \left(\begin{array}{c} |\gamma| + 2k + 2 \\ k + 1 \end{array} \right) ((k+1)!/(|\beta|\ldots|\beta+k+1|))\rho^{|\gamma|} \le$$

$$\le (\tilde{c})^{k+1} 2^{|\gamma|+2k+2}\rho w(\beta, k)\rho^{|\gamma|}$$

Thus estimation $(**)$ is proved by induction; w is defined below.

9^0. Consider $\omega^0 + F\omega^1 + (F)^2\omega^2 + \ldots$ as a series of $(\partial F/\partial x_1, \ldots, \partial F/\partial x_n)$ and F (more precisely – as μ such series). The coefficient before $(F)^k(F_x)^\gamma$ is estimated by $max_\lambda |d_{\gamma j}^k| \le \rho(2\tilde{c})^k (2\rho)^{|\gamma|} w(\beta, k)$, where $w = k!/(|\beta|\ldots|\beta + k|)$ has at most a power growth rate for $k \to \infty$ and β fixed and for $Re\beta > 0$ we have $0 < w < 1$. Hence, the series is convergent for $|F| < 2\tilde{c}$, $|\partial F/\partial x_j| < 2\rho$, $j = 1, \ldots, n$. This proves the lemma for $-\beta \notin N$.

10^0. For $\beta = 0$ equation (2.2.) has the evident solution $\sum_{j-1}^\mu \theta_j\varphi_j dx = \mathcal{L}[\psi dx]$ ($\mathcal{L}$ is defined in 2^0), $H = (\psi dx - \mathcal{L}[\psi dx])/F$. Let $\beta = -k, k \in N$. Present the form ψdx as

$$\psi dx = \Phi_2 dx + \Phi_1 dx + (F)^k \psi^k dx \quad , \quad \Phi_2 dx = (\Phi_2^0 + F\Phi_2^1 + \ldots + (F)^{k-1}\Phi_2^{k-1})dx$$

$$\Phi_1 dx = \Omega^0 + F\Omega^1 + \ldots + (F)^{k-1}\Omega^{k-1},$$

the forms $\Phi_2^j dx, \Omega^j$ being defined as in $2^0 - 4^0$. Set $\omega = \omega^0 + F\omega^1 + \ldots + (F)^{k-1}\omega^{k-1} + \omega^* + \tilde{\omega}$, $\theta = \theta^* + \tilde{\theta}$, where (ω^*, θ^*) is the solution to equation (2.2.) or (2.3.) with right-hand side $\Phi_1 dx$ and $(\tilde{\omega}, \tilde{\theta})$ (or $(\tilde{\omega}, \tilde{\theta}, \xi)$) is its solution with right-hand side $(F)^k\psi^k dx$. Set $(F)^k\psi^k dx = (F)^{k+1}\tilde{\psi}^k dx + (F)^k\mathcal{L}[\psi^k dx]$ and set $\tilde{\omega} = (F)^k\tilde{\omega}^0 + \tilde{\omega}^*$ where $d\tilde{\omega}^0 = \tilde{\psi}^k dx$. Then one verifies readily that $(\omega^* + \tilde{\omega}^*, \theta^* + \tilde{\theta})$ or $(\omega^* + \tilde{\omega}^*, \theta^* + \tilde{\theta}, \xi)$ is the solution to equation (2.2.) or (2.3.) with right-hand side $\Phi_1 dx + (F)^k\mathcal{L}[\psi^k dx]$ which is a form of the kind $(\sum_{i,j} c_{ij} F^i \varphi_j)dx$. This proves the lemma for $\beta = -k, k \in N$.

3.3   Preparation for the algorithm.

Consider equation

$$\beta dF_0 \wedge \omega + F_0 d\omega + (\sum_{j=1}^{\mu} \theta_j \varphi_j) dx = F_0 (\sum_{j=1}^{\mu} c_j(\lambda, \lambda') \varphi_j) dx \qquad (3.3.1.)$$

Lemma 3.3.1. *There exists $b_0 > 0$ such that for $Re\beta > b_0$ equation (3.3.1.) has a holomorphic solution $(\omega(x, \lambda, \lambda'), \theta(\lambda, \lambda'))$ which*

1) is equal to $\sum_{j=1}^{\mu} c_j(\lambda, \lambda')(\omega^j, \theta^j)$, where (ω^j, θ^j) is the solution to (3.3.1.) with right hand-side $F_0 \varphi_j dx$;

2) every component of the $(n-1)$-forms $\omega^1, \dots, \omega^{\mu}$ is a polynomial of $x_1, \dots, x_n$ (the set $\mathcal{M}$ of monomials being fixed for $Re\beta > b_0$) with coefficients depending on β; the choice of $\mathcal{M}$ is described in 1^0 of the proof of the lemma;

3) the solution (ω, θ) can be found algorithmically (the algorithm is described in 2^0 of the proof of the lemma);

4) the solution to (3.3.1.) is equal to $(\omega^0/\beta, \theta^0) + (\omega^1, \theta^1)$ where the components of ω^0, ω^1 are polynomials whose monomials belong to $\mathcal{M}$; (ω^0, θ^0) is the solution to

$$dF_0 \wedge \omega^0 + (\sum_{j=1}^{\mu} \theta_j^0 \varphi_j) dx = F_0 (\sum_{j=1}^{\mu} c_j(\lambda, \lambda') \varphi_j) dx \qquad (3.3.2.)$$

and $max_{\partial\Pi} |\theta_j^1| = O(1/\beta)$, $j = 1, \dots, \mu$, $|\omega_{j,k,i}| = O(1/\beta), j = 1, \dots, n; i = 1, \dots, \mu; k \in \mathcal{M}$ where Π is any fixed polydisc in $O\lambda\lambda'$ where c_j are defined and holomorphic and $\omega = \sum_{i=1}^{n} \sum_{k \in \mathcal{M}} \sum_{j=1}^{\mu} c_j \omega_{j,k,i} x^k dx/dx_i$, where $\omega_{j,k,i}$ depend only on β. The choice of (ω^0, θ^0) is given by the algorithm of finding the normal forms $\sum_{j=1}^{\mu} \theta_j^0 \varphi_j$ of the polynomials $F_0 \varphi_j$ with respect to a Gröbner basis of the ideal $\mathcal{J}$. The choice of (ω^1, θ^1) needs not to be unique, but one can fix a way of defining and algorithmically finding it for all $Re\beta > b_0$, which is assumed in 4).

Corollary 3.3.2. *There exists $b_1 \geq max(b_0, 1)$ such that for $Re\beta \geq b_1$ equation (2.2.) with $\lambda = 0$ has a holomorphic solution. The choice of b_1 depends only on the choice of θ^1 in 4) of the lemma and on the polydisc in $Ox\lambda\lambda'$ on which the right hand-side of (2.2.)is defined. We have $\omega = \sum_{i=0}^{\infty} (F)^i \omega_i$ where the $(n-1)$-forms ω_i are polynomials of one and the same set of monomials (namely, $\mathcal{M}$) whose coefficients are $O(1/i)$. Lemma 3.3.1. and Corollary 3.3.2. remain correct if the right hand-sides of (3.3.1.) and (3.3.2.) are polynomials of one and the same fixed set of monomials.*

Proof: 1^0. Define the linear operator $\mathcal{T} : C^{\mu} \to C^{\mu}$ by $\mathcal{T}(c_1, \dots, c_{\mu}) = (\theta_1^0, \dots, \theta_{\mu}^0)$, see (3.3.2). Set $\mathcal{T}(\theta^0) = \theta^1, \mathcal{T}(\theta^1) = \theta^2$ etc. Then $\theta^{n-1} = 0$ for any $(c_1, \dots, c_{\mu})$, i.e. $\mathcal{T}^n = 0$. Really, $(\sum_{j=1}^{\mu} \theta_j^{n-1} \varphi_j) dx = F_0^n (\sum_{j=1}^{\mu} c_j \varphi_j) dx - \beta dF_0 \wedge (\omega^{n-2} + F\omega^{n-3} + \dots + F^{n-2}\omega^0)$. The right-hand side belongs to $\mathcal{J}$

$(F_0^n \in \mathcal{J},$ see [BrSk]$)$. Hence $\theta_j^{n-1} = 0, j = 1, \ldots, \mu$.

2^0. For Reβ sufficiently large the solution (ω, θ) to (3.3.1) with right hand-side $(\sum_{i,j} c_{ij} F_0^i \varphi_j)dx$ can be found iteratively. Let for $(x, \lambda, \lambda') \in \Pi$ the following inequality holds: $|F| \le \rho$ $(\rho > 0)$. Set $\omega^* = \sum_{i=0}^{\infty} F_0^i \omega^{*i}$ where $(\omega^{*i}, \theta^{*i})$ is defined by 4) of the lemma solution to equation

$$(\beta + i)dF_0 \wedge \omega^{*i} + F_0 d\omega^{*i} + (\sum_{j=1}^{\mu} \theta_j^{*i} \varphi_j)dx = F_0(\sum_{j=1}^{\mu} c_{(i+1)j} \varphi_j)dx \qquad (3.3.3.)$$

(for $i => \ge 1$; for $i = 0$ the right-hand side is $(\sum_{j=1}^{\mu} c_{0j} \varphi_j + F_0 \sum_{j=1}^{\mu} c_{1j} \theta_j)dx$). Set $\omega = \omega^* + \tilde{\omega}, \theta = \theta^{*0} + \tilde{\theta}$. Then $(\tilde{\omega}, \tilde{\theta})$ is the solution to (2.2.) with right hand-side $(\sum_{i=1}^{\infty} \theta_j^{*i} F_0^i \varphi_j)dx$. One can write $\theta^{*i} = \mathcal{T}(c_{(i+1)1}, \ldots, c_{(i+1)\mu}) + Q_i(c_{(i+1)1}, \ldots, c_{(i+1)\mu})$ where $\|Q_i\| = O(1/(\beta + i))$ for $i \to \infty$, see 4) of the lemma.

3^0. Iterate the mapping $\Delta : \omega \mapsto \tilde{\omega}*, \theta \mapsto \tilde{\theta}$. For Re$\beta$ sufficiently large the norm of the right-hand-side of (3.3.1.) after Δ^n decreases $c > 1$ times. Really $\|(\mathcal{T} + Q_i) \circ \ldots \circ (\mathcal{T} + Q_{i+n-1})\| = O(1/(\beta + i))$, due to $\mathcal{T}^n = 0$. Hence, for Reβ sufficiently large the iterations of Δ are convergent and provide the solution of (2.2.) for $\lambda = 0$. The presentation of ω as a series with the properties claimed follows from 4) of the lemma.

Proof of the lemma: 1^0. 1) is evident. Let φ_μ be the greatest among the monomials φ_j. Present the leading monomial of $F_0\varphi_\mu - \text{lm}(F_0\varphi_\mu)$ – and all monomials which are less than it as $\sum_{j=1}^{n}(\partial F_0/\partial x_j)\omega_j + \sum_{j=1}^{\mu} \theta_j \varphi_j$ where the polynomials ω_j and the numbers θ_j can be found algorithmically (one finds a Gröbner basis of $\mathcal{J}$ and the normal form $\sum \theta_j \varphi_j$ of $\text{lm}(F_0\varphi_\mu)$ with respect to it). Let $\mathcal{M}_0^j$ be the set (for fixed j) of all monomials participating in some of these ω_j. To find $\mathcal{M}$ we are going to extend the sets $\mathcal{M}_0^j$ in such a way that the union of monomials which participate in all the above mentioned polynomials $(\partial F_0/\partial x_j)\omega_j$ should be the same as (or contain) the one of the polynomials $F_0 \partial \omega_j/\partial x_j$. Denote the desired extensions by $\mathcal{M}^j$ and set $\mathcal{M}_0^j/x_k = \{\varphi = x_1^{m_1} \ldots x_n^{m_n} | x_k \varphi \in \mathcal{M}_0^j\}$. To achieve our aim it suffices to have $\mathcal{M}^j/x_j = \mathcal{M}^k/x_k$. Therefore we set $\mathcal{M}^j = \mathcal{M}^j + \cup_{i \ne j}((x_j \mathcal{M}_0^j/x_i)\backslash \mathcal{M}_0^j)$. The sets $\mathcal{M}^j$ are the components of the set $\mathcal{M}$.

2^0. To prove 3) and 4), set in (3.3.1.) $\omega' = \beta\omega$:

$$dF_0 \wedge \omega' + \alpha F_0 d\omega' + (\sum_{j=1}^{\mu} \theta_j \varphi_j)dx = F_0(\sum_{j=1}^{\mu} c_j \varphi_j)dx \quad , \quad \alpha = 1/\beta \qquad (3.3.3.)$$

Equations (3.3.1.) – (3.3.3.) can be interpreted as linear systems of equations: 1) the unknowns y are the numbers θ_j and the coefficients before the

monomials of the sets $\mathcal{M}^j$ 2) the right hand-side are the coefficients of $F_0 \sum c_j \varphi_j$
3) the system is obtained by comparing the coefficients before the equal mono-
mials from the right and from the left. System (3.3.2.) has a solution for any
right hand-side $\varphi = lm(F_0\varphi_\mu)$ or $\varphi < lm(F_0\varphi_\mu)$. Set $m = \#\{\varphi|\varphi \le lm(F_0\varphi_\mu)\}$.
Hence, there exists a subset $\tilde{\mathcal{M}} \subset \mathcal{M}$, $\#\tilde{\mathcal{M}} = m - \mu$ such that the determinant
of system (3.3.2.) composed of its coefficients before $y \in \tilde{\mathcal{M}}$ is not zero. Hence,
one can set $y = 0$ for $y \in \mathcal{M}\backslash\tilde{\mathcal{M}}$ and find unique $y \in \tilde{\mathcal{M}}$ which give a solution
to (3.3.2.). Hence, the same is true for α sufficiently small in (3.3.3.). It is clear
that the unknowns (y, θ) obtained in this way as a solution to (3.3.3.) depend
analytically on α for α sufficiently small. This proves 4).

3^0. The definition of (ω^1, θ^1), see 4), depends on the choice of the set $\tilde{\mathcal{M}}$ in
2^0. Choosing different $\tilde{\mathcal{M}}$, we can obtain different pairs (ω^1, θ^1).

3.4   The case $\mathrm{Re}\beta >> 0$.

Consider (2.2.) with right hand-side a series of type C, see Lemma 3.2.1. Its
solution can be presented as a sum $(\omega^0 + \tilde{\omega}, \theta^0 + \tilde{\theta})$ where $\omega^0 = \sum_{i=0}^{\infty}(F)^i\omega_{0i}$,
ω_{0i} being polynomial in x $(n-1)$-forms with uniformly in (i, λ) restricted coef-
ficients, of a fixed set of monomials of x; $(\tilde{\omega}, \tilde{\theta})$ is a solution to (2.2.) with right
hand-side $-(\sum_{i=0}^{\infty} \sum_{j=1}^{\mu} \theta_{0,i+1,j}(F)^i\varphi_j)dx + \beta d(F - F_0) \wedge \omega^0 + (F - F_0)d\omega^0)$
Really, ω_{0i} can be found as solutions to

$$(\beta + i)dF_0 \wedge \omega_{0i} + F_0 d\omega_{0i} + (\sum_{j=1}^{\mu} \theta_{0ij}\varphi_j)dx = F(\sum_{j=1}^{\mu} c_{ij}\varphi_j)dx \qquad (3.4.1.)$$

In fact, they can be presented as $\omega_{0i} = \omega_{00i} + \lambda_1\omega_{01i} + \ldots + \lambda_\mu\omega_{0\mu i}$ where
ω_{0ji} (ω_{00i}) is a solution to last equation with right hand-side $\lambda_j\varphi_j(\sum c_{ij}\varphi_j)dx$
(with right hand-side $F_0(\sum c_{ij}\varphi_j)dx$).
    Present $(\beta+i)d(F - F_0)\omega_{0i}+(F - F_0)d\omega_{0i}$ as a series of type B, see Proposi-
tion 3.1.5. A series of type B is also a series of type $(*)$, see the proof of Lemma
3.2.1. For this series for λ sufficiently small we have $\max_\lambda|d_{\gamma j}| \le \tilde{C}(\tilde{\rho})^{|\gamma|}$,
where the constants $\tilde{C} > 0, \tilde{\rho} > 0$ can be taken arbitrarily small, see Propo-
sition 3.1.1. Let $\tilde{\rho} < \rho$, where ρ is from 1^0 of the proof of Lemma 3.2.1.
Each $\Phi^i dx = (F)^i((\beta + i)d(F - F_0) \wedge \omega_{0i} + (F - F_0)d\omega_{0i})$ can be presented
as $\tilde{\Phi}_1 dx + \tilde{\Phi}_2 dx$, $\tilde{\Phi}_1$, $\tilde{\Phi}_2$ are as Φ_1, Φ_2 from Lemma 3.1.1. and the coefficients
of $\tilde{\Phi}_1$ satisfy the inequalities

$$|c_{ij}| < \tilde{\rho}(2\tilde{c}')^k \qquad , \qquad \tilde{c}' = c_0\tilde{\rho} + c(\tilde{\rho})^2$$

(this follows from 9^0 of the proof of Lemma 3.1.1.), i.e. the radius of convergence
of each such series is at least $1/\tilde{\rho}$; the radius of convergence of Φ_1, see Lemma
3.1.1., is at least $1/\rho < 1/\tilde{\rho}$. Hence, the sum of all $\Phi^i dx$ is equal to $\Phi_1^* dx + \Phi_2^* dx$,
Φ_1^*, Φ_2^* are as Φ_1, Φ_2 from Lemma 3.1.1., with $|c_{ij}^*| \le C^*(\rho)^i$ where C^* can

be chosen arbitrarily small if λ is restricted to a sufficiently small polydisc (Proposition 3.1.1.).

The operator mapping the functions c_{ij} into the functions θ_{0ij} is the sum of a nilpotent operator with constant coefficients (this is operator T, see the proof of Corollary 3.3.2.), an operator with a small norm (contracting) for $\mathrm{Re}\beta$ large (it gives θ_j^1, see 4) of the lemma) and an operator with a small norm for $|\lambda|$ small. Hence, for $|\lambda|$ small and $\mathrm{Re}\beta$ large the iterations $(\omega, \theta) \mapsto (\tilde{\omega}, \tilde{\theta})$ are convergent, the consecutive ω^0 are series as described above and $|\omega_{0i}^{(k)}| \to 0$ for $k \to \infty$ uniformly in i (k is the number of the iteration). This proves the versality theorem for $\mathrm{Re}\beta \geq b_1$, ($b_1$ is defined in Corollary 3.3.2.).

3.5   The general case.

If β is not resonant, then the proof of the versality theorems is finished as in **3.4.** If it is, then for some i equation (3.4.1.) can't be solved. For these i we first present the right hand-side of (3.4.1.) as $F_0(\sum_{j=1}^{\mu} c_{ij}\varphi_j)dx + (F - F_0)(\sum_{j=1}^{\mu} c_{ij}\varphi_j)dx$, then present the second term as $\Phi_1 dx + \Phi_2 dx$, see Lemma 3.2.1.; $\Phi_1 dx$ (which is $O(\lambda)$) is added to the right-hand-side of (2.3.) for the next iteration. If equation (3.4.1.) with right-hand-side $F_0(\sum_{j=1}^{\mu} c_{ij}\varphi_j)dx$ still can't be solved, then using the algorithm from **3.6.** we find which c_{ij} can be left so that the equation could be solved. For the others write $c_{ij}F_0\varphi_j = c_{ij}F\varphi_j dx + c_{ij}(F_0 - F)\varphi_j dx$, present the second term as $\Phi_1 dx + \Phi_2 dx$, $\Phi_1 = O(\lambda)$ as in Lemma 3.2.1. and add $\Phi_1 dx$ to the right-hand-side of (2.3.) for the next iteration. The term $c_{ij}F\varphi_j dx$ is left in the normal form as a resonant term. *So the basic task of the algorithm in 3.6. is to define for which c_{ij} equation (3.4.1.) (or, equivalently, equation (3.3.1.), as was just shown) has a solution.*

3.6   The algorithm

In **3.3.** (see Lemma 3.3.1. and 1^0 of its proof) we defined the sets $\mathcal{M}^j$ of monomials of each component of the $(n-1)$-form ω where (ω, θ) is a solution to (3.3.1.). Let their coefficients y be the unknown variables. Equation (3.3.1.) can be written as a system $Ay = b$ with a matrix $A = A_0 + \beta A_1$, A_0, A_1 being constant, the vector b being defined by the coefficients of $F_0\varphi_j (1 \leq j \leq \mu)$. We first solve (3.3.1.) for each j separately, i.e. we consider the set of rows of A in which the corresponding element of b is $\neq 0$. For each minor $M \subset A$ with this set of rows we find its determinant $\det M = f_M(\beta)$ where f_M is a polynomial. We find the zeros of f_M for all such M. The roots of the reduced Bernstein's polynomial are among them. Present the solution $y(\beta)$ by Kramer's rule, for each choice of M. To this end we find a non-degenerate minor $N \subset A$ of greatest size, $M \subset N$, the variables y which are not multiplied by the elements of N are set to be zeros and the rest are defined by Kramer's rule.

Define for a fixed β the linear space T_β as $\{t = \theta_1\varphi_1 + \ldots + \theta_\mu\varphi_\mu | \theta_j \in C, \exists q \in$

$\Omega^{n-1}, q \in \mathcal{O}(x) : t = \beta dF_0 \wedge q + F_0 dq$. Suppose that for a fixed β for some j we can't solve (3.3.1.) with right hand-side $F_0 \varphi_j$ using the method above. Then we must check whether there exists a vector $t \in T_{\beta+1}$ such that some linear combination of these 'bad' $F_0 \varphi_j$ is equal to t. Hence, we must find the space T_β. For every $F_0 \varphi_j$ we find all vectors θ such that (ω, θ) is solution to (3.3.1.) with right hand-side $F_0 \varphi_j$ by the method above, i.e. we find all vectors θ which are obtained for some minor $N \subset A$ whose determinant is not identically in β equal to 0. Then we exclude these θ which for the given β are not defined, i.e. for their corresponding N det$N = 0$; det$N$ is a polynomial of β. The set of the θ which remains is denoted by S_β^0. Set $T_\beta^0 = \{t_1 - t_2 | t_1, t_2 \in S_\beta^0\}$. Then for any $t \in T_{\beta+1}^0$ find by the method above all the θ such that there exists (ω, θ) – a solution to (3.3.1.) with right hand-side $F_0 t dx$.

The set of these θ is denoted by S_β^1. Set $T_\beta^1 = \{t_1 - t_2 | t_1, t_2 \in S_\beta^1\}$. Define in the same way T_β^2, $T_\beta^3 \ldots$ and set $T_\beta = T_\beta^0 \cup T_\beta^1 \cup \ldots$. This union is finite because dim$T_\beta \leq \mu$. The criterion to stop looking for the next T_β^k is: let $\beta + k$ be non-resonant and let $T_\beta^{k+1} = 0$; then stop. The criterion is based on the fact that if we cut off in (*) with $\lambda = 0$ the sufficiently high powers of F_{0x} in the right hand-side, then the vector θ of the solution (ω, θ) to $(2.2.)|_{\lambda=0}$ changes little. Really, the coefficients of $(F_{0x})^\gamma$ grow at most like the terms of a geometric progression and their influence upon θ decreases like $1/|(\gamma/n)!|$; this follows from 3^0 of the proof of Corollary 3.3.2.. Hence, T_β is generated by such t for which the corresponding q are polynomials of x. The fact that equation (3.3.1.) can be solved for β non-resonant follows from [Var2].

R E F E R E N C E S :

[Arn1] Arnold V.I. Singularities in differential calculus. Itogi Nauki, Sovremennye Problemy Matematiki, Moscow, VINITI, vol. 22, pp; 3 - 55 (in Russian).

[Arn2] Arnold V.I. Classification of Poisson structures on the plane. Trudy Seminara im. I.G. Petrovskogo 1985, vol. 12, pp. 1 - 15.

[AVG] Arnold V.I., Varchenko A.N., Guseyn-Zade S.M. Singularités des applications différentiables. 1-ère partie: Classification des points critiques, des caustiques et des fronts d'onde. Moscou, Ed. Mir, 1986.

[Ber1] Bernstein I.N. The possibility of analytic continuation of f_+^λ for certain polynomials f. Functional Analysis and its Applications 1968, vol. 2, No 1.

[Ber2] Bernstein I.N. Continuation of generalised functions with respect to a parameter.Functional Analysis and its Applications 1972,vol. 6,No 4.

[BG] Bernstein I.N., Gelfand S.I. Meromorphness of the function P^λ. Functional Analysis and its Applications 1969, vol. 3, No 1.

[Bjo] Bjork I.E. I.N.Bernstein functional equation, local case. Preprint - Catholic University, Nijmengen, November 1975.

[BrSk] Briançon J., Skoda H.Sur la clôture intégrale d'un idéal de germes de fonctions holomorphes en un point de C^n. C.R.Acad. Sci. Paris, Ser.A, 278 (1974), 949-951.

[Bu1] Buchberger B. A criterion for detecting unnecessary reductions in the construction of Gröbner bases, Proc; Eurosam 79, Lecture Notes in C.S. 72 (1979).

[Bu2] Buchberger B. Note on the complexity of constructing Gröbner-Bases, in Proc. Eurocal 83, Lecture Notes in C.S. (1983).

[GSh] Gelfand I.M., Shilov G.E. Generalised functions and actions upon them, vol. 1. Fizmathgiz, 1959 (in Russian).

[GSt] Guillemin V., Sternberg S. Geometric asymptotics. Providence (R.I.), Amer. Math. Soc., 1977, XVIII.

[Gi1] Givental' A.B. Lagrangian varieties with singularities and irreducible sl_2-modules. Russian Mathematical Surveys 1983, vol. 38, No 6, pp. 121 - 122.

[Gi2] Givental' A.B. Twisted Picard-Lefschetz formulas. Functional Analysis and its Applications 1988, vol. 22, No 1, pp. 10 - 18.

[Kos1] Kostov V.P. Versal deformations of differential forms of degree α on a line. Functional Analysis and its Applications 1984, vol. 18, No 4, pp. 335 - 337.

[Kos2] Versal deformations of differential forms of real degree on the real line. Mathematics of the USSR - Izvestiya,1991,vol. 37,No 3,pp. 525–537.

[Lan1] Lando S.K. Normal forms of the degrees of a volume form. Functional Analysis and its Applications 1985, vol. 19, No 2, pp. 146 - 148.

[Lan2] Lando S.K. Deformations of differential forms. Candidate of the phys.-math. sciences dissertation, Moscow, Moscow State University, 1985.

[Mal] Malgrange B. Le polynôme de Bernstein d'une singularité isolée. Springer Lecture Notes in Math. 1974, vol. 459, pp. 98 - 119.

[Var1] Asymptotic Hodge structure in vanishing cohomologies. Mathematics of the USSR – Izvestiya 1982, vol. 18, No 3, pp. 469 – 512.

[Var2] Varchenko A.N. Local classification of volume forms in the presence of a hypersurface. Functional Analysis and its Applications 1985, vol. 19, No 4. pp. 269 – 276.

[Var3] Varchenko A.N. Gauss-Manin connection of an isolated singular point and Bernstein polynomial. Bull. Sc. Math., 2-e série, 1980, vol. 104, pp. 205 – 223.

V.P.Kostov
Université de Nice – Sophia Antipolis,
Laboratoire de Mathématiques, U.R.A. du C.N.R.S. No 168,
Parc Valrose,
06108 Nice Cedex 02, France
e-mail kostov@hera. unice.fr.

S.K.Lando
Institute of New Technologies,
11 Kirovogradskaya,
113587 Moscow, USSR.
e-mail: postmaster@globlab.msk.su, fax: (7095) 3150808.

Computing subfields: Reverse of the primitive element problem

D. Lazard [*]   A. Valibouze [*]

Abstract.   We describe an algorithm which computes all subfields of an effectively given finite algebraic extension. Although the base field can be arbitrary, we focus our attention on the rationals.

This appears to be a fundamental tool for the simplification of algebraic numbers.

Introduction

Many algorithms in computer algebra contain subroutines which require the use of algebraic numbers. Computing with them is especially important when polynomial systems of equation have to be solved. As an example let us consider the cyclic 7th–roots of unit, which are the solutions of the following system [4, 1]:

$$a + b + c + d + e + f + g = 0$$
$$ab + bc + cd + de + ef + fg + ga = 0$$
$$abc + bcd + cde + def + efg + fga + gab = 0$$
$$abcd + bcde + cdef + defg + efga + fgab + gabc = 0$$
$$abcde + bcdef + cdefg + defga + efgab + fgabc + gabcd = 0$$
$$abcdef + bcdefg + cdefga + defgab + efgabc + fgabcd + gabcde = 0$$
$$abcdefg = 1.$$

Some of the solutions of this system are of the form $(a, b, c, 1/c, 1/b, 1/a, 1)$ where a (and also b) is a root of the following polynomial, and b and c are expressed as polynomials of degree at most 11 in a [6, 3, 10].

$$P(x) = \quad x^{12} + 9x^{11} + 3x^{10} - 73x^9 - 177x^8 - 267x^7 - 315x^6 - 267x^5$$
$$-177x^4 - 73x^3 + 3x^2 + 9x + 1.$$

Both from an intuitive and a computational point of view it is much more suitable to represent the root a of P by the "nested" system:

$$x^2 - 3x - 3 = 0,$$

[*]Partially supported by PRC–GDR Mathématique–Informatique

$$y^3 + y^2 - 2y - 1 = 0,$$
$$a^2 - (xy - 1)a + 1 = 0.$$

This much simpler representation has also the advantage that it gives some additional information on the Galois group of the polynomial P.

Generalizing this example, we can say that the simplification of algebraic numbers consists of two problems. The first one is to decompose the extension in which the algebraic numbers are expressed in subextensions of degree as low as possible; the second, is to choose generators for these subextensions which have a simple minimal polynomial and on which the algebraic numbers to be expressed have a simple form.

We consider here only the first problem, but we have to mention the function polred in the system PARI [2] which appears to be the best known solution for the second problem.

For solving the first problem, we give in this paper an algorithm for computing all subextensions of a given algebraic field extension $K \longrightarrow L$.

As a byproduct, the algorithm could also output the subfields of low degree and index of the extensions of low degree of L. However, we have to emphasize that the algorithm does not compute in general the Galois group nor the Galois closure, not even when the Galois closure may be obtained as a tower of extensions of low degrees.

We know of two previous algorithms ([5, 15], the latter seeming found by S. Landau) for finding the subfields of an extension. It is rather difficult to decide which of these three algorithms is the most efficient, because their most time consuming steps are factorization of very different shape. Moreover, for our algorithm, we need only the factors of low degrees as output of the factorization; thus the timings may very different when using specially designed factorization routines instead of the standard ones.

1   Symmetric resolvents

The notion of resolvent was introduced by Lagrange [9] for studying the solutions of an equation. We recall here the general definition of a resolvent which is needed for our algorithm. The idea is to transform an equation by means of an appropriate function in order to obtain some information about its roots. In this paper the appropriate functions will be symmetric polynomials.

Throughout the paper, p will be an irreducible univariate polynomial of degree n over a field K, which is generally the field of the rationals. The set of its roots in an algebraic closure of K will be denoted by $\alpha = \{\alpha_1, \ldots, \alpha_n\}$.

Let $X = \{x_1, \ldots, x_n\}$ be a set of n indeterminates. The symmetric group S_n acts naturally on $K(X)$ by permuting the elements of X. This action leads to the following definition of a resolvent. (The notation is the same as in [7].)

Definition 1 *Let H be a subgroup of S_n and let f be an element of $K(x_1,$ $\ldots,x_n)$. The set of functions $f(x_{s(1)},\ldots,x_{s(n)})$, for s in H, is called the orbit of f under the action of H and is denoted by $H(f)$.*

Definition 2 *Let f be an element of $K(x_1,\ldots,x_n)$ and p be a univariate polynomial, the roots of which are $\alpha_1,\ldots,\alpha_n$. The (general) resolvent of p by f, denoted by $f_*(p)$, is the polynomial whose roots are the elements of the orbit of f evaluated at the roots of p:*

$$f_*(p)(y) = \prod_{h \in S_n(f)} (y - h(\alpha_1,\ldots,\alpha_n)).$$

In this case f is called a transformation function.

The irreducible factors over K of a general resolvent are polynomials in y which are called *irreducible resolvents*.

When a function depends on exactly k variables we say that its *arity* is k. If a transformation function of arity k is symmetric on these k variables we say that the resolvent is *k-symmetric*. We will recall in Section 5 a method for computing resolvents, which is very simple in the case of symmetric ones. It is easy to see that when the transformation function is a polynomial the resolvent may also be obtained by using resultants (see [7, 9] and also [11], where Soicher uses them to compute linear resolvents).

2   Finding equations for subfields

In this section we give the main theorem of this paper. It gives a necessary condition on the symmetric resolvents, for the existence of a field between K and $K(a)$. This condition leads to our algorithm for computing all subfields of index k in $K(a)$.

In what follows $\mathrm{Irr}(b, K)$ will denote the minimal polynomial of the algebraic number b over the field K.

Theorem 1 *Let p be an irreducible polynomial over K of degree n and a be one of its roots. If L is a subfield of $K(a)$ of index k, then any k-symmetric resolvent $s_*(p)$ has a factor of degree n/k which is a power of an irreducible polynomial (over K) which has a root in L. More precisely, there exists a polynomial q and an irreducible polynomial h of degree d over K with a root in L such that $s_*(p) = qh^{n/(kd)}$*

Proof. Let $a = \alpha_1,\ldots,\alpha_k$ be the conjugates of a over L. The element $b := s(\alpha_1,\ldots,\alpha_k)$ is in L. This symmetric function can be written by means of the elementary symmetric functions of $\alpha_1,\ldots,\alpha_k$ which are, up to their sign, the coefficients of the minimal polynomial of a over L. Thus, $\mathrm{Irr}(b, K)$ has a degree d dividing $n/k = [L : K]$ and we have the following inclusion:

$$K \xrightarrow{d} K(b) \xrightarrow{n/(kd)} L \xrightarrow{k} K(a).$$

We have to prove that $\mathrm{Irr}(b, K)^{n/(kd)}$ divides $s_*(p)$, i.e., that b is a root of $s_*(p)$ of multiplicity at least $n/(kd)$. The symmetric group S_n acts on the sets of k conjugates of a, and thus acts also on the roots of $s_*(p)$. The multiplicity of b is $\mu := [Stab_{S_n}(b) : Stab_{S_n}(\{\alpha_1, \ldots, \alpha_k\})]$ (we recall that the roots of p are all distinct). On the other hand, let $H := Stab_G(\{\alpha_1, \ldots, \alpha_k\})$ where G is the Galois groups of p over K; it is easy to verify that the map of homogeneous spaces $H/Stab_G(b) \longrightarrow Stab_{S_n}(\{\alpha_1, \ldots, \alpha_k\})/Stab_{S_n}(b)$ is injective; this implies that μ is larger than the index of H in $Stab_G(b)$. By Galois theory we know that $Stab_G(b)$ and H are the automorphism groups of the splitting field of $K(b)$ and L; thus

$$[Stab_G(b) : H] = [L : K(b)] = n/(kd).$$

Hence we have $\mu \geq n/(kd)$. $\Diamond$

Remark 1 *When looking for subfields of index k of $K(a)$, the following kinds of factorization may occur for the k-symmetric resolvent $s_*(p)$:*

(a) $s_*(p)$ has a simple factor of degree n/k. Lemma 3 of Section 3 shows that this factor is the minimal polynomial of a generator of a subfield of index k of $K(a)$.

(b) $s_*(p)$ has an irreducible factor of degree n/k of multiplicity greater than 1. This factor defines an extension of K, but further computations are needed to decide whether this extension has a conjugate included in $K(a)$.

(c) $s_*(p)$ has a factor of degree n/k which is a power of an irreducible polynomial h of degree dividing n/k. This situation may correspond to a subfield of index k, which has itself a subfield with h as defining polynomial; further computations are needed for deciding if the subfield of index k exists and for finding it.

(d) $s_*(p)$ does not have a factor of degree n/k which is a power of an irreducible polynomial. This implies that $K(a)$ does not have a subfield of index k.

Our algorithm mainly consists of applying this remark to the elementary symmetric functions. We recall that the i-th *elementary symmetric function* of k variables, denoted $e_i(x_1, \ldots, x_k)$, is the sum of the elements of the orbit of the monomial $x_1 \ldots x_i$ under the action of the symmetric group S_n. It is also the coefficient of t^i in the polynomial $\prod_{j=1}^n (t + x_j)$.

We present now our algorithm for computing subfields; beforehand, we introduce a data structure for representing the fields which appear as a tower of simple algebraic extensions embedded in $K(a)$:

Definition 3 *Let a be an algebraic number over K. We say that the list*

$$[[h_1, b_1, g_1], [h_2, b_2, g_2], \ldots, [h_r, b_r, g_r]],$$

is a descriptive chain of an intermediate field L between K and $K(a)$ if we have a sequence of field extensions:

$$K \xrightarrow{h_1} K(b_1) \xrightarrow{h_2} K(b_1, b_2) \xrightarrow{h_3} \cdots \xrightarrow{h_r} L = K(b_1, b_2, \ldots, b_r) \xrightarrow{g_r} K(a)$$

such that $h_i = Irr(b_i, K(b_1, \ldots, b_{i-1}))$ and $g_i = Irr(a, K(b_1, \ldots, b_i))$, for $i = 1, \ldots, r$.

The length of this chain is r. The descriptive chain of K is the empty list $[\,]$.

Algorithm Eqnfield(p, k)

Input : **p** an irreducible univariate polynomial over K
        k an integer dividing the degree of **p**
        a a root of **p** appearing implicitly as the variable of **p**
Output : A list of descriptive chains representing all subfields of index **k**
        in $K(\mathbf{a})$;
Begin
        Eqnfield2(p, [ ], k, 1)
End.

Algorithm Eqnfield2(p, F, k, i)

Input :   **p** an irreducible polynomial over K
        a a root of **p** appearing implicitly as the variable of **p**
        k an integer dividing the degree of **p**
        F a descriptive chain representing a field between K and
        $K(\mathbf{a})$ of index greater than **k** in $K(\mathbf{a})$
        i the index of the symmetric function to use, which is also
        the depth of the recursion
Output :   A list of descriptive chains representing all extensions of the
        field defined by **F**, of index **k** in $K(\mathbf{a})$
Begin
    sol := empty
    n := deg(p)
    s := i-th elementary symmetric function over **k** variables
    if i > k then return empty   {*see Remark 2*}
    R := s_*(p)   {*see Section 5*}
    for all distinct irreducible factors **h** of **R** of degree dividing **n/k**
        d:=deg(h)
        m:=multiplicity(h,R)
        if d=1 and m≥n/k then   {*case (c)*}

```
              sol := append(sol, Eqnfield2(p, F, k, i+1))
          elseif d=n/k and m=1 then                        {case (a)}
              b := a root of h in K(a), expressed as a polynomial in a
                                                     {see Section 4}
              sol := cons(endcons([h,b,Irr(a,F(b))], F), sol)
          elseif d=n/k and m>1 then                        {case (b)}
              for all root b of h such that b ∈ K(a) do
                                              {see Sections 3 and 4}
                  sol := cons(endcons([h,b,Irr(a,F(b))], F), sol)
          elseif d<n/k and m≥n/(kd) then                   {case (c)}
              for all root b of h such that b ∈ K(a) do
                                              {see Sections 3 and 4}
                  g := Irr(a,F(b))
                  sol := append(sol,
                    Eqnfield2(g, endcons([h, b, g], F), k, i+1))
          else               {h does not define a subextension of index k}
       return sol
End.
```

Remark 2 When a field of index greater than **k** is found, another symmetric function is tried, over the field generated by the preceding ones. The following lemma ensures that if a field of index **k** in $K(a)$ exists, it is eventually found.

If one would try each new symmetric function with the same base field, instead of increasing it as in **Eqnfield**, the computations would probably be faster, but it may arise that none of the elementary symmetric function generates over K the subfield of index **k** to be computed.

Lemma 1 *If L is a subfield of index k in $K(a)$ and $a = \alpha_1, \ldots, \alpha_k$ are the conjugates of a over L, then L is generated over K by the k elementary symmetric functions in $\alpha_1, \ldots, \alpha_k$.*

Proof. The elementary symmetric functions of $a = \alpha_1, \ldots, \alpha_k$ are (up to sign) the coefficients of $\mathrm{Irr}(a, L)$. Thus they are in L. The field they generate is clearly of index k in $K(a)$, and is equal to L.   ◇

3   Testing inclusion between fields

Assume that a k-symmetric resolvent $s_*(p)$ has an irreducible factor h of degree $m < n = \deg(p)$. By a permutation of the roots $\alpha = \{\alpha_1, \ldots, \alpha_n\}$ of p, we may choose a root $a = \alpha_1$ of p and a root b of h such that $b = s(\alpha_1, \ldots, \alpha_k)$.

We have the following diagram of field extensions

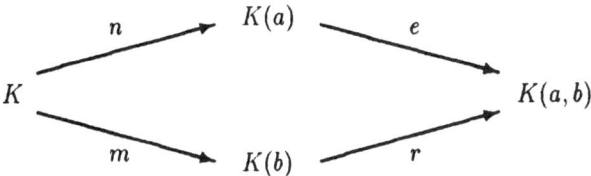

where the labels of the arrows represent the degrees of the extensions.

Lemma 2 *Let s be a symmetric function of arity $k < n$ and h an irreducible factor of the resolvent $s_*(p)$. If h is a simple factor of $s_*(p)$, then the degree r of the extension $K(a,b)/K(b)$ is less than or equal to k.*

Proof. Let σ be a K-automorphism of the splitting field of $K(a,b)$ and $\alpha'_1, \ldots, \alpha'_k$ be the images of $a = \alpha_1, \ldots, \alpha_k$ by σ. If b is fixed by σ we have $s(\alpha'_1, \ldots, \alpha'_k) = b = s(\alpha_1, \ldots, \alpha_k)$; since b is a simple root of $s_*(p)$ we have $\{\alpha'_1, \ldots, \alpha'_k\} = \{\alpha_1, \ldots, \alpha_k\}$. Thus the orbit of a by the automorphisms of the splitting field of $K(a,b)$ which fix b may have at most k values; since this orbit is the set of the conjugates of a over $K(b)$, the value k bounds the degree r of the extension $K(a,b)/K(b)$. $\Diamond$

Lemma 3 *Under the hypotheses of previous lemma, if the degree of the irreducible resolvent h is $m = n/k$, then $K(b)$ is a subfield of $K(a)$ and $r = k$.*

Proof. We have $ne = mr$, where e is the degree of the extension $K(a,b)/K(a)$. The previous lemma shows that $r \leq k$; thus $r \leq k = n/m = r/e$; hence $e = 1$ and $r = k$. $\Diamond$

Lemma 4 *Let p be an irreducible polynomial over $K[x]$ of degree n and a be a root of p. Let h be another irreducible polynomial over $K[x]$ of degree $m < n$.*

(i) If h has a linear factor $x - b$ in $K(a)[x]$ then $b \in K(a)$ and this factor gives an expression of b as a polynomial in a.

(ii) If p has an irreducible factor of degree n/m over $K[x]/(h)$ then h has a root b in $K(a)$ and the irreducible factor of p is $Irr(a, K(b))$, the minimal polynomial of a over $K(b)$.

Proof. Cases (i) and (ii) imply $e = 1$ in the previous diagram. $\Diamond$

Remark 3 *In case (i) (resp. (ii)) we can obtain $Irr(a, K(b))$ (resp. $Irr(b, K(a)))$ by using linear algebra. This will be discussed in the next section.*

Remark 4 *Dixon [5] tests such an inclusion by means of a factorization of a polynomial of degree nm over Q.*

4 Embedding subfields

At several steps of algorithm Eqnfield, we know of two extensions of K, given by the minimal polynomials (p and h) of some generators (a and b); we have seen in Section 3 how to test if the second extension may be embedded in the first one, and this gives the minimal polynomial $p_a = \mathrm{Irr}(a, K(b))$ or $p_b = \mathrm{Irr}(b, K(a))$.

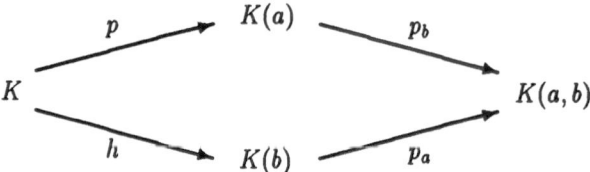

We show here how to compute one of these minimal polynomials from the other. This can not be done by a second factorization, not only because of its cost, but also because we would have to choose the good factor among several ones.

Thus, we proceed by linear algebra, in the following way. As the problem is symmetric in a and b, we only need to describe how to compute p_a from p_b.

More precisely, we want a polynomial of degree r in a, with coefficients in $K(b)$; this means that these coefficients are univariate polynomials in b of degree strictly less than m. We take the coefficients of these polynomials in b as unknowns, i.e. we introduce mr new indeterminates u_{mi+j} and write:

$$p_a(x) = x^r + \sum_{i=0}^{r-1} U_i(b)x^i \quad \text{where} \quad U_i(b) = \sum_{j=0}^{m-1} u_{mi+j} b^j.$$

The polynomial p_a has to vanish for $x = a$. Computing $p_a(a)$ in $K(a)(b)$, i.e reducing the powers of b by p_b and the powers of a by p, we get a polynomial which is linear in the u_i and has a degree less than n in a and less than e in b. The fact that a is a root of p_a implies that the $en = mr$ coefficients of the monomials in $a^i b^j$ are zero. This gives a square linear system; a solution of it gives a value for the u_i and, thus, for the coefficients of p_a. Therefore, the linear system has a unique solution.

Remark 5 *Suppose that we have $m < n$ and that we want to compute both p_a and p_b. We need one factorization over an algebraic extension. It is usually better to compute first p_a by factoring p over $K(b)$ and then p_b by linear algebra. In fact, the complexity of the available factorization routines depend more on the degree of the field extension than on the degree of the polynomial to factor. However, if p_b is factorized first, we only need the factors of degree 1, and this is probably faster if the factorization routine is customized in order to compute only these factors.*

Remark 6 *When p_a and p_b are known, it is easy to convert the representation of an element in $K(a)(b)$ (or in $K(a)$, if $e = 1$) to its representation in $K(b)(a)$ or conversely: it suffices to reduce by p_a and h or by p_b and p.*

5   Computing the symmetric resolvents

The most natural way for computing general resolvents consists of using symmetric functions, but, for avoiding an exponential increase of the representation of the expressions, it is necessary to contract them and to keep only one term by orbit.

In this section, we describe briefly an algorithm for computing a k-symmetric resolvent of a univariate polynomial p of degree n. It appears in [14] and is improved for the particular case studied here.

The degree of a k-symmetric resolvent is $\binom{n}{k}$; its coefficients may easily be deduced by Girard–Newton formulae from the *power functions* of the roots of the resolvent.

We recall that the i-th power function on a finite set A is the sum $\sum_{a \in A} a^i$. A *monomial form* on A is the sum of the orbit of a monomial under the action of S_n; more precisely, if I a k-uplet of positive integers (the exponent of the monomial) then $M_I(A) = \sum_{J \in S_n(I)} A^J$ is the corresponding monomial form. Generally we choose for I the *partition* (i.e., a decreasing sequence of integers) of the orbit. For example $M_{(2,1)}(x,y) = x^2 y + xy^2$, the elementary symmetric functions, e_i are $M_{(1,\ldots,1)}$ and the i-th power function, p_i, is $M_{(i,0,\ldots)}$. Any symmetric polynomial is a linear combination of monomial forms, and any monomial form may be expressed as a polynomial in the power functions or in the elementary symmetric functions. For example $M_{(2,1)} = e_1 e_2 - 3e_3 = p_2 p_1 - p_3$.

With these notations we can describe the computation of the resolvent. Let $s(x_1, \ldots, x_k)$ be a symmetric function used as transformation function for computing the resolvent. Since s is symmetric, we express it on the base of monomial forms $M_I(x_1, \ldots, x_k)$. Let r be an integer between 1 and the degree $\binom{n}{k}$ of the resolvent; using the product formula for symmetric functions [12], we can expand s^r on the same base, in order to obtain

$$s^r(x_1, \ldots, x_k) = \sum_{I \in E} c_I M_I(x_1, \ldots, x_k), \tag{1}$$

where E is some set of partitions, and the c_I are integers.

For computing the r-th power function of the roots of the resolvent, we have to compute the sum q of the elements of the orbit of $s^r(x_1, \ldots, x_k)$ under S_n (see definition 2). As s is symmetric, the expression of q is easily deduced from (1):

$$q(X) = \sum_{I \in E} \binom{n - lg(I)}{k - lg(I)} c_I M_I(x_1, \ldots, x_k), \tag{2}$$

where $\lg(I)$ denotes the *length* of the partition I (the number of non-zero parts of I).

From this, the r-th power function of the roots of $s_*(p)$ is obtained by specializing the variables of q as the roots of p, i.e. by expressing q or the monomial forms appearing in it as polynomials in the symmetric functions of the roots of p, which are (up to the sign) its coefficients.

Remark 7 *When the transformation function is the sum $x_1 + \cdots + x_k$, the polynomial q is easily obtained by the multinomial formula: in (2) the set E is the set of all partitions $I = (i_1, \ldots, i_k)$ of length at most k such that $i_1 + \cdots + i_k = r$ and the coefficient c_I of I is the multinomial coefficient $c_I = \binom{r}{i_1, \ldots, i_k}$. In this case one may also use the algorithm of Soicher [11] because s is linear.*

Remark 8 *Formula (2) avoids the computation of the action of the symmetric group. But an algorithm is needed for computing the expression of q in function of the coefficients of p without expanding monomial forms. Such an algorithm has been implemented in Macsyma [13].*

6   An illustrating example

As an example, we show in this section how algorithm `Eqnfield` works on the polynomial P of degree 12 given in the introduction. We call a one of its roots.

This polynomial is clearly reciprocal and $d := a + 1/a$ generates a subfield of index 2. This element d is the sum of two conjugates of a; thus its minimal polynomial may be obtained by factoring the resolvent by $x_1 + x_2$. But we already know a generator of the subfield; thus its minimal polynomial may more easily be obtained by computing the minimal linear relation between the powers of d in $Q(a)$. This minimal polynomial is

$$p(d) = d^6 + 9d^5 - 3d^4 - 118d^3 - 180d^2 - 3d + 43.$$

$Q(a)$ has a subfield of index 3, not contained in $Q(d)$, but, from now on, we restrict ourselves to the subfields of $Q(d)$. Since the degree of a subfield divides the degree of the field, the only possible values for k are 2 and 3. For $k = 3$ and $s = x_1 + x_2 + x_3$ we have:

$$\begin{aligned}
factor(s_*(p)) \;=\; & (x^2 + 9x + 15) \\
& (x^6 + 27x^5 + 204x^4 + 27x^3 - 3681x^2 - 4698x + 13581) \\
& (x^6 + 27x^5 + 246x^4 + 720x^3 - 972x^2 - 5454x + 2241) \\
& (x^6 + 27x^5 + 246x^4 + 846x^3 + 729x^2 - 1107x - 1161).
\end{aligned}$$

Thus, by case (a) of Remark 1, $Q(d)$ has exactly one subfield of degree 2 generated by, say, $b_0 := (\sqrt{21} - 9)/2$, such that $Irr(b_0, Q) = x^2 + 9x + 15$. This subfield has a simpler generator $b := (3 - \sqrt{21})/2 = -b_0 - 3$ with minimal polynomial

$h_1 := \mathrm{Irr}(b, Q) = x^2 - 3x - 3$. This minimal polynomial has smaller coefficients; we have preferred it on $x^2 - 5x + 1$ (which generates the same field) because it gives simpler results in the computations which follow.

Similarly, for $k = 2$ we obtain one subfield $F = Q(c_0)$ of index 2 in $Q(d)$, such that $\mathrm{Irr}(c_0, Q) = x^3 + 9x^2 + 6x - 43$; if we set $c_1 := c_0 - 3$ we obtain a new generator of F such that $\mathrm{Irr}(c_1, Q) = x^3 - 21x - 7$. A simpler generator is $c = (c_1 - 1)/3$ with minimal polynomial $h_2 := \mathrm{Irr}(c, Q) = c^3 + c^2 - 2c - 1$; it is better because the powers of c generate the ring of the integers of F and also because this generator leads to a very simple final result. Note that $c = \omega + 1/\omega$ where ω is a primitive 7-th root of unity.

Now, as 2 and 3 are relatively prime, we have $Q = Q(b) \cap Q(c)$; thus $Q(b, c)$ is of degree 6 and equal to $Q(d)$. We shall "compute" this equality by expressing d as a function of b and c, and also b and c as functions of d.

For this purpose we first compute the minimal polynomials of d over $Q(b)$ and $Q(c)$. These polynomials are respectively obtained by the factorizations:

$$
\begin{aligned}
factor(p, h_1) \;=\; & (d^3 + (b+3)d^2 + (-4b-3)d - 17b - 14) \\
& (d^3 + (-b+6)d^2 + (4b-15)d + 17b - 65), \\
factor(p, h_2) \;=\; & (d^2 + (-3c+2)d - 3c^2 - 3c + 1) \\
& (d^2 + (-3c^2 + 8)d + 3c - 2) \\
& (d^2 + (3c^2 + 3c - 1)d + 3c^2 - 8).
\end{aligned}
$$

We find several polynomials of the same degree which correspond to the fact that $Q(b)$ and $Q(c)$ are Galois extensions and have several generators with the same minimal polynomial. In both cases we choose the first factor, which is the simplest; let us call it h_3 and h_4, respectively.

This gives the following diagram of extensions, where the labels are the polynomials defining the extensions.

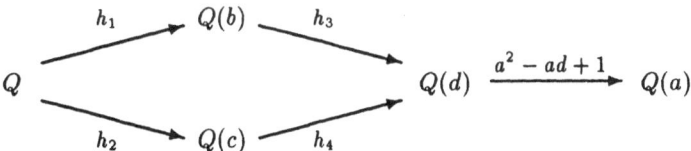

Now, having computed the minimal polynomial of each extension of this diagram, the algorithm of Section 4 may compute the expressions of b and c as polynomials of degree 5 in d or of degree 12 in a.

For computing the expression of d in term of b and c, we may reduce this expression of c by $h_3 = \mathrm{Irr}(d, Q(b))$ and $h_1 = \mathrm{Irr}(b, Q)$, to obtain a polynomial c_{val} in b and d. Reducing similarly the monomials $b^i c_{val}^j$ for $i = 0, 1$ and $j = 0, 1, 2$, we get the expression in $Q(d)$ of the elements of the canonical basis of $Q(b, c)$. The expression of d as a linear combination of these basis elements may be found by solving a linear system, as in Section 4. This gives the expression

$d = bc - 1$, which completes the simplification of a given in the introduction.

Remark 9 The only part of the above computation which is not completely algorithmic is the choice of the generators of $Q(b)$ and $Q(c)$.

7   Using known facts

It arises frequently that some information on the subfields or on the Galois group is directly available or easily detected. This is the case (as above) for reciprocal polynomials and for decomposable polynomials of the form $h(g(x))$, for example the even polynomials of the form $p(x^2)$.

Incidentally, the present paper may be viewed as a generalization of decomposition algorithms (see [8]) not only because any decomposition defines a subfield, but also because we compute a decomposition of the form $h(g(x)) \equiv 0 \bmod p(x)$.

For all of above cases we know a generator of a subfield, $a + 1/a$, $g(a)$ or a^2, where a is a root of the initial polynomial. When a generator of a subfield is known, there is no need to compute a resolvent nor to factorize: the minimal polynomial of the generator of the initial field over the intermediate one may be computed by linear algebra, with the method of Section 4.

Sometimes, an automorphism of the field $Q(a)$ is known. This is the case for reciprocals and even polynomials. When these arise, a symmetric function of the images of the field generator by the powers of this automorphism gives a generator of a subfield, the index of which is generally the order of the automorphism.

For example, in the case of a reciprocal polynomial, the automorphism is $x \rightarrow 1/x$, and $a + 1/a$ is the result of the symmetric function $x_1 + x_2$ applied to a and its image $1/a$; in this case, the symmetric function $x_1 x_2$ gives 1 which generates the trivial extension.

With even polynomials, the automorphism is $x \rightarrow -x$. Taking the sum as a symmetric function leads to a trivial result, but the symmetric function $-x_1 x_2$ gives a^2 which is usually taken for simplifying even polynomials.

For the above polynomial of degree 12, an automorphism of order 3 was known[3], which consists of replacing a by the second component b of the solution of cyclic 7th roots problem. The decomposition described in the introduction was first found by using this automorphism, without computing resolvents.

This possibility of using known automorphisms or known generators of subfields is in practice very important, because it avoids the computation of the resolvent and any factorization, which may be rather time consuming.

References

[1] Baeckelin, J. and Fröberg, R (1991). How we proved that there are exactly 924 cyclic 7–roots.*Proceedings of ISSAC'91* (S.W. Watt ed.). ACM Press (New–York), 103–111.

[2] Batu, C., Bernadi, D., Cohen, H. and Olivier, M. (1991). *User's Guide to PARI-GP*. Available by anonymous ftp from math.ucla.edu (128.97.4.254).

[3] Björck G. and Fröberg, R., (1989), A faster way to count the solutions of inhomogeneous systems of algebraic equations, with applications to cyclic n-roots, *Reports, Matematiska Institutionen, Stockholms Universitet, 1989- No 7.*

[4] Davenport, J.H. (1987). Looking at a set of equations. *Technical report 87-06*, University of Bath.

[5] Dixon, J.D. (1990). Computing subfields in algebraic number fields. *J. Austral. Math. Soc. (serie A)* **49**, 434-448.

[6] Faugère, J.C., Gianni, P., Lazard, D. and Mora, T. (1989). Efficient Computation of Zero–dimensional Gröbner Bases by Change of Ordering. *Submitted to J. Symb. Comp.* Technical Report LITP 89-52.

[7] Giusti, M., Lazard, D. and Valibouze, A. (1988). Algebraic transformations of polynomial equations, symmetric polynomials and elimination. *Symbolic and Algebraic Computation, International Symposium ISSAC '88* (P. Gianni, ed.), Lect. Notes in Comp. Sc. **358**, 309–314

[8] Kozen, D and Landau, S (1989). Polynomial decomposition algorithms. *J. Symb. Comp.* **7**, 445–456.

[9] Lagrange, J.L., (1770-1771) Réflexions sur la résolution algébrique des équations. *Nouveaux Mémoires de l'Académie royale des Sciences et Belles-Lettres de Berlin.*

[10] Lazard, D. (1992). Solving zero–dimensional algebraic systems. *J. Symbolic Computation* **13**, 117–131.

[11] Soicher, L. (1981). *The computation of the Galois groups.*. Thesis, Concordia University, Montreal (Quebec, Canada, 1981).

[12] Valibouze, A. (1987). Fonctions symétriques et changements de bases. *European Conference on Computer Algebra, Leipzig, GDR, 1987.* (J.H. Davenport, ed.) Lect. Notes in Comp. Sc. **378**.

[13] Valibouze, A. (1989). Symbolic computation with symmetric polynomials, an extension to Macsyma. *Computers and Mathematics (1989, MIT, Cambridge, Mass.).* Springer-Verlag.

[14] Valibouze, A. (1989). Résolvantes et fonctions symétriques. *Proc. of the ACM-SIGSAM 1989 Intern. Symp. on Symbolic and Algebraic Computation, ISSAC'89 (Portland, Oregon)*. ACM Press, 390-399.

[15] Zippel , Book to appear.

D. Lazard, A. Valibouze
LITP, Université P. et M. Curie,
4 place Jussieu,
F-75252 Paris Cedex 05
e-mail: dl@litp.ibp.fr, avb@litp.ibp.fr

Applications of the Eisenbud-Levine's theorem to real algebraic geometry

A. Lęcki   Z. Szafraniec

Introduction

Let $f : (R^n, 0) \to (R^p, 0)$ be the germ of an analytic mapping. The fibre $f^{-1}(0)$ is locally homeomorphic to a cone, with vertex 0. The base L of the cone is the intersection of $f^{-1}(0)$ with a small sphere S_ϵ centred at 0. Investigation of topology of L is one of the most crucial aims of singularity theory.

In this paper we shall present some recent results which express topological invariants of L in terms of algebraic invariants of appropriate local algebras. The first author has written a computer program which is able to calculate effectively these invariants. The program is based on the Eisenbud & Levine's theorem which gives an algebraic method for computing local topological degrees. We shall give examples of calculations.

The paper is organized as follows. In section 1 we introduce the notion of the diagram of initial exponents and we recall its properties. We also formulate the Eisenbud & Levine's theorem and we give a brief description of the computer program. In section 2 we show how to study the topology of L in case of isolated singularity, in section 3 we investigate homogeneous polynomials. In section 4 we shall explain how to compute the number of branches of a curve having an isolated singular point at 0. References include papers not mentioned in the text but important to the subject.

1   Algorithm for the local topological degree

Let $G = (G_1, ..., G_m) : (R^m, 0) \to (R^m, 0)$ be a polynomial mapping. If G has an isolated zero at the origin then $deg(G)$ will denote the topological degree of the mapping $x \to G(x)/\|G(x)\|$ from a small sphere centered at the origin to the unit sphere in R^m. In this section we shall briefly explain how to check whether $0 \in R^m$ is algebraically isolated in $G^{-1}(0)$ and how to compute $deg(G)$. The method is based on the Eisenbud & Levine's theorem (see [8]) which gives an algebraic formula for $deg(G)$:

Theorem 1.1
Let I be the ideal in $R[[x]] = R[[x_1, ..., x_m]]$ generated by $G_1, ..., G_m$. Let $J \in R[[x]]/I$ denote the residue class of the Jacobian of the map G. Assume that $dim R[[x]]/I$ is finite. Then

(i) $J \neq 0$,

(ii) if $\phi : R[[x]]/I \to R$ is a linear functional such that $\phi(J) > 0$ and B_ϕ is the symmetric bilinear form on the vector space $R[[x]]/I$ defined by $B_\phi(f, g) = \phi(fg)$ for $f, g \in R[[x]]/I$, then the form B_ϕ is non-singular and $deg(G) = signature(B_\phi)$. $\square$

Now we shall explain how to investigate the algebra $R[[x]]/I$. The diagram of initial exponents $N(I)$, introduced by Hironaka (cf. [2],[3],[12]), is a subset of N^m defined as follows: If $\beta = (\beta_1, ..., \beta_m) \in N^m$, put $|\beta| = \beta_1 + ... + \beta_m$. The lexicographic (from the right) ordering on $(m + 1)$-tuples $(\beta_1, ..., \beta_m, |\beta|)$ induces a total ordering on N^m. Let $f = \Sigma a_\beta x^\beta \in R[[x]]$, let $supp(f) = \{\beta : a_\beta \neq 0\}$, let $\nu(f)$ denote the smallest element of $supp(f)$, by $\alpha(f)$ we denote $a_\beta x^\beta$, where $\beta = \nu(f)$. Then $N(I)$ is defined as $\{\nu(f) : f \in I, f \neq 0\}$. Clearly, $N(I) + N^m = N(I)$. It follows that there is a smallest finite subset $V(N(I))$ of $N(I)$ such that $N(I) = V(N(I)) + N^m$. Let $\#V(N(I)) = s$. Then any $g_1, ..., g_s \in I$ such that $V(N(I)) = \{\nu(g_1), ..., \nu(g_s)\}$ are called vertices of the diagram. We associate to vertices $g_1, ..., g_s$ the following decomposition of N^m : Set $\Delta_1 = \nu(g_1) + N^m, .., \Delta_i = (\nu(g_i) + N^m) - \Delta_1 \cup ... \cup \Delta_{i-1}, i = 1, ..., s$. Put $\Delta = N^m - \Delta_1 \cup .. \cup \Delta_s = N^m - N(I)$.

Theorem 1.2 (Grauert-Hironaka formal division algorithm [2],[3],[12])

For every $f \in R[[x]]$, there exist unique $r, h_1, ..., h_s \in R[[x]]$ such that $\nu(g_i) + supp(h_i) \subset \Delta_i$ for $i = 1, ..., s, supp(r) \subset \Delta$, and $f = r + h_1 g_1 + ... + h_s g_s$. $\square$

Corollary 1.3
$dim R[[x]]/I = \#\Delta$. In particular, if $\#\Delta$ is finite then

(i) $0 \in R^m$ is algebraically isolated in $G^{-1}(0)$,

(ii) monomials $x^\beta, \beta \in \Delta$, form a basis in $R[[x]]/I$. $\square$

To formulate the algorithm we need some more definitions.

If $f, g \in R[x]$ then $S(f, g) = (f\alpha(g) - g\alpha(f))/g.c.d.(\alpha(f), \alpha(g))$, where $g.c.d.(x^\alpha, x^\beta) = x^\gamma$ and $\gamma_i = min(\alpha_i, \beta_i)$, $i = 1, .., m$. Let $mindeg(f) = min\{|\beta| : \beta \in supp(f)\}$ and $maxdeg(f) = max\{|\beta| : \beta \in supp(f)\}$. Let $E(f) = maxdeg(f) - mindeg(f)$. If T is a finite subset of $R[x]$ and $f \in R[x]$ such that $\nu(f) \in \nu(T)$ then by $ch(f, T)$ we denote an arbitrary polynomial $h \in T$ satisfying: $\nu(h)$ divides $\nu(f)$ and for any $g \in T$ if $\nu(g)$ divides $\nu(f)$ then $E(h) \le E(g)$.

Our algorithm is divided into two parts:

PART 1: *Computing vertices of the diagram.*

The algorithm for computing vertices of the diagram, so called "the tangent cone algorithm", is analogous to the algorithm presented in [14] [15]. Roughly speaking it looks:

(A) $C := \{G_1, ..., G_m\}$ $R := \emptyset$ $V := \emptyset$

(B) $f := minC$ (i.e. $\nu(f) \le \nu(g)$ for all $g \in C$)

(C) if $\nu(f) \in \nu(V)$

        then

(C1) $C := C - \{f\} \cup \{S(f, ch(f, R))\}$

(C2)   if $E(f) < E(ch(f, R))$ then $R := R \cup \{f\}$

        else

(C3) $C := C - \{f\} \cup \{S(f, v) : v \in V\}$

(C4) $V := V \cup \{f\}$

(C5) $R := R \cup \{f\}$

        end if

(D) if $C \neq \emptyset$ then goto (B)

(E) END

Using the method presented in [5],[14] we can reduce the number of reduction in the point (C3). It is enough to add to the set C only $\{S(f, v') : v' \in V'\}$, where V' is a subset of V such that for any $v \in V$ there is $v' \in V'$ such that $l.c.m.(\nu(f), \nu(v'))$ divides $l.c.m.(\nu(f), \nu(v))$.

The above algorithm always stops in the point (E) after finite number of loops. Then the family V is the family of vertices of the diagram generated by $G_1, .., G_m$. If Δ is finite then we can continue computation in part 2.

PART 2: *Computing a form and its signature.*

Let E denote the linear space spanned by monomials $x^\beta, \beta \in \Delta$.

Step 1. Using the Grauert-Hironaka formal division algorithm express the residue classes of J and of all monomials $x^\alpha x^\beta$, where $\alpha, \beta \in \Delta$, as elements of E (We shall use the same symbols to denote these elements).

Step 2. Fix the linear form $\phi : E \longrightarrow R$; $\phi(\alpha(J)) = 1$; $\phi(x^\beta) = 0$ for $\beta \in \Delta$ and $\beta \neq \nu(J)$.

Step 3. Put $a_{\alpha,\beta} = \phi(x^\alpha x^\beta)$ for $\alpha, \beta \in \Delta$.

Step 4. $deg(G) := signature[a_{\alpha,\beta}]$.

2 Isolated singularity

Let $f : (R^n, 0) \longrightarrow (R, 0)$ be the germ of an analytic function, let L be the intersection of $f^{-1}(0)$ with a small sphere centered at the origin, let $\chi(L)$ denote the Euler characteristic of L. From [1],[16] $\chi(L)$, and consequently $dim_{Z_2}(H_*(L, Z_2))$, is even.

In this section we shall assume that f has an isolated critical point at 0. In this case L is a smooth compact orientable $(n-2)$-manifold. Let $\nabla f = (\partial f/\partial x_1, ..., \partial f/\partial x_n) : (R^n, 0) \longrightarrow (R^n, 0)$ denote the gradient of f. Hence 0 is isolated in $(\nabla f)^{-1}(0)$.

Clearly, if n is odd then $\chi(L) = 0$. Arnold and Wall (see [22]) have proved

Theorem 2.1

$\chi(L) = 2(1 - deg(\nabla f))$, if n is even. $\square$

Example

Let $f(x, y, z, w) = x^6 + x^2 w^3 - 2xyz^2 w + yz^5 + 3y^3 z + w^7$. The computer has calculated that $deg(\nabla f) = -2$, and then $\chi(L) = 6$.

Let $\mu = dim R[[x]]/(\partial f/\partial x_1, ..., \partial f/\partial x_n)$ be the Milnor number of f.

Wall [22] has proved

Theorem 2.2

If f has an algebraically isolated singularity, i.e. μ is finite, then

$1/2 \, dim_{Z_2}(H_*(L, Z_2)) \equiv \mu + 1 \quad \mod 2.$ □

Example

Let $f(x, y, z) = x^{10} + y^8 + z^9 - 2x^8yz + 3xy^5z^3 - 5y^5z^5 + x^6y^6$. The computer has calculated that $\mu = 504$, and then L consists of an odd number of circles.

3   Homogeneous polynomials

Let $f : R^n \to R$ be a homogeneous polynomial of degree $d \geq 3$. It is proper to add that in this section we shall not assume that f has an isolated singularity. Set either $k = d + 2$ if d is even or $k = d + 1$ if d is odd. Put $g_1 = f - (x_1^k + ... + x_n^k)/k$, $g_2 = -f - (x_1^k + ... + x_n^k)/k$. We have (see [19],[20])

Theorem 3.1

Functions g_1, g_2 have algebraically isolated singularity at 0. Moreover,
$\chi(L) = 2 - (deg(\nabla g_1) + deg(\nabla g_2) + \chi(S^{n-1}))$
$\chi(L)/2 \equiv 1 + \mu + \chi(S^{n-1})/2 \qquad \mod 2 \quad$ if d is odd,
$\chi(L)/2 \equiv 1 + (\mu + (-1)^{(d/2-1)n})/2 \quad \mod 2 \quad$ if d is even,
where μ is the Milnor number of g_1. □

Example

Let $f(x, y, z) = 3x^2z^3 + 2xyz^3 + 2y^2z^3 - x^3y^2 + x^5 - 4y^5$. The computer has calculated that $\mu = 65$, $deg(\nabla g_1) = deg(\nabla g_2) = -1$. So $\chi(L) = 2$.

Let $L_{reg} = \{x \in L : L$ is a topological $(n - 2)$-manifold in some neighbourhood of $x\}$, and let $L_{sing} = L - L_{reg}$. Clearly, the number $\#L_{sing}$ is an important topological invariant of L. If the restricted function $f|S^{n-1}$ is Morse then $\#L_{sing}$ is finite.

Let $F_i : (R \times R^n, 0) \to (R, 0)$ be the polynomial given by
$F_i(\lambda, x) = \lambda x_i + \partial f/\partial x_i, \quad i = 1, ..., n$.
Define polynomials $\Psi_i, \Delta_i : (R \times R^n, 0) \to (R, 0), \quad i = 1, 2, 3 :$
$\Psi_1 = (\lambda^2 + x_1^2 + ... + x_n^2)/2,$
$\Psi_i = \lambda + (-1)^i (x_1^{k-2} + ... + x_n^{k-2})/(k-2), \quad i = 2, 3,$
$\Delta_i = Jacobian(\Psi_i, F_1, ..., F_n), \quad i = 1, 2, 3.$
Set $H_i = (\Delta_i, F_1, ..., F_n) : (R \times R^n, 0) \to (R \times R^n, 0)$. We have (see [21])

Theorem 3.2

$f|S^{n-1}$ is Morse if and only if $0 \in R \times R^n$ is isolated in $H_1^{-1}(0)$. If $f|S^{n-1}$ is Morse then $0 \in R \times R^n$ is isolated in $H_i^{-1}(0)$, $i = 2, 3$, and
$\#L_{sing} = deg(H_2) - deg(H_3)$ for even d,
$\#L_{sing} = 2deg(H_2)$ for odd d. $\square$

Example

Let $f(x, y, z) = 3x^2 z + 2xyz + y^2 z - x^3 - 2y^3 - xy^2$. Then $d = 3$ and $k = 4$.

The computer has calculated that 0 is isolated in $H_1^{-1}(0)$, and then $f|S^{n-1}$ is a Morse function. The computer has also calculated that $deg(H_2) = +1$. So L_{sing} consists of two points.

4   Curves having an isolated singular point

Let $f = (f_1, ..., f_{n-1}) : (R^n, 0) \to (R^{n-1}, 0)$ be an analytic mapping. We shall say that $f^{-1}(0)$ has an isolated singular point at 0 if 0 is isolated in $\{x \in f^{-1}(0) : rank[Df(x)] < n - 1\}$. Let $\Omega = (x_1^2 + ... + x_n^2)/2$, let $\Delta = Jacobian(\Omega, f_1, ..., f_{n-1})$, and let $H = (\Delta, f_1, ..., f_{n-1}) : (R^n, 0) \to (R^n, 0)$. We have (see [21])

Theorem 4.1

$f^{-1}(0)$ has an isolated singular point at 0 if and only if 0 is isolated in $H^{-1}(0)$. $\square$

Clearly, if $f^{-1}(0)$ has an isolated singular point at 0 then $dim f^{-1}(0) \leq 1$, and then L is finite. Fukuda et al.[9],[10],[11] have proved

Theorem 4.2

If $f^{-1}(0)$ has an isolated singular point at 0 then $\#L = 2deg(H)$. $\square$

Example

Let $f_1 = x^4 + 3xyz^3 - 2yz^4$, $f_2 = y^3 - x^2z^2 + y^3z^2 + z^5$. The computer has calculated that $deg(H) = +3$, so L consists of six points.

References

[1] R.Benedetti, J.-J.Risler, Real algebraic and semi-algebraic sets, Hermann, Paris (1990).

[2] E. Bierstone, P.D. Milman, Relations among analytic functions, Ann.Inst. Fourier 37 (1987), 187-239.

[3] J. Briançon, Weierstrass préparé à la Hironaka, Asterisque 7-8 (1973), 67-73.

[4] J.W. Bruce, Euler characteristic of real varieties, Bull. London Math.Soc. 22 (1990), 547-552.

[5] B. Buchberger, A criterion for detecting unnecessary reduction in the construction of Gröbner bases, Proc. EUROCAM '79' , Lecture Notes in Compt. Sciences 72 (1979), 3-21.

[6] F. Cucker, L.M. Pardo, M. Raimondo, T. Recio, M.-F. Roy, On the computation of the local and global analytic branches of a real algebraic curve, Lect. Notes in Compt. Sci. 356 , Springer, Berlin-New York (1989).

[7] J.Damon, On the number of branches for real and complex weighted homogeneous curve singularities, Topology 30 (1991), 223-229.

[8] D. Eisenbud, H.I. Levine, An algebraic formula for the degree of a C^∞ - map germ, Ann. of Math. 106 (1977) 19-44.

[9] T. Fukuda, K. Aoki, W.Z. Sun, On the number of branches of a plane curve germ, Kodai Math. Journal 9 (1986), 178-187.

[10] T. Fukuda, K. Aoki, T. Nishimura, On the number of branches of the zero locus of a map germ $(R^n, 0) \rightarrow (R^{n-1}, 0)$, Topology and Computer Science : Proceedings of the Symposium held in honor of S.Kinoshita, H.Noguchi and T.Homma on the occasion of their sixtieth birthdays (1987), 347-363.

[11] T. Fukuda, K. Aoki, T.Nishimura, An algebraic formula for the topological types of one parameter bifurcations diagrams, Archive for Rational Mechanics and Analysis 108 (1989), 247-265 .

[12] A. Galligo, Théorème de division et stabilité en géométrie analytique locale, Ann. Inst. Fourier 29 (1979), 107-184.

[13] J. Montaldi, D. van Straten, One-forms on singular curves and the topology of real curve singularities, Topology 29 (1990), 501-510.

[14] T. Mora, An algorithm to compute the equation of tangent cones, Proc. EUROCAM '82' , Lect. Notes in Computer Sciences 144 (1982), 158-165.

[15] G. Pfister, The tangent cone algorithm and some applications to local algebraic geometry, in "Effective Methods in Algebraic Geometry" , ed. T. Mora, C. Traverso, Boston (1991), 401-409.

[16] D.Sullivan, Combinatorial invariants of algebraic spaces, Proc. of Liverpool Sing. Symposium I , Lecture Notes 192, Springer Verlag (1971)

[17] Z. Szafraniec, On the number of branches of an 1-dimensional semianalytic set, Kodai Math. Journal 11 (1988), 78-85.

[18] Z. Szafraniec, The Euler characteristic of algebraic complete intersections, J. reine angew. Math. 397 (1989), 194-201.

[19] Z. Szafraniec, On the Euler characteristic mod 2 of real projective hypersurfaces, Bull. Polish Acad. Sci. 37 (1989), 103-107.

[20] Z. Szafraniec, Topological invariants of weighted homogeneous polynomials, Glasgow Math. J. 33 (1991), 241-245.

[21] Z. Szafraniec, On the number of singular points of a real projective hypersurface, Math. Annalen 291 (1991), 487-496.

[22] C.T.C. Wall, Topological invariance of the Milnor number mod 2 , Topology 22 (1983), 345-350.

Andrzej Lęcki, Zbigniew Szafraniec
Uniwersytet Gdanski
Instytut Matematyki
ul. Wita Stowsza 57
80–952 GDANSK, Poland

Applications of Algebraic Geometry to Computer Vision

S.J. Maybank

1 Introduction

There is an increasing interest in applications of algebraic geometry to computer vision. There are at least three possible reasons for this: i) certain vision problems naturally involve polynomial equations; ii) with the increase in available computing power it is easier to implement algorithms which stay close to the geometry underlying vision; and iii) algebraic geometry may in future provide methods for assessing the stability of algorithms against small perturbations in the data.

Applications of algebraic geometry to reconstruction, camera calibration and model based vision are discussed. The book [8] is a good general introduction to computer vision.

2 Reconstruction

Let two or more images of the same scene be taken from different viewpoints. Two features in different images are said to correspond if they are the projections of the same feature in the scene. If enough pairs of corresponding features are given then the shape of the scene and the relative positions of the cameras can be calculated. The most commonly used features are points [12,17,20,21]. More recently reconstruction methods have been developed for straight lines [11,19,22] and for general image curves [3]. Reconstruction from images always involves an unknown scale factor. There may be additional ambiguities depending on the available data [1,5,7,13,14,17].

The map from space to the image is assumed to be projective linear. The unique point at which the map is not defined is called the optical centre of the camera. If the full calibration (interior orientation) of each camera is known then the images can be rectified to yield the images that would have been obtained by projection onto a unit sphere. If the full calibration is unknown then reconstruction is still possible but more data is required and the ambiguities in the reconstruction are more numerous [17]. Quantities associated with the second or third images are often but not always indicated by superscripts ' or

" attached to the corresponding quantities defined for the first image. It is assumed that each image is taken by a different camera. If the camera calibration is known then an equivalent assumption is that the images are taken by a single moving camera.

2.1 Equations underlying reconstruction with known camera calibration

It is assumed that the images are obtained by projection onto the unit sphere. The projection of a point x in space to the surface of a sphere centred at the origin is $x \mapsto x/\|x\|$. Let x project to q (q') in the first (second) image. The coordinates of q and q' are measured in coordinate frames attached to each camera. The displacement of the second camera from the first consists of a translation a followed by a rotation R^{T} in SO(3). Let the coordinates of x in the first camera coordinate frame be pq. It follows that $p'q' = R(pq - a)$. The distances p and p' are eliminated to give

$$q'^{\mathsf{T}} R(q \times a) = 0 \tag{1}$$

Let T_a be the antisymmetric matrix defined such that $T_a q = q \times a$ and let E be the so called essential matrix defined by $E = RT_a$. Then (1) takes the form

$$q'^{\mathsf{T}} Eq = 0 \tag{2}$$

Any non-zero 3×3 matrix which is the product of an orthogonal matrix and an antisymmetric matrix is by definition an essential matrix. Reconstruction from image point correspondences can be based on (2), as shown in [12]. If $q_i \leftrightarrow q'_i$ for $1 \leq i \leq n$ is a set of n given image correspondences then a 3×3 matrix E is found such that $|q'^{\mathsf{T}}_i Eq_i|$ is small for $1 \leq i \leq n$. The rotation R and the translation a are then recovered from the decomposition $E = RT_a$. In practice there are severe difficulties due to instability [21]. Small perturbations in the data $q_i \leftrightarrow q'_i$ cause large perturbations in the estimate of the displacement $\{R, a\}$.

2.2 Properties of essential matrices

In this subsection E is a non-zero 3×3 matrix.

Let U and V be orthogonal matrices and let λ be a non-zero scalar. Then E is an essential matrix if and only if one of λE, E^{T} and UEV is an essential matrix [17]. Let E be a real essential matrix with symmetric part S and antisymmetric part T_c, ie. $E = S + T_c$. Then the eigenvalues λ_i of S satisfy $\lambda_1 + \lambda_2 = \lambda_3$. If e_3 is the eigenvector of S corresponding to λ_3 then $e_3.c = 0$.

Let E be a real matrix and let σ_1, σ_2, σ_3 be the singular values of E. It is shown in [9] that E is an essential matrix if and only if $\sigma_1 = \sigma_2 > 0$ and $\sigma_3 = 0$.

A rank two matrix E is an essential matrix if and only if [2]

$$EE^T E = \frac{1}{2} \text{trace}(EE^T)E \qquad (3)$$

Let $E = [e_1|e_2|e_3]$ be a real matrix, and let $e_i.e_j \neq 0$ for all i, j. Then E is an essential matrix if and only if [9]

$$(e_1.e_2)^{-1}e_3 + (e_2.e_3)^{-1}e_1 + (e_3.e_1)^{-1}e_2 = 0$$

Each matrix E defines a point in $\mathbf{P}^8$ obtained by writing the entries of E as a single vector. The essential matrices comprise a rational algebraic variety $\mathcal{M}$ in $\mathbf{P}^8$ defined by the polynomial equations (3). The dimension of $\mathcal{M}$ is five, and the degree of $\mathcal{M}$ is ten. The degree of $\mathcal{M}$ is by definition the number of points contained in the intersection of $\mathcal{M}$ with a general linear subspace of $\mathbf{P}^8$ of codimension five. The variety $SO(3) \times \mathbf{C}^3$ is an étale cover of $\mathcal{M}$ of degree two under the map $(R, a) \mapsto RT_a$. Let the image correspondences $q_i \leftrightarrow q_i'$ be given for $1 \leq i \leq 5$. Then there in general exactly ten essential matrices E such that

$$q_i'^T E q_i = 0 \qquad (1 \leq i \leq 5) \qquad (4)$$

There exist sets of five image point correspondences for which all ten essential matrices are real. The figure ten is obtained by showing that the number of solutions to (4) is equal to the degree of $\mathcal{M}$ [2,17].

2.3 Reconstruction from image line correspondences

The correspondences between two images of a set of lines do not place any constraints on the relative position of the cameras. For reconstruction three images are required [11,19,22]. Let the displacements of the second and third cameras relative to the first camera be $\{R, a\}$ and $\{S, b\}$ respectively. Let l be a line in space projecting to lines with normals n, n' and n'' in the three images. It is assumed that $l.m = 0$. Let q' (q'') be a general point on the second (third) image line. It follows that

$$p'q' = R(l + tm - a) \qquad\qquad p''q'' = S(l + tm - b) \qquad (5)$$

It follows from the definitions of n' and n'' that $n'.q' = n''.q'' = 0$. The equations (5) thus yield

$$n'^T R(l + tm - a) = 0 \qquad\qquad n''^T S(l + tm - b) = 0 \qquad (6)$$

The equations (6) are true for all t thus

$$
\begin{aligned}
n'^T Rm &= 0 & n'^T R(l - a) &= 0 \\
n''^T Sm &= 0 & n''^T S(l - b) &= 0
\end{aligned}
\qquad (7)
$$

The vector m is parallel to $R^\mathsf{T} n' \times S^\mathsf{T} n''$ and normal to n thus

$$(R^\mathsf{T} n' \times S^\mathsf{T} n'').n = 0 \tag{8}$$

The vector l is parallel to $n \times m$ or equivalently parallel to $n \times (R^\mathsf{T} n' \times S^\mathsf{T} n'')$. The equation

$$(n'^\mathsf{T} Rl)(n''^\mathsf{T} Sb) - (n''^\mathsf{T} Sl)(n'^\mathsf{T} Ra) = 0$$

obtained from (7) yields the following equation on substituting $n \times (R^\mathsf{T} n' \times S^\mathsf{T} n'')$ for l,

$$(n''^\mathsf{T} Sb)[(n^\mathsf{T} S^\mathsf{T} n'')(n'.n') - (n^\mathsf{T} R^\mathsf{T} n')(n'^\mathsf{T} RS^\mathsf{T} n'')]$$
$$- (n'^\mathsf{T} Ra)[(n^\mathsf{T} S^\mathsf{T} n'')(n'^\mathsf{T} RS^\mathsf{T} n'') - (n^\mathsf{T} R^\mathsf{T} n')(n''.n'')] = 0 \tag{9}$$

If six triples of corresponding lines are given then the six equations (8) can in principle be solved for R and S. There remain six linear equations in a, b, of which (9) is the prototype. The equations can be solved for a, b provided the 18 image lines satisfy a single algebraic constraint.

2.4 Reconstruction in the absence of camera calibration

Reconstruction is still possible if the full camera calibration is unknown. Reconstruction without the full calibration tends to be simpler mathematically, but more data are required and there is in general a wider range of unavoidable ambiguity in the reconstruction.

The epipolar transformation is the fundamental geometric construction underlying reconstruction from image point correspondences [8]. Let two images of the same scene be obtained by cameras with optical centres at the points o and a respectively. The epipole p in the first image is the projection of the line $\langle a, b \rangle$ to the first image, and the epipole p' in the second image is the projection of $\langle a, b \rangle$ to the second image. The image lines containing p (or p') are the epipolar lines. Let Π be a plane in the pencil with axis $\langle o, a \rangle$. Then Π projects to a line l in the first image and to a line l' in the second image. The projection $\Pi \mapsto l$ establishes a homography between the pencil of planes and the epipolar lines in the first image. Similarly there is a homography between the pencil of planes and the epipolar lines in the second image. The two homographies $l \overline{\wedge} \Pi \overline{\wedge} l'$ yield the epipolar transformation $l \overline{\wedge} l'$. Two reconstructions are regarded as identical if they are based on the same epipolar transformation, even if the reconstructions do not give the same set of points in $\mathbf{P}^3$.

If $q_i \leftrightarrow q_i'$ for $1 \leq i \leq n$ is a set of corresponding image points then for reconstruction it is sufficient to find p, p' such that

$$\langle p, q_i \rangle \overline{\wedge} \langle p', q_i' \rangle \qquad (1 \leq i \leq n) \tag{10}$$

(If the full camera calibration is known then there is an additional constraint: if $l_1 \overline{\wedge} l_1'$ and $l_2 \overline{\wedge} l_2'$ then the angle between l_1, l_2 is equal to the angle between

l_2, l_2'.) If $n = 7$ in the uncalibrated case then there are in general exactly three pairs p, p' such that (10) holds. If $n = 6$ then p lies on a cubic plane curve c [20] and the correspondence $p \leftrightarrow p'$ establishes a birational equivalence between c and c'.

The equations (10) have an equivalent formulation

$$q_i'^\mathsf{T} F q_i = 0 \qquad\qquad (1 \le i \le n) \qquad\qquad (11)$$

where F is a non-zero 3×3 matrix. The epipoles are obtained by solving (11) for F subject to the condition $\det(F) = 0$. The epipoles p and p' are given by $Fp = 0$ and $p'^\mathsf{T} F = 0$. The matrix F is analogous to the essential matrix. The analogue of the variety $\mathcal{M}$ of essential matrices is the variety $\mathcal{D}$ in $\mathbf{P}^8$ of 3×3 matrices with determinant zero.

2.5   Ambiguity

For certain special configurations of image data the reconstructions compatible with the data show a wider range of ambiguity than is found in the general case. The term 'ambiguous reconstruction' is reserved for these special cases. In reconstruction from image point correspondences ambiguity is only possible if the points in space lie on a critical surface [7,13,14,17,20]. In reconstruction from image line correspondences ambiguity is only possible if the lines in space belong to a critical line congruence [1].

2.5.1   Critical surfaces

It is assumed that the camera calibration is known. Let o be the optical centre of the first camera and let a, b be the two possible optical centres for the second camera. The reconstruction is ambiguous only if the points x in space lie a quadric surface containing o, a and b. The quadric may have a singular point or it may split into a plane pair. Let ψ be the reconstructed quadric when the second camera is at a. The quadric ψ is subject to two cubic polynomial constraints which arise from the camera calibration. Let coordinates be chosen with the origin at o and let the equation of ψ be $x^\mathsf{T} M x + l.x = 0$. The first constraint is

$$\det\left(M - \frac{1}{2}\mathrm{trace}(M)I\right) = 0 \qquad\qquad (12)$$

where I is the 3×3 identity matrix. If (12) is satisfied then there exist vectors m, n such that

$$M = \frac{1}{2}(m \otimes n + n \otimes m) - (m.n)I$$

The vectors m, n are associated with the points $(m, 0)$, $(n, 0)$ in the plane at infinity. Let τ_ψ be the unique rigid involution of ψ that fixes $(m, 0)$ and $(n, 0)$. Then the second constraint on ψ is obtained from the condition $\tau_\psi(a).l = 0$.

Let ϕ be the critical surface obtained when the second camera is at b. Then $\psi \cap \phi = c \cup l$ where c is a horopter curve and l is the line through o which cuts c twice. If ψ is fixed then b and ϕ can be chosen such that $\tau_\psi(c) = c$ and $\tau_\psi(a) = b$. A horopter curve is a twisted cubic which cuts the plane at infinity at three points n, i_n, j_n such that i_n, j_n are the points of contact of the tangents drawn from n to the absolute conic.

2.5.2   Critical line congruences

Reconstruction from triples of corresponding lines is ambiguous if and only if the lines in space belong to a critical line congruence Ψ. The lines in $\mathbf{P}^3$ are parameterised by the points of the Klein quadric hypersurface Ω_4 in $\mathbf{P}^5$. A line congruence is by definition an algebraic variety of dimension two contained in Ω_4. The order of Ψ is the number of lines of Ψ through a general point of $\mathbf{P}^3$. The degree of Ψ is the number of lines of Ψ contained in a general plane of $\mathbf{P}^3$. Let l be a general line of $\mathbf{P}^3$. The rank of Ψ is the number of points x of l at which l is coplanar with two of the lines of Ψ passing through x. The sectional genus of Ψ is the genus of the ruled surface formed by the lines of Ψ which intersect l. It is shown in [1] that a critical line congruence Ψ has order three, degree three, rank five and sectional genus five.

3   Camera Calibration

The camera calibration allows the calculation of the angles between the different rays entering a camera. In order to calculate an angle it is sufficient to know the calibration and the coordinates of the image points to which the rays project. There are several different ways of describing the calibration. For example, if the camera is at a known position and orientation in $\mathbf{P}^3$ then the calibration is determined by the 3×4 matrix M giving the projection from $\mathbf{P}^3$ to the image plane. The absolute conic is used to specify the camera calibration in a way which is independent of the position and orientation of the camera [5,17]. The absolute conic Ω is the unique conic invariant under the rigid motions of $\mathbf{P}^3$. It is contained in every sphere in $\mathbf{P}^3$. The image w of Ω is independent of the position and orientation of the camera; it depends only on the camera calibration. Conversely, if w is given then the camera calibration is determined. To show that w determines the camera calibration, let h_1, h_2 be two rays entering the camera and let q_1, q_2 be the projections of h_1 and h_2 to the image. Let h_1 and h_2 intersect the plane at infinity at the points p_1, p_2 and let the line $\langle p_1, p_2 \rangle$ intersect Ω at i, j. The angle θ between h_1 and h_2 is given by the Laguerre formula

$$\theta = \frac{1}{2i} \log(\{p_1, p_2; i, j\}) \tag{13}$$

Let i_p, j_p be the points in the image at which $\langle q_1, q_2 \rangle$ cuts w. The cross ratio on the right-hand side of (13) is invariant under projection to the image thus

$$\{p_1, p_2; i, j\} = \{q_1, q_2; i_p, j_p\}.$$

3.1 The Kruppa equations

The Kruppa equations link the camera calibration to the epipolar transformation associated with a displacement of the camera. Let each pencil of epipolar lines be parameterised by the points of the line $z = 0$. A general epipolar line l_t in the first image has an equation $p \times (1, t, 0)$ and the corresponding epipolar line $l'_{t'}$ in the second image has an equation $p' \times (1, t', 0)$. The coordinate t' is related to t by a bilinear transformation, $t' = (at + b)/(ct + d)$. The line l_t is tangent to w if and only if $l'_{t'}$ is tangent to w. Let D be the matrix of the dual conic of w. Then l_t and $l'_{t'}$ are tangent to w if and only if the following two equations are satisfied,

$$
\begin{aligned}
(p \times (1, t, 0))^\top D(p \times (1, t, 0)) &= 0 \\
(p' \times (1, t', 0))^\top D(p' \times (1, t', 0)) &= 0
\end{aligned}
\tag{14}
$$

On expanding the equations of (14), substituting $(at+b)/(ct+d)$ for t and then clearing fractions two quadratic equations in t are obtained,

$$A_2 t^2 + A_1 t + A_0 = 0 \qquad\qquad A'_2 t^2 + A'_1 t + A'_0 = 0 \tag{15}$$

The coefficients A_i, A'_i are homogeneous linear polynomials in the entries of D. The roots of the first equation of (15) coincide with the roots of the second equation. It follows that

$$A_1 A'_2 - A_2 A'_1 = 0 \qquad A_2 A'_3 - A_3 A'_2 = 0 \qquad A_3 A'_1 - A_1 A'_3 = 0 \tag{16}$$

The Kruppa equations (16) are homogeneous of degree two in the entries of D. Only two of the equations are independent. The camera calibration is described by five independent parameters. The camera calibrations compatible with the epipolar transformations obtained from two independent camera motions are parameterised by an algebraic curve of genus four. In principle the camera calibration is determined uniquely by the epipolar transformations obtained from three camera motions [16].

Kruppa used (16) to show that if the camera calibration is known then there are at most 11 reconstructions compatible with five image correspondences [10]. Each reconstruction can be obtained from one of the intersections of two plane curves of degree six. Kruppa showed that 25 of the intersections are parasitic, in that they do not yield reconstructions. In fact one of the remaining eleven intersections is also parasitic [5].

4 Model Based Vision

Let x_i for $1 \leq i \leq n$ be set of points at known relative positions in space and let q_i for $1 \leq i \leq n$ be a set of image points. The q_i are recognised as a projection of the x_i if there is a camera position and orientation relative to the x_i such that x_i projects to q_i for $1 \leq i \leq n$. The x_i are usually easily identified points on an object of known shape, for example the nose and wing tips of an aircraft. The method can be extended to include more general features such as straight lines and curves.

The results of the mathematical theory of invariants suggest that model based object recognition is possible without first estimating the position and orientation of the camera relative to the object. Let I be a function of the x_i invariant under projection to the image. For example I could be the cross ratio of four collinear points. The value $I(q_i)$ is computed from the image data and compared with the value $I(x_i)$ computed from the model. If $I(q_i) \neq I(x_i)$ then the image points q_i are not the projections of the points x_i. Current applications of invariant theory to model based vision are described in a book due to appear soon [18].

4.1   Algebraic invariants

The cross ratio of four collinear points is invariant under projection to an image. A set of n coplanar points possesses $2(n - 4)$ independent invariants under projection. Four points on a non singular conic define a cross ratio and hence a j-invariant. Two coplanar conics c_1 and c_2 have exactly two independent invariants under collineation, namely the j-invariants of $c_1 \cap c_2$, firstly as points of c_1 and secondly as points of c_2. Let a single image of two known conics be obtained by an uncalibrated camera. In [15] it is shown that the locus of the optical centre is a curve contained in the intersection of two surfaces, one of degree 12 and the other of degree 24. In [6] a system for object recognition based on the invariants of coplanar pairs of conics is described. The conics are required to be covariants of the image data.

4.2   Differential invariants

Let $t \mapsto x(t)$ be a smooth parameterised plane curve c. There exist functions $t \mapsto \Theta_3(t)$ and $t \mapsto \Theta_8(t)$ which are invariant under collineation and which determine c up to a collineation [23]. All the invariants of c are functions of Θ_3 and Θ_8. The calculation of Θ_3 and Θ_8 involves derivatives of c to order seven. If c contains identifiable points then the so called semi-differential invariants can be constructed from derivatives of lower order. The application of differential invariants to model based vision is impeded by the near impossibility of obtaining accurate estimates of the derivatives.

References

1. Buchanan T. 1992 Critical sets for reconstruction using lines. *Proc. 2nd European Conf. on Computer Vision, ECCV92.*

2. Demazure M. 1988 Sur deux problèmes de reconstruction. *Technical Report No. 882, INRIA, Rocquencourt, France.*

3. Faugeras O.D. 1990 On the motion of 3D curves and its relationship to optical flow. *Proc. 1st European Conf. on Computer Vision, ECCV90*, Lecture Notes in Computer Science 427, Springer-Verlag.

4. Faugeras O.D. Luong Q.-T. & Maybank S.J. 1992 Camera self-calibration: theory and experiments. *Proc. 2nd European Conf. on Computer Vision, ECCV92.*

5. Faugeras O.D. & Maybank S.J. 1990 Motion from point matches: multiplicity of solutions. *International J. Computer Vision* 4, 225-246.

6. Forsyth D., Mundy J., Zisserman A., Coelho C., Heller A. & Rothwell C. 1991 Invariant descriptors for 3-D object recognition and pose. *IEEE Trans. Pattern Analysis and Machine Intelligence* 13, 971-991.

7. Hofmann W. 1950 Das Problem der "Gerfährlichen Flächen" in Theorie und Praxis. *Dissertation, Fakultät für Bauwesen der Technischen Hochschule München, München, FR Germany. Published in Reihe C, No. 3 der Deutschen Geodetischen Kommission bei der Bayerischen Akademie der Wissenschaften, München 1953.*

8. Horn B.K.P. 1986 *Robot Vision.* Cambridge, Massachusetts: The MIT Press.

9. Huang T.S. & Faugeras O.D. 1980 Some properties of the E matrix in two view motion estimation. *IEEE Trans. Pattern Analysis and Machine Intelligence* 11, 1310-1312.

10. Kruppa E. 1913 Zur Ermittlung eines Objektes zwei Perspektiven mit innere Orientierung. *Sitz-Ber. Akad. Wiss., Wien, math. naturw. Kl., Abt. IIa.* 122, 1939-1948.

11. Liu Y. & Huang T.S. 1988 Estimation of rigid body motion using straight line correspondences. *Computer Vision, Graphics, and Image Processing* 44, 35-57.

12. Longuet-Higgins H.C. 1981 A computer algorithm for reconstructing a scene from two projections. *Nature* 293, 133-135.

13. Longuet-Higgins H.C. 1988 Multiple interpretations of a pair of images of a

surface. *Proc. Royal Soc. London, Series B* 227, 399-410.

14. Maybank S.J. 1990 The projective geometry of ambiguous surfaces. *Phil. Trans. Royal Soc. London, Series A* 332, 1-47.

15. Maybank S.J. 1991 The projection of two non-coplanar conics. In *Proc. First DARPA-ESPRIT Joint Workshop on Applications of Invariant Theory to Computer Vision, Reykjavik, 25-28 March 1991.*

16. Maybank S.J. & Faugeras O.D. 1992 A theory of self-calibration of a moving camera. *Accepted by International J. Computer Vision.*

17. Maybank S.J. 1992 *Theory of Reconstruction from Image Motion.* Springer-Verlag, to appear.

18. J. Mundy & A. Zisserman (eds.) 1992 *Applications of Invariance in Computer Vision.* To appear, MIT Press.

19. Spetsakis M.E. & Aloimonos J. 1990 Structure from motion using line correspondences. *International J. Computer Vision* 4, 171-183.

20. Sturm R. 1869 Das Problem der Projectivität und seine Anwendung auf die Flachen zweiten Grades. *Math. Annalen* 1, 533-573.

21. Tsai R.Y. & Huang T.S. 1984 Uniqueness and estimation of three-dimensional motion parameters of rigid objects with curved surfaces. *IEEE Trans. Pattern Analysis and Machine Intelligence* 6, 13-27.

22. Weng J., Huang T.S. & Ahuja N. 1992 Motion and structure from line correspondences: closed-form solution, uniqueness, and optimization. *IEEE Trans. Pattern Analysis and Machine Intelligence* 14, 318-336.

23. Wilczynski E.J. 1906 *Projective Differential Geometry of Curves and Surfaces.* Leipzig: Teubner.

S.J. Maybank
GEC-Marconi Hirst Research Centre, East Lane,
Wembley, Middlesex HA9 7PP, UK

Disproving Hibi's Conjecture with CoCoA
or
Projective Curves with bad Hilbert Functions

G. Niesi   L. Robbiano *

Introduction.

In this paper we show how to combine different techniques from Commutative Algebra and a systematic use of a Computer Algebra System (in our case mainly CoCoA (see [G-N] and [A-G-N])) in order to explicitly construct Cohen-Macaulay domains, which are standard k-algebras and whose Hilbert function is "bad". In particular we disprove a well-known conjecture by Hibi.

To be more precise, we recall that a ring A is called a *standard k-algebra*, or simply *standard*, if k is a field and A is a finitely generated k-algebra, which is generated by its forms of degree 1 (see [S$_1$]). To every such a ring A a numerical function H_A is associated, namely the function $H_A : \mathbf{N} \longrightarrow \mathbf{N}$, which is defined by $H_A(r) := dim_k A_r$ for every $r \in \mathbf{N}$. Here it should be noted that A can be represented as the quotient of $R := k[X_0, \ldots, X_n]$ modulo a homogeneous ideal I, hence $dim_k A_r \leq dim_k R_r = \binom{n+r}{r}$. Such a function is called the *Hilbert function* of A. It is well-known (see [A-M]) that H_A can be encoded in the power series $\mathcal{P}_A(\lambda) := \sum_r H_A(r)\lambda^r \in \mathbf{Z}[[\lambda]]$, which is called the Hilbert-Poincaré series (or simply the Poincaré series) of A. This series is rational of type $\mathcal{P}_A(\lambda) = \frac{Q_A(\lambda)}{(1-\lambda)^d}$, with $Q_A(1) \neq 0$; moreover $Q_A(\lambda) = \sum h_i(A)\lambda^i \in \mathbf{Z}[\lambda]$ and if δ is its degree, then the integral vector $\mathbf{h}(A) := (h_0(A), h_1(A), \ldots, h_\delta(A))$ is called the *h-vector of A*. It turns out that all the information of H_A can be encoded in $(\mathbf{h}(A), d)$; in particular d is the dimension and $\sum_i h_i(A) = Q_A(1)$ is the multiplicity of A. An efficient algorithm for the computation of $\mathcal{P}_A(\lambda)$ is described in [B-C-R] and implemented in CoCoA.

In the paper [Hi] Hibi defines an h-vector $(h_0, h_1, \ldots, h_\delta)$ to be *flawless* if
   i) $h_0 \leq h_1 \leq \cdots \leq h_{[\delta/2]}$ and
   ii) $h_i \leq h_{\delta-i}$ for every i such that $0 \leq i \leq [\delta/2]$
and he states the following
CONJECTURE: The h-vector $\mathbf{h}(A) := (h_0(A), h_1(A), \ldots, h_\delta(A))$ of a standard Cohen-Macaulay domain is flawless (see [Hi], Conjecture 1.4).

* The paper was partly supported by Consiglio Nazionale delle Ricerche.

The conjecture was supported by some empirical evidence and the fact that the statement is true under some additional hypotheses (see [Hi], Theorem 3.1).

The main goal of this paper is to construct *explicit examples of standard Cohen-Macaulay domains, whose h-vector is not flawless*. To this end we use a technique introduced by Galligo in [G], which yields sets of points in the projective space with the Uniform Position Property. A good deal of freedom in choosing equations allows us to construct projective coordinate rings of sets of points, whose h-vector "has a flaw". Then we lift these sets to irreducible reduced rational projective curves in $\mathbf{P}_{\mathbf{C}}^4$, which turn out to be projectively Cohen-Macaulay. Their coordinate rings are the desired counterexamples.

This paper is largely inspired by the work of Galligo (see [G]) and by the calculations of some Galois groups, which were shown to the second author by G. Scheja, during a short visit to the University of Tübingen in June 1991. To both we are largely indebted.

§1. Preliminaries.

In this section we recall all the definitions and results, which we need later.

Definition. Let k be an infinite field, A a standard k-algebra, i.e. a graded k-algebra which is finitely generated by its linear forms and let $\mathcal{P}_A(\lambda) = \frac{Q_A(\lambda)}{(1-\lambda)^d}$, with $Q_A(1) \neq 0$, be its Poincaré series. Let $\delta := deg(Q_A(\lambda))$ and $Q_A(\lambda) := \sum_{i=0}^{\delta} h_i(A)\lambda^i$.

Then $\mathbf{h}(A) := (h_0(A), h_1(A), \ldots, h_\delta(A))$ is called the h-vector of A. Sometimes it is denoted by $(h_0, h_1, \ldots, h_\delta)$, if there is no ambiguity about A.

The following facts are part of the folklore and are recalled only for the sake of completeness. The non explained terminology is part of the basic literature in Commutative Algebra and Algebraic Geometry (see for instance [A-M] and [Hart]).

Lemma 1.1. *Let $k \subset F$ be fields, A a standard k-algebra, $A_F := A \otimes_k F$. Then $\mathbf{h}(A_F) = \mathbf{h}(A)$.*

Proof. Indeed $\mathcal{P}_{A_F}(\lambda) = \mathcal{P}_A(\lambda)$ •

Proposition 1.2. *Let k be an infinite field and A a standard k-algebra of dimension d. Then there exist d linear forms $L_1, \ldots, L_d$, such that $dim(A/(L_1, \ldots, L_d)) = 0$. If moreover A is Cohen-Macaulay (C-M), then $L_1, \ldots, L_d$ is a regular sequence in A.*

Corollary 1.3. *Let A be a Cohen-Macaulay standard k-algebra over a field k, let $L_1, \ldots, L_d$ be a maximal regular sequence of linear forms in A and denote by $B := A/(L_1, \ldots, L_d)$. Then $\mathbf{h}(A) = \mathbf{h}(B)$.*

Proof. Indeed $\mathcal{P}_B(\lambda) = (1-\lambda)^d \mathcal{P}_A(\lambda)$ •

Let now $S := k[X_0, X_1, \ldots, X_n]$, $\mathfrak{M}$ a maximal homogeneous relevant ideal in S and $K := K_0(S/\mathfrak{M})$ its associated field, i.e. the field of homogeneous fractions of degree 0 of $S/\mathfrak{M}$. The scheme $Proj(S/\mathfrak{M})$ has a unique point, whose associated local ring is $K_0(S/\mathfrak{M})$. If $X_0 \notin \mathfrak{M}$, then we can dehomogenize $\mathfrak{M}$ with respect to X_0 and we get a maximal ideal $\mathfrak{m}$ in $R := k[X_1, \ldots, X_n]$ (this can be done by putting $X_0 = 1$). It turns out that $K \cong R/\mathfrak{m}$. Moreover every generic linear change of coordinates yields the following shape of $\mathfrak{m}$

$$\mathfrak{m} = (f(X_1), X_2 - g_2(X_1), X_3 - g_3(X_1, X_2), \ldots, X_n - g_n(X_1, \ldots, X_{n-1}))$$

It is clear that $K \cong k[X]/(f(X))$, hence $deg(f(X)) = dim_k K$.

Definition. In the above described situation we say that $f(X)$ represents $\mathfrak{M}$.

Definition. Let $\mathfrak{M}$ be a maximal homogeneous relevant ideal in S, K its associated field and $d = dim_k K$. We say that $\mathfrak{M}$ is G-symmetric if the Galois group $Gal_k(K)$ is the full symmetric group Σ_d.

Corollary 1.4. *Let $\mathfrak{M}$ be a maximal homogeneous relevant ideal in S, K its associated field and $f(X)$ a polynomial of degree d representing $\mathfrak{M}$. Then $\mathfrak{M}$ is G-symmetric if and only if $Gal_k(f(X)) = \Sigma_d$*

Now we recall a well-known criterion

Theorem 1.5. *Let $f(X) \in \mathbb{Z}[X]$, $d := deg(f(X))$ and assume that*
a) *$f(X)$ is irreducible*
b) *There exist two prime numbers p_1, p_2 such that, if we denote by $f_i(X)$ the residue classes of $f(X)$ modulo p_i, $i = 1, 2$, then*
  b_1) *$f_1(X)$ decomposes as the product of a linear factor and an irreducible factor of degree $d - 1$.*
  b_2) *$f_2(X)$ decomposes as the product of an irreducible factor of degree 2 and irreducible factors of odd degrees.*
  Then $Gal_{\mathbb{Q}}(f(X)) = \Sigma_d$.

Proof. See [W] and [S-S] •

Definition. Let F be an infinite field and let $E \subset \mathbb{P}_F^n$ be a finite set of reduced points. Let I be the homogeneous ideal of $F[X_0, \ldots, X_n]$ which defines E. We say that E is G-symmetric if there exists a subfield $k \subseteq F$ and a maximal homogeneous relevant ideal $\mathfrak{M}$ in $S := k[X_0, \ldots, X_n]$ such that $I = \mathfrak{M}F[X_0, \ldots, X_n]$ and $Gal_k(K_0(S/\mathfrak{M})) = \Sigma_d$

Theorem 1.6. *Let C be a projective irreducible reduced curve in $\mathbb{P}_{\mathbb{C}}^n$. Then the generic hyperplane section of C is a G-symmetric set.*

Proof. See [G], Proposition 13. The proof given for a smooth curve in $\mathbb{P}^3$ works as well in general •

Theorem 1.7. *Let* $E \subset \mathbf{P}^n_F$ *be a G-symmetric set of points and* $A := F[X_0, \ldots, X_n]/I$ *its coordinate ring. Let* $\mathbf{h}(A) := (h_0, h_1, \ldots, h_\delta)$ *be the h-vector of A. Then the following inequalities hold:*

$$h_0 + h_1 + \cdots + h_i \leq h_{\delta-i} + \cdots + h_{\delta-1} + 1$$

for every $i = 1, \ldots, [\delta/2]$

Proof. It is proved in [G] that if E is G-symmetric, then it has the Uniform Position Property (UPP), i.e. the Hilbert function of its subsets depends only on their cardinality. Now if E is UPP, then every subset has the Cayley Bacharach (CB) property (see [G-K-R] for definitions and properties), and the conclusion follows again by [G-K-R] •

§2. The construction of the counterexamples.

Now we are ready to use the above described machinery in order to produce standard Cohen Macaulay domains with bad Hilbert functions. All the following computations have been done using CoCoA 1.7b on a Macintosh.

The first step is to produce polynomials $f(X) \in \mathbf{Z}[X]$ of degree d, such that $Gal_{\mathbf{Q}}(f(X)) = \Sigma_d$. This is not difficult, since the "generic" one has this property; however we want polynomials, which are not too dense, since we want to use them to make further computations.

Lemma 2.1. *The polynomial* $f(X) := X^{18} - X - 1$ *is such that*

$$Gal_{\mathbf{Q}}(f(X)) = \Sigma_{18}.$$

Proof. The polynomial $f(X)$ is irreducible. Moreover $f(X) = (X^2 - X - 1)(X^{13} - X^{12} + X^{11} + X^8 - X^7 - X^6 - X^5 + X^4 + X^3 - X^2 + 1)(X^3 - X^2 + 1)$ mod 3 and $f(X) = (X + 2)(X^{17} - 2X^{16} - X^{15} + 2X^{14} + X^{13} - 2X^{12} - X^{11} + 2X^{10} + X^9 - 2X^8 - X^7 + 2X^6 + X^5 - 2X^4 - X^3 + 2X^2 + X + 2)$ mod 5. The conclusion follows from theorem 1.5 •

Corollary 2.2. *Let* I *be the ideal of* $\mathbf{Q}[X, Y, Z]$ *generated by* $(X^{18} - X - 1, Y - g(X), Z - h(X, Y)$ *with* $g(X) \in \mathbf{Q}[X]$ *and* $h(X, Y) \in \mathbf{Q}[X, Y]$. *Let* $\mathfrak{M} := {}^h I$ *i.e the homogeneization of* I *with respect to a new indeterminate* W. *Then* $\mathfrak{M}$ *is a G-symmetric maximal relevant ideal in* $\mathbf{Q}[X, Y, Z, W]$.

Proof. Clearly $\mathbf{Q}[X, Y, Z]/I \cong \mathbf{Q}[X]/(X^{18} - X - 1)$ and this is a field, hence I is a maximal ideal of $\mathbf{Q}[X, Y, Z]$. Consequently $\mathfrak{M}$ is a maximal relevant ideal in $\mathbf{Q}[X, Y, Z, W]$. The conclusion follows from Corollary 1.4 and Lemma 2.1 •

Now the homogeneization of I is computed via a Gröbner basis computation with respect to an ordering which is degree-compatible and the leading term ideal of I and of ${}^h I$ are generated by the same elements, hence they have the same h-vector. The key point is now that we are *totally free* in the choice of $g(X)$ and $h(X,Y)$. We use again CoCoA and again we are lucky, because we do not need many experiments. Namely

Example 2.3. *Let I be the ideal of $\mathbf{Q}[X,Y,Z]$ generated by $(X^{18} - X - 1, Y - X^3, Z - XY)$ and let $\mathfrak{M}$ be its homogeneization with respect to the new indeterminate W. Then $A := \mathbf{Q}[X,Y,Z,W]/\mathfrak{M}$ is a standard $\mathbf{Q}$-algebra, which is a Cohen-Macaulay domain and whose h-vector is $(1,3,5,4,4,1)$. It satisfies the inequalities of Theorem 1.7, but it is not flawless.*

Proof. Let $A := \mathbf{Q}[X,Y,Z,W]/\mathfrak{M}$. The computation shows that $\mathfrak{M} = (XY - ZW, \quad X^3 - YW^2, \quad X^2Z - Y^2W, \quad Y^3 - XZ^2, \quad Y^2Z^3 - XW^4 - W^5, \quad YZ^4 - X^2W^3 - XW^4, \quad Z^5 - X^2W^3 - YW^4)$ and that $\mathcal{P}_A(\lambda) = \frac{(1+3\lambda+5\lambda^2+4\lambda^3+4\lambda^4+\lambda^5)}{(1-\lambda)}$. Moreover A is a domain and it is Cohen-Macaulay, since it is 1-dimensional and W is a non zero divisor modulo $\mathfrak{M}$ •

This is already a counterexample to Hibi's conjecture!

However the fact that A is a domain heavily relies on the special ground field. Namely if we replace $\mathbf{Q}$ with $\mathbf{C}$ (it suffices to replace it with the decomposition field of $f(X)$), then $\mathbf{C}[X,Y,Z,W]/\mathfrak{M}\mathbf{C}[X,Y,Z,W]$ is a reduced $\mathbf{C}$-algebra, hence it is the coordinate ring of a G-symmetric set of 18 points in $\mathbf{P}^3_\mathbf{C}$. Its h-vector is still the same (see Lemma 1.1), but it is no more a domain.

So the final part is devoted to find a standard algebra whose h-vector is not flawless and which is a "geometric" domain, i.e the fact that it is a domain is not affected by any extension of the base field. The idea is to "lift" our previous example to an irreducible reduced curve in $\mathbf{P}^4_\mathbf{C}$. A small deformation of our equation $f(X)$ does the trick. Namely

Example 2.4. *Let $\mathfrak{p}$ be the ideal of $\mathbf{C}[X,Y,Z,T]$ generated by $(X^{18} - X - 1 - T, Y - X^3, Z - XY)$ and let $\mathfrak{P}$ be its homogeneization with respect to the new indeterminate W. Then $A := \mathbf{C}[X,Y,Z,T,W]/\mathfrak{P}$ is a standard $\mathbf{C}$-algebra, which is a Cohen-Macaulay domain and whose h-vector is $(1,3,5,4,4,1)$, hence it is not flawless.*

Proof. It is clear that $\mathbf{C}[X,Y,Z,T]/\mathfrak{p} \cong \mathbf{C}[X]$, hence $\mathfrak{p}$ is a prime ideal, hence $\mathfrak{P}$ is a prime ideal. It defines a projective rational curve in $\mathbf{P}^4_\mathbf{C}$. The actual computation yields the following minimal system of generators for $\mathfrak{P}$.

$$\mathfrak{P} = (XY - ZW, \quad X^3 - YW^2, \quad X^2Z - Y^2W, \quad Y^3 - XZ^2, \quad Y^2Z^3 + XW^4 + TW^4 + W^5, \quad YZ^4 + X^2W^3 + XTW^3 + XW^4, \quad Z^5 + X^2TW^2 + X^2W^3 + YW^4).$$

The computation of the Poincaré series yields
$$\mathcal{P}_A(\lambda) = \frac{(1+3\lambda+5\lambda^2+4\lambda^3+4\lambda^4+\lambda^5)}{(1-\lambda)^2}.$$ It remains to prove that A is a Cohen-Macaulay ring. For, it is enough to show that W,T is a regular sequence $mod\ \mathfrak{P}$. Indeed W is a non zero divisor $mod\ \mathfrak{P}$, since it is the

homogeneizing indeterminate. If we compute the quotient modulo W, we get $\mathbf{C}[X,Y,Z,T]/(XY,X^3,X^2Z,Y^3-XZ^2,Y^2Z^3,YZ^4,Z^5)$. Hence clearly T does not divide zero •

We conclude the paper with some remarks.

Remark. Example 2.4 disproves Hibi's Conjecture. So it is interesting to check that it does not fit with the special class described by Hibi in [Hi] Theorem 3.1. There it is required that the associated order ideal of monomials is pure. Without going too much in to the details, we check that in our case the associated order ideal of monomials is

$$(1,X,Y,Z,X^2,XZ,Y^2,YZ,Z^2,XZ^2,Y^2Z,YZ^2,Z^3,XZ^3,Y^2Z^2,$$
$$YZ^3,Z^4,XZ^4).$$

Its maximal elements are X^2,Y^2Z^2,YZ^3,XZ^4 whose degrees are 2,4,4,5; therefore the order ideal of monomials is not pure.

Remark. One can construct similar examples to 2.4. For instance if we consider the ideals $(X^{22}-X-1-T,Y-X^3,Z-XY)$, $(X^{26}-X-1-T,Y-X^3,Z-XY)$, $(X^{30}-X-1-T,Y-X^3,Z-XY)$ and carry over the same construction as in 4.2, we see that:

the Galois group of $X^{22}-X-1$ is Σ_{22}. The primes who do the trick as in Lemma 2.1 are 29 and 107. The corresponding h-vector is (1,3,5,4,4,4,1).

The Galois group of $X^{26}-X-1$ is Σ_{26}. The primes who do the trick as in Lemma 2.1 are 19 and 67. The corresponding h-vector is (1,3,5,4,4,4,4,1).

The Galois group of $X^{30}-X-1$ is Σ_{30}. The primes who do the trick as in Lemma 2.1 are 5 and 53. The corresponding h-vector is (1,3,5,4,4,4,4,4,1).

All of them are not flawless.

References

[A-G-N]   Armando, E., Giovini, A., Niesi, G., *CoCoA User's Manual v. 1.5*, (1991), Dipartimento di Matematica, Università di Genova.

[A-M]   Atiyah, M. F., Macdonald, I. G., *Introduction to Commutative Algebra*, Addison-Wesley (1969).

[B-C-R]   Bigatti, A., Caboara, M., Robbiano, L., *On the computation of Hilbert-Poincaré series*, Applicable Algebra in Engineering, Communications and Computing (1990), To appear.

[G]   Galligo, A., *Exemples d'ensembles de Points en Position Uniforme*, Proceedings of MEGA-90 (1991), Birkhauser.

[G-K-R]   Geramita, A., Kreuzer, M., Robbiano, L., *Cayley-Bacharach schemes and their canonical modules*, Trans. Amer. Math. J. (1991), To appear.

[G-M-R]   Geramita, A.V., Maroscia, P., Roberts, L.G., *The Hilbert function of a reduced k-algebra*, J. London Math. Soc. **28** (1983), 443–452.

[G-M]   Geramita, A.V., Migliore, J.C., *Hyperplane sections of a smooth curve in* $\mathbf{P}^3$, Comm. Algebra **17** (1989), 3129–3164.

[G-N]   Giovini, A., Niesi, G., *CoCoA: a user-friendly system for commutative algebra*, Proceedings of DISCO-90, Lecture Notes in Computer Sciences **429** (1990), Springer-Verlag.

[Ha]   Harris, J., *The genus of space curves*, Math. Ann. **249** (1980), 191–204.

[H-E]   Harris, J. (with the collaboration of D. Eisenbud), *Curves in projective space* in *Sém. de Mathématiques Supérieures, Université de Montreal* (1982), .

[Hart] Hartshorne, R., *Algebraic Geometry*, Springer (1977).

[Hi]   Hibi, T., *Flawless O-sequences and Hilbert functions of Cohen-Macaulay integral domains*, J. Pure Appl. Algebra **60** (1989), 245–251.

[Ra]   Rathmann, J., *The uniform position principle for curves in characteristic* p, Math. Ann. **276** (1987), 565–579.

[R$_1$] Robbiano, L., *Introduction to the theory of Gröbner bases* in *The Curves Seminar at Queen's, vol. V, Queen's Papers in Pure and Appl. Math.* 80 (1988), Queen's University, Kingston B1–B29.

[R$_2$] Robbiano, L., *On the theory of Hilbert functions*, Queen's Papers in Pure and Appl. Math., Kingston Canada **85** (1990), Vol VII.

[S-S]   Scheja, G., Storch, U., *Lehrbuch der Algebra*, B. G. Teubner Stuttgart (1988).

[S$_1$] Stanley, R, *Hilbert functions of graded algebras*, Adv. in Math. **28** (1978), 57–83.

[S$_2$] Stanley, R, *On the Hilbert function of a graded Cohen-Macaulay domain*, (1990), preprint, Massachusetts Inst. of Technology, Cambridge.

[W]   van der Waerden, B.L., *Algebra, Vol I*, Ungar (1970).

Gianfranco Niesi, Lorenzo Robbiano
Dipartimento di Matematica dell'Università di Genova
Via L.B. Alberti 4
16132 Genova ITALY

Counting real zeros
in the multivariate case

P. Pedersen   M.-F. Roy   A. Szpirglas

Abstract.   In this paper we show, by generalizing Hermite's theorem to
the multivariate setting, how to count the number of real or complex points of
a discrete algebraic set which lie within some algebraic constraint region. We
introduce a family of quadratic forms determined by the algebraic constraints
and defined in terms of the trace from the coordinate ring of the variety to
the ground field, and we show that the rank and signature of these forms are
sufficient to determine the number of real points lying within a constraint region.
In all cases we count geometric points, which is to say, we count points without
multiplicity. The theoretical results on these quadratic forms are more or less
classical , but forgotten too, and can be found also in [3].

We insist on effectivity of the computation and complexity analysis : we
show how to calculate the trace and signature using Gröbner bases, and we
show how the information provided by the individual quadratic forms may be
combined to determine the number of real points satisfying a conjunction of
constraints. The complexity of the computation is polynomial in the dimension
as a vector space of the quotient ring associated to the defining equations. In
terms of the number of variables, the complexity of the computation is singly
exponential. The algorithm is well parallelizable.

We conclude the paper by applying our machinery to the problem of effectively calculating the Euler characteristic of a smooth hypersurface.

Introduction

Let R be any real closed field, and C its algebraic closure. We want to count
points of a discrete real affine algebraic variety $V(f_1, \ldots, f_s)$ defined by some
number of real polynomial equations:

$$f_1(\bar{x}) = 0, \ldots, f_s(\bar{x}) = 0; \qquad f_i \in K[\bar{x}]; \qquad \bar{x} = (x_1, \ldots, x_n).$$

where K is a subfield of R.

We denote by $V_C(I)$ the C-points and by $V_R(I)$ the R-points of such a variety. The fundamental assumption we shall make is that the C-points form a finite (hence discrete) set, or equivalently, that the coordinate ring $A = K[\bar{x}]/I$ is a finite dimensional K-vector space, where $I = (f_1, \ldots, f_s) \subset K[\bar{x}]$ is the ideal generated by the given polynomials. This implies that $A_R = R[\bar{x}]/I$ (resp. $A_C = C[\bar{x}]/I$) is a finite dimensional R (resp. C)-vector space.

We shall consider constraint regions of the form

$$P = \{\bar{x} \in R^n \; : \; h_1(\bar{x})\varepsilon_1, h_2(\bar{x})\varepsilon_2, \ldots, h_k(\bar{x})\varepsilon_p\},$$

where $\varepsilon_i \in \{> 0, < 0, = 0\}$.

The constraint polynomials h_j are taken in $K[\bar{x}]$.

1   Definition of the quadratic form

The finite dimensional K-vector space $A = K[\bar{x}]/I$ is also a commutative K-algebra, hence for any $f \in A$ we may consider the vector space endomorphism induced by multiplication by f, which we denote by $m_f \in End_K(A)$. This defines a homorphism $m : A \longrightarrow End_K(A)$, so that $m_f m_g = m_{fg}$. We could also consider the corresponding homomorphism of A_R and A_C which correspond to extending scalars.

Since A is finite dimensional, we have a trace $Tr : End_K(A) \longrightarrow K$, which we may apply to m_f. This is nothing but the trace of the matrix representing the action of multiplication by f on A. In this way we may define a symmetric bilinear form $B : A \otimes A \longrightarrow K$

$$B(f,g) = Tr(m_f \cdot m_g) = Tr(m_{fg}),$$

In a given basis $\{v_i\}$ for A, the matrix for B is just

$$B_{i,j} = (Tr(m_{v_i v_j})).$$

Degeneracy of the form arises from multiplicity in the variety V, as we shall see shortly. For each polynomial h, we construct an associated bilinear form as follows:

$$B_h(f,g) = B(hf,g) = Tr(m_{fgh}).$$

and its associated quadratic forms Q_h.

This quadratic form has been studied by Hermite in the univariate and bivariate case ([9],[8]). See also [2].

2   The main theorem

We now have sufficient vocabulary to state our main theorem:

Theorem 2.1 . — *Given a field K, a finite affine algebraic variety V, defined by the ideal I generated by $(f_1, \ldots, f_s)$, where $f_i \in K[\bar{x}]$, R a real closed field such that $K \subset R$ and C its algebraic closure, and given one other polynomial $h \in K[\bar{x}]$, then*

$$\sigma(Q_h) = \#\{\bar{x} \in V_R(I) \ : \ h(\bar{x}) > 0\} - \#\{\bar{x} \in V_R(I) \ : \ h(\bar{x}) < 0\},$$
$$\rho(Q_h) = \#\{\bar{x} \in V_C(I) \ : \ h(\bar{x}) \neq 0\},$$

where σ denotes the signature and ρ the rank of the associated quadratic form.

Before proving the main theorem, we would like to motivate the discussion by considering the univariate case. So suppose $f(x) \in K[x]$ is of degree d, then a basis for $C[x]/(f)$ is provided by the monomials $\{1, x, x^2, \ldots, x^{d-1}\}$. If $f(x) = \prod_{i=1}^{s}(x - \alpha_i)^{e_i}$ is a factorization over C, then since the various factors are co-maximal, the quotient ring has the following structure:

$$A_C = C[x]/(f) \cong \prod_{i=1}^{s} C[x]/(x - \alpha_i)^{e_i}.$$

The image of h in $C[x]/(x - \alpha)^{e_i}$ will consist of the first e_i terms of the Taylor expansion of h about α_i, and the isomorphism corresponds to Hermite interpolation of h at the roots of f. We may consider the isomorphism as $h(x) \mapsto [h_{\alpha_1}, h_{\alpha_2}, \ldots, h_{\alpha_s}]$, where each h_{α_j} consists of the first e_j terms of the Taylor expansion at α_j.

We have the following result: if, for any $i \in \{1, \cdots, s\}$, we have $h_{\alpha_i} = 0$, then for any $g \in C[x]$, $(gh)_{\alpha_i} = 0$. This is because for every derivative of order r $(r \leq (e_i - 1))$: $(gh)^{(r)}(\alpha_i) = \sum \binom{j}{r} g^{(j)}(\alpha) h^{(r-j)}(\alpha_i) = 0$. Therefore every multiplication operator has a block structure which depends only on f. Each block may be put in a triangular form by the following choice of basis:

$$B_i = \{1, (x - \alpha_i), \frac{(x - \alpha_i)^2}{2!}, \ldots, \frac{(x - \alpha_i)^{e_i - 1}}{(e_i - 1)!}\}.$$

If we call these basis vectors $\{v_0, v_1, \ldots, v_{e_i-1}\}$, then an easy computation shows that

$$h \cdot v_k = \sum_{j \geq k} \binom{k}{j} h^{(j-k)}(\alpha) v_j,$$

so that the matrix of multiplication by h looks like

$$\begin{bmatrix} h(\alpha_i) & 0 & \cdots & 0 \\ h'(\alpha_i) & h(\alpha_i) & \cdots & 0 \\ h''(\alpha_i) & h'(\alpha_i) & \cdots & 0 \\ \vdots & \vdots & & \vdots \\ h^{(e_i-1)}(\alpha_i) & h^{(e_i-2)}(\alpha_i) & \cdots & h(\alpha_i) \end{bmatrix}.$$

It is evident that $h(\alpha_i)$ is an eigenvalue of algebraic multiplicity e_i. In particular,

$$Tr(m_h) = \sum_{i=1}^{s} e_i h(\alpha_i), \quad and \quad B(f,g) = \sum_{i=1}^{s} e_i f(\alpha_i)g(\alpha_i).$$

Now consider the matrix of B with respect to the union of the bases $\mathcal{B}_i = \{1, (x - \alpha_i), \frac{(x-\alpha_i)^2}{2!}, \ldots, \frac{(x-\alpha_i)^{e_i-1}}{(e_i-1)!}\}$,
$i = 1, \ldots, s$. In each subspace $C[x]/(x - \alpha_i)^{e_i}$ we have

$$m_{(x-\alpha_i)^k} = \begin{bmatrix} 0 & 0 & \cdots & 0 & \cdots & 0 \\ \vdots & & & \vdots & & \vdots \\ 1 & 0 & \cdots & 0 & \cdots & 0 \\ 0 & 1 & \cdots & 0 & \cdots & 0 \\ \vdots & & \ddots & 0 & \cdots & \vdots \\ 0 & 0 & \cdots & 1 & \cdots & 0 \end{bmatrix},$$

where the 1's fill the k-th lower diagonal. Therefore, $Tr(m_{(x-\alpha_i)^k} \cdot m_{(x-\alpha_i)^{k'}}) = 0$, unless $k = k' = 0$, in which case it equals e_i. Then matrix for B in the basis $\mathcal{B}_1 \cup \ldots \cup \mathcal{B}_s$ looks like

$$\begin{bmatrix} e_1 & 0 & \cdots & 0 & 0 & 0 & \cdots & 0 & \cdots \\ 0 & 0 & \cdots & 0 & \vdots & & & \vdots & \vdots \\ \vdots & & \ddots & \vdots & \vdots & & & \vdots & \\ 0 & 0 & \cdots & 0 & 0 & 0 & \cdots & 0 & \cdots \\ 0 & 0 & \cdots & 0 & e_2 & 0 & \cdots & 0 & \cdots \\ \vdots & & & \vdots & 0 & 0 & \cdots & 0 & \\ \vdots & & & \vdots & \vdots & & \ddots & \vdots & \\ 0 & 0 & \cdots & 0 & 0 & 0 & \cdots & 0 & \cdots \\ \vdots & & & \vdots & \vdots & & & \vdots & \ddots \end{bmatrix}.$$

Q_h is the quadratic form with p variables $x_0, x_1, \ldots, x_{p-1}$, defined by:

$$Q_h(x_0, x_1, \cdots, x_{p-1}) = \sum_{i=1,\ldots,p} h(\alpha_i)(x_0 + x_1\alpha_i + \ldots + x_{p-1}\alpha_i^{p-1})^2$$

It is clear that $Q_h(x_0, x_1, \cdots, x_{p-1})$ has coefficients in K, since the expression is symetric in the α_i's.

Denoting by $s(h)_k$, for $k = 0, \ldots, 2p - 2$ the sum $\sum_{i=1,\ldots,p} h(\alpha_i)\alpha_i^k$ one has:

$$Q_h(x_0, x_1, \cdots, x_{p-1}) = \sum_{\substack{k=0,\ldots,p-1 \\ j=0,\ldots,p-1}} s(h)_{k+j} x_k x_j$$

When $h = 1$, one notes Q the quadratic form Q_h; so

$$Q(x_0, x_1, \cdots, x_{p-1}) = \sum_{i=1,\ldots,p} (x_0 + x_1\alpha_i + \ldots + x_{p-1}\alpha_i^{p-1})^2$$

Denoting by s_k the Newton sum

$$\sum_{i=1,\ldots,p} \alpha_i^k$$

one has:

$$Q_h(x_0, x_1, \cdots, x_{p-1}) = \sum_{\substack{k=0,\ldots,p-1 \\ j=0,\ldots,p-1}} s_{k+j} x_k x_j$$

Let $\beta_1, ..., \beta_n$ be the distinct real zeroes of f , $m_1, ..., m_n$ their multiplicities $\gamma_1, \overline{\gamma_1}, \ldots, \gamma_m, \overline{\gamma_m})$ the complex distinct zeroes of f , $w_1, ..., w_n$ their multiplicities.

For $a \in C$, let y be the linear form on C^n defined by:

$$y(a, x) = x_0 + x_1 a + \ldots + x_{p-1} a^{p-1}$$

and let $b(a, x) = y(a, x)^2$ The quadratic form Q_h is equal to

$$\begin{aligned} Q_h(x_0, x_1, \cdots, x_{p-1}) &= \sum_{j=1,\ldots,n} m_j f(\beta_j) b(\beta_j, x) \\ &+ \sum_{h=1,\ldots,m} w_h(f(\gamma_h) b(\gamma_h, x) + Q(\overline{\gamma_h}) b(\overline{\gamma_h}, x)) \end{aligned}$$

Linear forms $y(\beta_j, x), y(\gamma_h, x)$, $y(\overline{\gamma_h}, x)$ are linearly independant (the zeroes are distinct and it is sufficient to consider a Vandermonde determinant). This gives the second statement of theorem 2.1 in the univariate case.

Writing $f(\gamma_h) = d_h^2$ and decomposing $d_h b(\gamma_h, x)$ under the form $d_h + id\prime_h$ with d_h et $d\prime_h$ real linear forms, it is clear that $f(\gamma_h)b(\gamma_h, x) + f(\overline{\gamma_h})b(\overline{\gamma_h}, x)$ is the difference of two squares of real linear forms. This concludes the univariate analysis. As we shall see, essentially every step we have made can be duplicated in the multivariate setting.

Proof of the main theorem We can prove theorem 2.1 by using standard theory of Artinian rings (see [AM 1969]). But, as H.Lombardi pointed it out, there is a proof which does not use this theory, and can be understood only with the help of "elementary" commutative algebra. It is this proof (in fact essentially the same as the classical one) which is given in the following section. The structure theory for finite dimensional algebras which we shall present was first developed by Stickelberger (see [SS 88]).

For a in C^n, we denote by $\mathcal{M}_a$ the maximal ideal of $C(\bar{x})$ defined by a, i.e:
$$\mathcal{M}_a = \{f \in C[\bar{x}] \; : \; f(a) = 0\} = (x_1 - a_1, \ldots, x_n - a_n)C[\bar{x}], \text{ where } a = (a_1, \ldots, a_n)$$

We need first the following theorem:

Theorem 2.2 . $-$ *If* $V_C(I) = \{\alpha_1, \cdots, \alpha_r\}$ *then:*

1. *there exists an integer k ($k > 0$) such that, for each i in $\{1, \cdots, r\}$, $\mathcal{M}_{\alpha_i}^{k+1} + IC = \mathcal{M}_{\alpha_i}^k + IC$;*

2. *$A_C \cong \prod_{i=1}^{r} A_C/(\bar{\mathcal{M}}_{\alpha_i})^k$, where for any α in C^n, $\bar{\mathcal{M}}_\alpha$ is the image of $\mathcal{M}_\alpha$ in A_C, i.e $\bar{\mathcal{M}}_\alpha = \mathcal{M}_\alpha + IC/IC$;*

3. *for each $i \in \{1, \cdots, r\}$, $A_C/(\bar{\mathcal{M}}_{\alpha_i})^k$ is a finite dimensional algebra over C, with unique maximal ideal $\tilde{\mathcal{M}}_{\alpha_i} = \bar{\mathcal{M}}_{\alpha_i}/(\bar{\mathcal{M}}_{\alpha_i})^k$.*

We prove in fact the following proposition, which gives theorem 2.2:

Proposition 2.3 *The following conditions are equivalent:*

a) A is finite dimensional over K.

b) $\#V_C(I)$ is finite.

Proof of proposition 2.3

- We have: $[\alpha \in V_C(I) \iff \mathcal{M}_\alpha \supset I]$. Note that, if $\alpha \neq \beta$ then $\mathcal{M}_\alpha$ and $\mathcal{M}_\beta$ are comaximal. So, if $\alpha_1, \cdots, \alpha_r$ are r distinct points of V_C, the ideals $\bar{\mathcal{M}}_{\alpha_i}$ and $\bar{\mathcal{M}}_{\alpha_j}$ are comaximal when $i \neq j$. The Chinese remainder theorem gives the following isomorphisms of C algebras:

$$\bigcap_{i=1}^{r} \bar{\mathcal{M}}_{\alpha_i} = \prod_{i=1}^{r} \bar{\mathcal{M}}_{\alpha_i} \text{ and } A_C/(\cap \bar{\mathcal{M}}_{\alpha_i}) \cong \prod_{i=1}^{r}(A_C/\bar{\mathcal{M}}_{\alpha_i}) \cong L^r. \text{ More pre-}$$

cisely, for h in $C[\bar{x}]$, if $\tilde{h}$ is the image of h in $A_C/(\cap\bar{\mathcal{M}}_{\alpha_i})$, then the image of $\tilde{h}$ by the isomorphism $A_C/(\cap\bar{\mathcal{M}}_{\alpha_i}) \xrightarrow{\sim} \prod_{i=1}^{r}(A_C/\bar{\mathcal{M}}_{\alpha_i}) \cong L^r$ is $(h(\alpha_1), \cdots, h(\alpha_r))$.

We proved a) $\Rightarrow$ b) and: $\#V_C(I) \leq dim_C A_C = dim_K A$.

- If $V_C(I)$ has exactly r points ($V_C(I) = \{\alpha_1, \cdots, \alpha_r\}$), Hilbert Nullstellensatz theorem implies that for each h in $C[\bar{x}]$ such that $h(\alpha_1) = \cdots = h(\alpha_r) = 0$, there exists an integer $p > 0$ with $h^p \in IC$. Denote by J the ideal $\bigcap_{i=1}^{r} \mathcal{M}_{\alpha_i} = \prod_{i=1}^{r} \mathcal{M}_{\alpha_i}$. Denote by $\{h_1, \cdots, h_l\}$ a system of generators of J deduced from the generators of each $\mathcal{M}_{\alpha_i}$.

  Then, for each $j \in \{1, \cdots, l\}$, there exists an integer $k_j > 0$ such that

  $h_j^{k_j} \in IC$. So we have: $J^k \subset IC$, with $k = \sum_{j=1}^{l} k_j$ and $\prod_{i=1}^{r} \mathcal{M}_{\alpha_i}^k \subset IC$.

  But, $\mathcal{M}_{\alpha_i}^k$ and $\mathcal{M}_{\alpha_j}^k$ are comaximal when $i \neq j$ and $C[\bar{x}]/J^k \cong \prod_{i=1}^{r} C[\bar{x}]/(\mathcal{M}_{\alpha_i})^k$.

  So, $A_C \cong \prod_{i=1}^{r} A_C/(\bar{\mathcal{M}}_{\alpha_i})^k$. This proves b) $\Rightarrow$ a), as:

  $C[\bar{x}]/(\mathcal{M}_{\alpha_i})$ is finite dimensional over C and $A_C \cong \prod_{i=1}^{r} A_C/(\bar{\mathcal{M}}_{\alpha_i})^k = \prod_{i=1}^{r} C[\bar{x}]/(< f_1, \cdots, f_s > +\mathcal{M}_{\alpha_i^k})$. And we get theorem 2.2.

□

We may express this decomposition directly in terms of localizations as $A_C = \prod_{\alpha \in V_C} A_\alpha$, where A_α is the localization of A_C at $\bar{\mathcal{M}}_\alpha$. Let $C[\bar{x}]_\alpha$ denote the localization of $C[\bar{x}]$ at the multiplicative subset $S_\alpha = C[\bar{x}] \setminus \mathcal{M}_\alpha$, then A_α is also canonically isomorphic to the quotient $C[\bar{x}]_\alpha/IC[\bar{x}]_\alpha$, and the ideals $\bar{\mathcal{M}}_\alpha$ for $\alpha \in V_C$ are precisely all the maximal ideals of A_C.

Definition 2.4 *The multiplicity of $V_C(I)$ at α is $e_\alpha = dim_C A_\alpha$.*

Now we consider the matrix of the multiplication operator m_h, where $h(\bar{x}) \in K[\bar{x}]$.

Lemma 2.5 *The subspaces A_α are invariant under m_h.*

Proof of lemma 2.5 Each A_α may be identified with the quotient $C[\bar{x}]_\alpha/IC[\bar{x}]_\alpha$. If $p/q \in A_\alpha$, with $q(\alpha)I \neq 0$, then $h \cdot (p/q) = (hp)/q$ is of the same form. □

Lemma 2.6 *Multiplication by $g(\bar{x}) = h(\bar{x}) - h(\alpha)$ is a nilpotent operator in A_α.*

Proof of lemma 2.6 We have $A_\alpha \cong A/\bar{\mathcal{M}}_\alpha^k$. A is a polynomial ring over an algebraically closed field, and $\bar{\mathcal{M}}_\alpha^k$ is an $\mathcal{M}_\alpha$-primary ideal whose variety consists of the single point $\alpha \in V_C$. Clearly $g(\alpha) = 0$, hence by the Hilbert Nullstellensatz there is some p such that $g^p \in \mathcal{M}_\alpha$, i.e. g is nilpotent on A_α. $\square$

Therefore multiplication by $g(\bar{x})$ is upper triangular in A_α, and the matrix of multiplication by $h(\bar{x})$ is of the form

$$\begin{bmatrix} h(\alpha) & * & \cdots & * \\ 0 & h(\alpha) & \cdots & * \\ \vdots & & \ddots & \vdots \\ 0 & 0 & \cdots & h(\alpha) \end{bmatrix}.$$

We see that $h(\alpha)$ is an eigenvalue of multiplicity e_α, where $e_\alpha = dim_C(A_\alpha)$ is the multiplicity of α.

Theorem 2.7 *If $I \subset K[\bar{x}]$ is an ideal defining a discrete variety, let $A = K[\bar{x}]/I$ and $A_C = C[\bar{x}]/I$, and $B_h : A \otimes A \longrightarrow K$ be the trace form associated with $h \in A$, then*

$$B_h(f,g) = \sum_{\alpha \in V_C(I)} e_\alpha h(\alpha)f(\alpha)g(\alpha).$$

Proof As we have seen, the multiplication operators have all block triangular representations which depend only on the structure of A_C. Products of such block triangular matrices remain block triangular, and the eigenvalues of the multiplication by $h(\bar{x})f(\bar{x})g(\bar{x})$ are the products $h(\alpha)f(\alpha)g(\alpha)$ with multiplicity e_α. $\square$

We observe that right hand side of the equation in the theorem above is a symmetric function of the coordinates of the C points of the variety, and hence lies in the field of coefficients K.

Suppose now that $B = \{\omega_0, \omega_1, \ldots, \omega_{p-1}\}$ is a monomial basis for the K-vector space A. We may always assume: $\omega_0 = 1$. Let us denote by $B_h(A/B)$ the symmetric matrix associated to the bilinear form B_h in the basis B. Its components are

$$(B_h)_{i,j} = Tr(m_{h\omega_i\omega_j}) = \sum_{\alpha \in V_C} e_\alpha h(\alpha)\omega_i(\alpha)\omega_j(\alpha).$$

We may proceed just as in the univariate case. Let

$$W = \begin{bmatrix} 1 & 1 & \cdots & 1 \\ \omega_1(\alpha_1) & \omega_1(\alpha_2) & \cdots & \omega_1(\alpha_p) \\ \omega_2(\alpha_1) & \omega_2(\alpha_2) & \cdots & \omega_2(\alpha_p) \\ \vdots & \vdots & & \vdots \\ \omega_{p-1}(\alpha_1) & \omega_{p-1}(\alpha_2) & \cdots & \omega_{p-1}(\alpha_p) \end{bmatrix},$$

and $\Delta_h = diag[h(\alpha_1), h(\alpha_2), \ldots, h(\alpha_p)]$. Then $Q_h = W\Delta_h W^t$. It is understood here that the point α_i is taken e_{a_i} times, so that the form has degeneracies exactly corresponding to the multiplicities of the variety.

The same argument that was used in the univariate case now applies. If we restrict the inputs I to be real, then the zeros occur in conjugate pairs, the trace form is a real form, and the rank and signature information are as claimed in the main theorem. This concludes the proof of the main theorem. □

Signature calculations. The quadratic form Q_h is given by the symmetric matrix $B_h(A/B)$. The signature of Q_h is equal to the difference between positive eigenvalues and negative eigenvalues of $B_h(A/B)$. The characteristic polynomial of $B_h(A/B)$ has all its roots real, so we can apply the following proposition, which is an application of Descartes' rule:

Proposition 2.8 *([11]) Let S be a real symetric $p \times p$ matrix; and $\chi_S(\lambda) = det(S - \lambda I)$ be the characteristic polynomial of S. Denote:*

$$\chi_S(\lambda) = a_0 + a_1\lambda + \cdots + a_{p-1}\lambda^{p-1} + (-1)^p \lambda^p,$$
$$D(S) = [a_0, a_1, \cdots, a_{p-1}, (-1)^p].$$

The number of sign variations in the sequence $D(S)$ is equal to the number of positive eigenvalues of S.

So, in order to evaluate $\sigma(Q_h)$, we need only , if $B_h(A/B)$ is known, to compute its characteristic polynomial.

3 Quadratic form calculations using Gröbner bases.

As we have just seen, in order to calculate the signatures $\sigma(Q_h)$, we must first know how to calculate traces of multiplication operators in $A = K[\bar{x}]/I$. We can do this using Gröbner bases [5].

Definition 3.1 *An admissible ordering of monomials is any ordering $<$ such that $\omega < \omega' \implies \eta\omega < \eta\omega'$, for any three monomials ω, ω', η, and such that $1 < \omega$ for any other monomial ω (so that $<$is a total ordering).*

Definition 3.2 *For any subset $S \subset K[\bar{x}]$, let $Init(S)$ denote the set of initial terms of S with respect to some admissible ordering of monomials, that is, extract the highest ordered monomial of each polynomial appearing in S.*

Definition 3.3 *A Gröbner basis of a polynomial ideal I is any set G of generators for I with the property that $Init(I) = (Init(G))$, where the right hand side refers to the ideal generated by the elements of $Init(G)$.*

Definition 3.4 *A polynomial P is reduced with respect to a Gröbner basis if and only if no monomial of P is multiple of an initial monomial in the Gröbner basis.*

Proposition 3.5 *For any polynomial P, there exists a polynomial P_1, reduced with respect to the Gröbner basis, such that $P - P_1 \in I$. The polynomial P_1 is called the reduction with respect to the Gröbner basis of the polynomial P.*

Theorem 3.6 *With respect to a Gröbner basis reductions are canonical, i.e. all the polynomials in any coset of I in $K[\bar{x}]/I$ have the same reduction.*

Proof of theorem 3.6   See [5]. □

This property makes it possible to calculate within the quotient rather easily. If $[f]$ denotes the reduction of f with respect to the Gröbner basis, then $[f+g] = [f] + [g]$, and $[fg] = [[f][g]]$.

Theorem 3.7 *If the ideal I is zero dimensional, then a basis for the vector space $K[\bar{x}]/I$ over K is provided by the set of monomials which appear earlier in the admissible ordering than any leading monomial of the Gröbner basis.*

Proof of theorem 3.7   See [5]. □

This collection of monomials is generally referred to as the *Gröbner staircase* because if one plots the exponent vectors of the leading monomials of the basis on an integer lattice, then the set of multiples, *i.e.* higher ordered, of these monomials correspond to positive orthants attached at the plotted points, and the finite residual subset of lattice points which are not multiples of any leading monomial of the basis forms a small staircase-like subset near the origin. The structure of this staircase will be of considerable interest to us.

Let $B = \{\omega_0, \omega_1, \ldots, \omega_{p-1}\}$ be the monomial basis determined by the Gröbner staircase. We will express the computation of the matrix components of the symmetric form B_h for any $h \in R[\bar{x}]$ in terms of the multiplication tables $\omega_i \omega_j$, $i, j \in [0..p-1]$, $x_l \omega_j$ $l \in [1..n]$, $j \in [0..p-1]$, and $x_l x_m$ $l, m \in [1..n]$. These tables are determined by the reductions with respect to the Gröbner basis of the corresponding monomials. The matrix of m_{ω_i} with respect to the basis B may be read off from the i-th row of the $\omega_i \omega_j$ table. Likewise, the matrix for m_{x_l} may be read off from the $x_l \omega_j$ table.

So that, knowing the K-algebra structure of A through the multiplication tables of monomials in a Gröbner staircase, we are able to compute the matrix $B_h(A/B)$ and the signature of Q_h .

4 Counting real points inside a real semi algebraic constraint region

Characteristic functions.

We can proceed in the following way to calculate the number of real points of a discrete variety which lie within any real algebraic constraint region. Suppose

$$P = \{\bar{x} \in V_R(I) \; : \; h_1(\bar{x}) > 0, h_2(\bar{x}) > 0, \ldots, h_k(\bar{x}) > 0\},$$

The characteristic function of P (in the case where the h_i have no zeros on $V_R(I)$) may be expressed as

$$\chi_P(\bar{x}) = 2^{-k} \prod_{i=1}^{k}(1 + sgn(h_i(\bar{x}))),$$

and if we expand this expression the result is

$$2^{-k} \sum_{J \subset [1..p]} \prod_{i \in J} sgn(h_i(\bar{x})).$$

Denote by $h_J(\bar{x})$: $h_J(\bar{x}) = \prod_{i \in J} h_i(\bar{x})$, then

$$\#(P) = \sum_{\alpha \in V_R(I)} \chi_P(\alpha) = 2^{-k} \sum_{J \subset [1..k]} \sum_{\alpha \in V_R(I)} sgn(h_J(\alpha)).$$

For any $J \subset [1..p]$,

$$\sum_{\alpha \in V_R(I)} sgn(h_I(\alpha)) = \#\{\alpha \in V_R(I) \; : \; h_I(\alpha) > 0\} - \#\{\alpha \in V_R(I) \; : \; h_I(\alpha) < 0\}$$

may be evaluated as $\sigma(Q_{h_I})$, and we are done. The same expression appears in [B 91], where it is considered as a "Pfister form" $<< h_1, h_2, \ldots, h_l >>$.

The characteristic function is an inefficient way to count the points, since it is exponential in the number k of constraint polynomials. But, there are numerous linear dependencies among the summands. These dependencies make it possible to do the calculation, indeed to calculate the number of real points within *every* sign-component defined by the h_i using a number of steps which depends on the total number of real points and is polynomial in the number of constraints. We turn to this question in the next section.

The Algorithm of Ben-Or, Kozen, and Reif.

Consider a family of polynomials in n variables $H = [h_1, \ldots, h_k]$, then the *sign-components* determined by H are the non-empty subsets of R^n determined by some conjunction of sign conditions

$$h_1(\bar{x}) \, \varepsilon_1 \, \wedge \, h_2(\bar{x}) \, \varepsilon_2 \, \ldots \, h_k(\bar{x}) \, \varepsilon_k,$$

where the ε_i are drawn from the set of relations $\{< 0, = 0, > 0\}$.

As in the univariate case, it is possible to make an efficient parallel calculation of exactly how many real zeros of I lie in each such component using the algorithm of Ben-Or, Kozen, and Reif (BKR for short), which we proceed to describe. This algorithm has been extensively described in various articles, see for example [4] [6] [12] ... The first idea is that there is a linear relation (*)

$$\begin{bmatrix} 1 & 1 & 1 \\ 0 & 1 & -1 \\ 0 & 1 & 1 \end{bmatrix} \begin{bmatrix} c_{=0}(F,h) \\ c_{>0}(F,h) \\ c_{<0}(F,h) \end{bmatrix} = \begin{bmatrix} \sigma(1) \\ \sigma(Q_h) \\ \sigma(Q_{h^2}) \end{bmatrix}.$$

The cardinalities on the left hand side refer to the number of real zeros of the family $F = (f_1, \ldots, f_s)$ where $h = 0$, $h > 0$, etc. The first row merely restates the obvious fact that the conditions $h = 0$, $h > 0$, and $h < 0$ are exhaustive; the second row expresses the content of our main theorem; and the last row follows from the observation that h^2 is positive and application of the main theorem.

Two such identities may be combined as follows, with the help of a tensor product

$$Ac = s, \quad A'c' = s' \implies (A \otimes A')(c \otimes c') = s \otimes s'.$$

The vector $c \otimes c'$ has components which count the number of zeros satisfying the conjunction of sign-conditions appearing in c and c', and the vector $s \otimes s'$ has components which are signatures of the corresponding products of polynomials from s and s'. Applied naively, this construction generates matrices which grow as 3^{2^l} after l combining steps. If there are k constraints to combine, then, after $\log(k)$ steps, one has a matrix of size 3^k, and nothing has been gained.

The second important idea in the BKR algorithm is that one may reduce the size of the tensored matrices after each step, because all the components of $c \otimes c'$ which are equal to 0 correspond to conjunctions which cannot already be verified by any point, hence there is no point pursuing them any further: those branches should be pruned from the computation tree. One deletes the columns of $A \otimes A'$ corresponding to such components, extracts a full rank square submatrix from the result, and then continues to operate with this reduced matrix. The result is matrices that never exceed $\#V_R$ in size.

These two ideas together produce a parallel algorithm which starts with $2k + 1$ signature calculations: $\sigma(1)$, and $\sigma(Q_{h_j}), \sigma(Q_{h_j^2})$ for $j = 1, \ldots, k$, and then simultaneously combines these pairwise in $\log(k)$ levels using signature calculations for some useful subset of the set of all possible products $\prod_{i \in I} h_i$, $I \subset [1..k]$. Each node of the computation tree involves the calculation of no more than $(\#V_R)^2$ new signatures before the current reduction (some of the signatures being already calculated). There are at most $2l - 1$ nodes in any binary tree with l leaves, so the total cost of the algorithm is $O(kr^2)$, where $r = \#V_R$. In fact being more precise it is possible to see that we are reduced at each node of the tree to signature calculations of the Hermite quadratic forms for products of at most $\log(r)$ polynomials h_j or squares of h_j.

5   Complexity

We shall use the following notations:

- $F = \{f_1, f_2, \cdots, f_s\}$ is a set of multivariate polynomials in $C[\bar{x}]$, with integer coefficients, which defines the ideal I

- $A = C[\bar{x}]/I$, which is of finite dimension p over C.

- $V_R(F) = \{x \in R^n, \ f_i(x) = 0, \ 1 \le i \le s\}$

- $r = \#(V_R(F))$

- $H = \{h_1, h_2, \cdots, h_k\}$ is a set of multivariate polynomials in $C[\bar{x}]$, with integer coefficients;

- for $\varepsilon = (\varepsilon_1, \cdots, \varepsilon_k) \in \{< 0, = 0, > 0\}^k$, $c_\varepsilon(F, H)$ will be define by:
  $c_\varepsilon(F, H) = \#(\{x \in V_R(F), h_i \varepsilon_i, \ for \ 1 \le i \le k\})$

In this section, we compute the complexity of the algorithm which gives the $c_\varepsilon(F, H)$ for any ε in $\{< 0, = 0, > 0\}^k$ in terms of complexity parameters which in a certain sense are "canonical" with respect to the sets F and H. In order to apply the BKR method, we have to determine, for each ε element of $\{< 0, = 0, > 0\}^k$, the number $c_\varepsilon(F, h)$, for some products h of elements in H. This will be done by applying the main theorem and using relation (*).

For any multivariate polynomial h, if $B_h : A \otimes A \to A$ is the bilinear form defined by $B_h(f, g) = trace(m_{fhg})$, where m_{fhg} is the matrix of the multiplication operator induced by fhg in A, with respect to a basis B, denote by $B_h(A/B)$ the matrix of this bilinear form with respect to the basis B.

Monomials and multiplication tables.

As a preprocessing, we construct a convenient basis $B = \{\omega_0, \omega_1, \cdots, \omega_{p-1}\}$ of linearly independent monomials of A deduced from a Gröbner basis of I. We compute also the multiplication table $MT(A, B)$ which gives the products, mod

I, $\omega_i\omega_j$, x_lx_m, $x_l\omega_i$ for any i and j in $\{0, \cdots, p-1\}$ and any l and m in $\{1, \cdots, n\}$ (i.e $MT(A, B)$ gives the coefficients of each of these products in A with respect to B). Note that $MT(A, B)$ is a triple index table; it is a matrix whose entries are vectors in Z^p, namely the coordinates in B of each basic product $\omega_i\omega_j$, x_lx_m and $x_l\omega_i$, for i and j in $\{0, \cdots, p-1\}$ and l and m in $\{1, \cdots, n\}$.

Then the matrices (with respect to B) of the operators m_{ω_i} are known for any i in $\{0, \cdots, p-1\}$; and the matrices (with respect to B) of the operators m_{x^α} are easy to compute for any monomial x^α.

Remark 1 . — If the entries of $MT(A, B)$ are not in Z (and this can happen if the leading monomials of the elements of the Gröbner basis have not their coefficients equal to 1), a change of variables has to be done: this will be a part of the preprocessing.

Remark 2 . — Since we are in a 0-dimensional situation, the complexity of this preprocessing is polynomial in the degree d and the number s of input polynomials and singly exponential in the number n of variables. More precisely, when the ideal I is 0-dimensional projective, or when a probabilistic method is used, the complexity of this preprocessing is $s^{0(1)}(dn)^{O(n)}$, its complexity for a deterministic algorithm in the affine 0-dimensional case is $s^{0(1)}(dn)^{O(n^2)}$ (for all this see [7]and [10]) . However it is well known from experiments that the size of the output of a Gröbner basis computation may be quite small compared to the intermediate calculations needed to compute it. It is only the size of this output that we need for our quadratic form and signature computations and we shall express our complexity results in terms of parameters describing the size of the Gröbner basis.

The complexity parameters we will consider as describing the size of the Gröbner basis are p, the dimension of the vector space A and c the length of the entries in $MT(A, B)$

The first part of this section is first devoted to the determination of the matrices m_{h_j} for any j in $\{1, \cdots, k\}$, with an evaluation of the complexity of this computation; then to the computation of the matrices m_h for h any product of elements in H.

In the second part of this section, we will apply the result of this first part in order to determine the matrix of B_h (and its trace), for any product h of elements in H, with the complexity of the computation.

Then the quadratic form Q_h is determined and we can compute the complexity of the computation of its signature $\sigma(Q_h)$.

In conclusion, we will summarize the preceeding parts by giving the algorithm which leads to $\sigma(Q_h)$, for any product h of elements in H, and giving its general complexity.

Part 1: Matrices of multiplication by polynomials.

Remark . — We will use here the following result: "if f and g are two multivariate polynomials, $m_f m_g = m_{fg}$ and $m_f + m_g = m_{f+g}$" in two different situations, which will cover the two cases we need in our context: the case "element of H" and the case "product of elements in H".

1^{rst} case: multivariate polynomial given by its canonical representation.

Let h be a multivariate polynomial. We say that h is known by its canonical representation when we write: $h(x) = \sum_{\alpha \in \mathcal{A}} a_\alpha x^\alpha$, with:

$$\alpha = (\alpha_1, \cdots, \alpha_n), |\alpha| = \sum_{i=1}^{n} \alpha_i, x^\alpha = \prod_{i=1}^{n} x_i^{\alpha_i}.$$

and $\mathcal{A} \subset N^n$ is the index set for the monomials in the canonical representation of h.

In this case, in order to compute m_h, we use:

$$m_h = \sum_{\alpha \in \mathcal{A}} a_\alpha \left(\prod_{i=1}^{n} m_{x_i}^{\alpha_i} \right)$$

The matrices m_{x_i} for $1 \leq i \leq n$ are known as soon as $MT(A, B)$ is known without any further computation.

So the natural complexity parameters for h in this computation, where h is given by its canonical representation are the following:

- the degree $d(h)$ of h

- $t(h) = \#(\mathcal{A})$ ($t(h)$ is the number of monomials in h)

- $c(h) = \left(\sum_{\alpha \in \mathcal{A}} a_\alpha^2 \right)^{\frac{1}{2}}$ ($c(h)$ is called the size of the polynomial h).

With these definitions, we have the

Proposition 5.1 *The complexity of the computation of m_h is polynomial in n, p, c, $d(h)$, $c(h)$, $t(h)$.*

The length of its entries is in $O\left(t(h) + c(h) + d(c + p) + np\right)$.

Proof We have the following lemma:

Lemma 5.2 *The length of the entries in $m_{x_i}^{\alpha_i}$ ($1 \leq i \leq n$) is in $O(\alpha_i(c+p))$. The complexity of its computation is in $O\left(\alpha_i p^3 (\alpha_i(c+p))^2\right)$.*

Proof of lemma 5.2 m_{x_i} is a $(p \times p)$ matrix; denote by $c_{l-1}(x_i)$ a bound for the length of the entries in $m_{x_i}^{l-1}$. We have: $c_l(x_i) = c_{l-1}(x_i) + c + p$; from which we obtain the first part of lemma. To compute $m_{x_i}^{\alpha_i}$, we need at most $\alpha_i p^3$ elementary arithmetical operations; this leads to the second part of lemma.

We deduce then:

- *The computation of each $a_\alpha m_{x^\alpha}$ (knowing each $m_{x_i}^{\alpha_i}$ for $1 \leq i \leq n$) is in*
$$O\left(dp^3\left(c(h) + d(c+p) + np\right)^2\right);$$

- *the length of its entries is in $O\left(c(h) + d(c+p) + np\right)$.*

Knowing each $m_{x_i}^{\alpha_i}$ for $1 \leq i \leq n$, we have to make at most $p^2 + np^3$ elementary arithmetical operations, so the number of elementary arithmetical operations in this computation is in $O\left(np^3\right)$.

Suppose that the length of the entries in $m_{x_1}^{\alpha_1}, \cdots, m_{x_{j-1}}^{\alpha_{j-1}}$ is in $O(J)$, then the length of the entries in $m_{x_1}^{\alpha_1}, \cdots, m_{x_j}^{\alpha_j}$ is in $O(J + \alpha_i(c+p) + p)$. Hence, the length of the entries in m_{x^α} is in $O\left(d(c+p) + np\right)$ and the length of the entries in $a_\alpha m_{x^\alpha}$ is in $O\left(c(h) + d(c+p) + np\right)$. So we get lemma 5.2.□

To obtain the complexity of the computation of m_h, we need to add the complexities found in lemma 5.2 to the complexity of the computation of $\sum_{\alpha \in \mathcal{A}} a_\alpha m_{x^\alpha}$, knowing each $a_\alpha m_{x^\alpha}$. But this leads to $t(h)p^3$ elementary arithmetical operations, with entries of length in $O\left(t(h) + c(h) + d(c+p) + np\right)$.

This ends the proof of proposition 5.1.□

2nd case: multivariate polynomials given by a product of canonically represented polynomials.

Let h be a product of l elements in H, i.e $h = \prod_{j=1}^{l} h_j$ with $h_j \in H$.

Note that a polynomial h_j can appear several times in the product (never, once or twice);

Denote by:

- $t(H)$ the first integer bigger than sup $\{t(h_j), 1 \le j \le k\}$;

- $c(H)$ the first integer bigger than sup $\{c(h_j), 1 \le j \le k\}$;

- $d(H)$ the first integer bigger than sup $\{degree(h_j), 1 \le j \le k\}$.

Remark . — Applying the proposition 5.1, with the above notations, we deduce that the length of the entries in m_{s_j} for $1 \le j \le l$ is in :

$$O\left(t(H) + c(H) + d(H)(c + p) + np\right)$$

We have: $m_h = \displaystyle\prod_{j=1}^{l} m_{h_j}$, so we get the

Proposition 5.3 *Knowing each m_{h_j} for $1 \le j \le l$, the complexity of the computation of m_h is polynomial in l and p. The length of its entries is in $O\left(l(t(H) + c(H) + d(H)(c + p) + np)\right)$.*

Proof of proposition 5.3 It is a direct consequence of proposition 5.1, knowing that a product of $(p \times p)$ matrices needs $O\left(p^3\right)$ elementary arithmetical operations.□

From proposition 5.3, we deduce the

Proposition 5.4

The complexity of the computation of m_h, where h is a product of l elements of H, is polynomial in n, p, c, $d(H)$, $c(H)$, $t(H)$ and l.

The length of its entries is in $O\left(l(t(H) + c(H) + d(H)(c + p) + np)\right)$.

Part 2: The quadratic form and its signature.

Let h be a product of l elements in H.

In order to compute the bilinear form B_h, i.e its matrix $B_h(A/B)$ relative to the basis B of A, if m_h is already determined, we have only to compute $trace(m_{\omega_i h \omega_j})$ for any (i, j) in $\{0, \cdots, p - 1\}$.

m_{ω_i} is known without any further computation by the knowledge of $MT(A, B)$. So, we get the

Proposition 5.5

The complexity of the computation of $m_{\omega_i h \omega_j}$ for any (i, j) in $\{0, \cdots, p - 1\}$ is polynomial in c, n, p, $c(H)$, l, and $t(H)$; the length of its entries is in $O\left(l(t(H) + c(H) + d(H)(c + p) + np)\right)$

Proof It is clear by applying proposition 5.3.□

We have then as an immediate consequence the

Proposition 5.6 *The complexity of the computation of $B_h(A/B)$ is polynomial in c, n, p, $c(H)$, l, and $t(H)$.*

The length of its entries is in $O\left(l(t(H) + c(H) + d(H)(c + p) + np)\right)$

We deduce the

Proposition 5.7 *When $B_h(A/B)$ is given, the complexity of the computation of $\sigma(Q_h)$ (the signature of Q_h) is polynomial in c, n, p, $t(H)$, $c(H)$, $d(H)$, and l.*

The length of the integers in the computation is in:
$$O\left(p^2 l(t(H) + c(H) + d(H)(c + p) + np)\right).$$

Proof The complexity of the computation of the characteristic polynomial of $B_h(A/B)$ (as well as of its rank) is polynomial in p and in the length of its entries.□

Part 3: The numbers $c_\varepsilon(F, H)$ and CRZ- algorithm.

We recall first the definition of the complexity parameters, and describe the CRZ- algorithm (for "Counting Real Zeros" algorithm) whose output is the set $CRZ(F, H) = \left\{ c_\varepsilon(F, H), \ \varepsilon \in \{< 0, = 0, > 0\}^k \right\}$ giving the numbers $c_\varepsilon(F, H)$ for any ε in $\{< 0, = 0, > 0\}^k$ for which $c_\varepsilon(F, H)$ is not equal to 0. Note that $c_\varepsilon(F, H)$ is the number of points in V which verify the condition $(H\varepsilon) = (\forall i \in \{1, .., k\}, \ h_i \varepsilon_i)$.

Complexity parameters.

- c is the first integer bigger than the length of all the entries of $MT(A, B)$;

- p is the dimension of A;

- r $(r \leq p)$ is the number of distinct points in V;

- n is the number of variables;

- $d(H)$ is equal to $sup\{degree(h_i), \ for \ 1 \leq i \leq k\}$;

- $c(H)$ is the first integer bigger than $sup\{c(h_i), \ for \ 1 \leq i \leq k\}$ (where $c(h_i)$ is the size of the polynomial h_i for $1 \leq i \leq k$);

- $t(H)$ is equal to $sup\{t(h_i),\ for\ 1 \leq i \leq k\}$ (where $t(h_i)$ is the number of monomials in the polynomial h_i for $1 \leq i \leq k$);

- k is equal to the number of polynomials in H.

Preparatory material $PM(F, H)$.

Definition 5.8 $PM(F, H) = \left\{m_{h_i},\ 1 \leq i \leq k\right\}$

The complexity of the computation of $PM(F, H)$ is in:

$O\left(k(t(H) + c(H) + d(H)(c + p) + np)\right)$

We have now all the tools for the computation of the numbers $c_\varepsilon(F, H)$.

Description of CRZ-algorithm.

First step: computation of $PM(F, H)$;

Second step: divided in $\log(k)$ steps, namely, for $1 \leq j < k$, knowing the $CRZ(F, H_J)$, J having j elements, compute the $CRZ(F, H_{J'})$, J' having j' elements $(J \subset J')$.

Remark . — For each of these $\log(k)$ steps we need to compute at most r^2 numbers $\sigma(Q_h)$, where h is a product of at most $\log(r)$ elements of H.

We have proved the theorem:

Theorem 5.9 *CRZ-algorithm has a complexity polynomial in n, c, p, $d(H)$, $c(H)$, $t(H)$, r and k.*

6 Calculating the Euler characteristic of a smooth hypersurface

The Euler characteristic may be determined using Morse theory. Suppose X_n is a Morse function on the smooth compact hypersurface V defined by an equation $f = 0$ in n variables, then the topological type of V depends on the indices of V at the critical points of X_n on V. At each critical point α of X_n of index k, one shows that, for any $\varepsilon > 0$ sufficiently small, the topological type of the level set $X_n^{-1}(\alpha + \varepsilon)$ is obtained from the topological type of the level set $X_n^{-1}(\alpha - \varepsilon)$ by adding a handle which is topologically a k-disc.

From this, one can obtain that the Euler characteristic of the set V is the alternated sum of the indices of V at the critical points of X_n.

More precisely, let $\mathcal{C}$ be the set of critical points of X_n on V. Let α be an element of $\mathcal{C}$, i_a the index of V at a that is $n - \sigma(f, \alpha)$ where $\sigma(f, \alpha)$ is the signature of the quadratic form with coefficients $\frac{\partial^2 f}{\partial X_i X_j}(\alpha)$. Then the Euler characteristic of V is $\sum_{\alpha \in \mathcal{C}} (-1)^{i_\alpha}$.

This computation can be done by applying our algorithm in the following particular case. The set $\mathcal{C}$ of critical points is defined by the 0-dimensional ideal $(f, (\frac{\partial f}{\partial X_i})_{i=1,\dots,n-1})$.

The list of polynomials whose signs need to be evaluated at the points of $\mathcal{C}$ are the principal minors of the $(n-1, n-1)$ matrix $M_{i,j} = \frac{\partial^2 f}{\partial X_i X_j}$. So, we get for each possible value of the index i the number of points of $\mathcal{C}$ with this index.

This leads to a single exponential computation (in terms of the dimension n) of the Euler characteristic.

Conclusion

An alternative for computing Hermite quadratic forms in the multivariate case appears in [11], where computations of multivariate symmetric functions were performed. The method we present here is more general since there is no need for the number of equations to be equal to the number of variables. The multivariate symmetric functions computations are replaced by traces of multiplication endomorphisms, that is by simple linear algebra.

Our complexity results are satisfactory since they are polynomial in the input size of a presentation of the quotient ring A: the dimension of A as a vector space and the coefficients size of the matrices of multiplication by monomials. If we have the point of view: "efficient computations", the algorithm is promising since it is obtained by a very easy computation from a Gröbner basis (for any compatible order).

References

[1] Atiyah, M. F., and I. G. Macdonald, *Introduction to Commutative Algebra*, (1969), Addison-Wesley, Reading, Massachusetts.

[2] Becker, E., *Sums of Squares and Trace forms in Real Algebraic Geometry*, Cahiers du Séminaire d'Histoire des Mathématiques, 2ème série **Vol 1** (1991), Université Pierre et Marie Curie.,

[3] Becker, E., Wörmann T., *On the trace formula for quadratic forms and some applications*, to appear in Proceedings of the Special Year in Real Algebraic Geometry and Quadratic Forms, University of Berkeley.

[4] Ben-Or M., Kozen D. , Reif J., *The complexity of elementary algebra and geometry*, J. of Computation and Systems **32** (1986), 251–264.

[5] Buchberger, B., *Gröbner: An algorithmic method in polynomial ideal theory, in Multidimensional Systems Theory, chapter 6*, (1985), N. K. Bose Ed., D. Reidel.

[6] Cucker F., Lanneau H., Mishra B., Pedersen P., Roy M.-F., *Real algebraic numbers are in NC*, To appear in Applicable Algebra in Engineering, Communication and Computing.

[7] Dickenstein A., Fitchas N., Giusti M., Sessa C., *The membership problem for unmixed polynomial ideal is solvable in single exponential time*, (1987), AAECC Toulouse.

[8] Hermite C., *Remarques sur le théorème de Sturm*, C. R. Acad. sci. Paris **36** (1853), 52–54.

[9] Hermite C., *Sur l'extension du théorème de M. Sturm à un système d'équations simultanées*, Oeuvres de Charles Hermite, Tome 3, (1969), 1–34.

[10] Lakshman Y. N., Lazard D., *On the complexity of zero-dimensional algebraic systems*, Proceedings of MEGA 90, Birkhäuser (1991), 217–226.

[11] Pedersen, P., *Counting Real Zeros, Thesis*, (1991), Courant Institute, New York University.

[12] Roy M.-F., Szpirglas A., *Complexity of computations with real algebraic numbers*, J. of Symbolic Computation **10** (1990), 39–51.

[13] Scheja, G. and U. Storch, *Lehrbuch der Algebra*, B. G. Teubner, Stuttgart (1988), **Band 2.**

P. Pedersen

Departement of Mathematics

Cornell University

Ithaca, New-York 14B53-7901, USA

Marie-Françoise Roy
IRMAR
Université de Rennes I
Campus de Beaulieu
35042 Rennes Cedex FRANCE

Aviva Szpirglas
CNRS URA 742
Université Paris-Nord
Avenue Jean-Baptiste Clément
93430 Villetaneuse FRANCE

Finding the number of distinct real roots of sparse polynomials of the form $p(x, x^n)$

D. Richardson

Abstract. Cylindrical decomposition and false derivatives are used to find the number of distinct real solutions of a polynomial with integral coefficients $p(x, x^n)$, where n is large. The computation time depends only polynomially on the *size* of the problem, where this is defined to be the maximum of d, $log(n)$ and $log(C)$, where d is the total degree of $p(x, y)$, assumed to be small relative to n, and C is the maximum of the absolute values of the coefficients.

Introduction

Let $p(x, y)$ be a polynomial with total degree d and integral coefficients whose absolute value is bounded by C. Define the *size* of $p(x, x^n)$ to be the maximum of d, $log(n)$ and $log(C)$. A method is given below to find the number of distinct real solutions of $p(x, x^n) = 0$ in a time which depends polynomially on the size. It is thought that this algorithm will be useful when the size is not very large, but n is large. So if polynomial $q(x)$ is sparse and of high degree, the algorithm could be used if we could find a polynomial of small degree, $p(x, y)$, so that $p(x, x^n) = q(x)$ for some n.

Suppose $p(x, y)$ has been given. Assume $p(x, x^n)$ is not identically zero. (This can be checked in a time polynomial in size.) We can factorize $p(x, y)$ in a time polynomial in size, and apply our algorithm to the factors. We can therefore assume that $p(x, y)$ is irreducible.

Let the interval in which we are looking for roots of $p(x, x^n)$ be called the domain of the problem. Following a suggestion of an anonymous referee, we can make changes of variable

$$x \to -x$$

$$x \to (1/x)$$

$$x \rightarrow (-1/x)$$

which will reduce our original problem with domain $(-\infty, \infty)$ to four problems with domain $[1, \infty)$. After clearing of denominators the new polynomials have size at most double that of the original polynomial.

¿From now on, we are trying to find roots of $p(x, x^n)$ in $[1, \infty)$. Once again, the intention is to decompose the domain into finitely many subdomains in which the problem is, in some sense, simple.

Let $C = \{(x, y) \mid p(x, y) = 0 \land x \geq 1 \land y \geq 1\}$.

Let $(\gamma_1, ..., \gamma_k)$ be a list of points in $[1, \infty)$. We will say this is a complete decomposition for $p(x, x^n)$ if the following four conditions hold.

1) $1 = \gamma_1 < \gamma_2 < ... < \gamma_k = \infty$.

2) In each subinterval (γ_i, γ_{i+1}), C consists of finitely many non intersecting graphs of functions $y_{i1}(x), ..., y_{in_i}(x)$ which are all continuous and totally defined over (γ_i, γ_{i+1}). These will be called the branches of C over the interval.

3) For each branch of C, $y(x)$, we have $p_y(x, y(x)) \neq 0$ on (γ_i, γ_{i+1}).

4) For each branch of C, $y(x)$, there is at most one point w in $[\gamma_i, \gamma_{i+1}]$ so that $w^n = y(w)$; and if there is such a point in (γ_i, γ_{i+1}), i.e. in the interior of the interval, the tangents to the two curves at the point of intersection are not equal.

The roots of $p(x, x^n)$ can now be found in two stages.

I) Decomposition: Find a complete decomposition $(\gamma_1, ..., \gamma_k)$ for $p(x, x^n)$ in $[1, \infty)$.

II) Location: Locate the graph of $y = x^n$ relative to C. This may be done as follows. For each point γ in the decomposition, find the set of points $Y_\gamma = \{(\gamma, y) \mid p(\gamma, y) = 0 \land y \geq 1\}$. Order Y_γ by the size of the y coordinate, and place (γ, γ^n) correctly in this ordered set. We can compute, using [Roy 1987], how many connected curves there are in C above each interval (γ_i, γ_{i+1}), and we can discover how the points in adjacent Y_γ sets are connected to each other by these curves. Since the graph of $y = x^n$ is guaranteed not to cross any of the branches more than once, we can then find how many intersections there are in each interval. Thus we can find the number of distinct real roots.

The main idea which is used in stage I) of this algorithm is the false derivative. This is defined as follows.

Let $f(x)$ be a differentiable function and let $f*(x)$ be a continuous function on an interval I. We will say that $f*(x)$ is a *false derivative* of $f(x)$ on I if the sign of $f*(x)$, (either negative, zero, or positive), is the same as the sign of $f'(x)$ at every root of $f(x)$ on I.

It is shown below that such a false derivative $f*(x)$, although it may be simpler than a true derivative, has the useful property that its roots separate the roots of the original function $f(x)$. Therefore false derivatives can be used for root finding. False derivative techniques have been developed to find roots and solve other problems for non algebraic functions containing terms, such as e^x, defined by first order algebraic differential equations. See [Richardson, 1991]. It seems that these techniques can also apply to algebraic functions, in this case by regarding x^n as a solution of $y' = ny/x$, rather than as the result of n-fold multiplication.

In the second stage of the algorithm we have real algebraic numbers α and β, both greater than one, and of fairly small degree and height, and we wish to decide between the three possibilities, $\alpha^n > \beta$, $\alpha^n = \beta$, and $\alpha^n < \beta$, where n is relatively large. We do this by comparing $nlog(\alpha)$ and $log(\beta)$, and applying a beautiful result of Mignotte and Waldschmidt about logarithmic forms.

After the example below, the next two sections describe the two stages of the algorithm in turn.

Example

Let $p(x, x^n) = x^{300} - 10x^{203} + 15x^{104} - x^6 + 3$.

We wish to find roots of $p(x, x^n)$ for $x \geq 1$. Taking $n = 100$, replace x^n by y to get $p(x, y) = y^3 - 10y^2x^3 + 15yx^4 - x^6 + 3$.

As is usual in cylindrical algebraic decomposition, we first eliminate y between $p(x, y)$ and $p_y(x, y)$. Let $r_1(x) = resultant(p, p_y, y)$. The degree of $r_1(x)$ is 15.

$$r_1(x) = 4000x^{15} + ... + 243$$

The real roots of $r_1(x)$ are [-0.83739..., 0.824871..., 5.638864...]. So far we have $(\alpha_1, \alpha_2, \alpha_3) = (1, 5.638864..., \infty)$

On each of the two subintervals C is a stack of non intersecting graphs. Since 5^{100} is so large, only the first interval, (α_1, α_2) is interesting.

We need to consider the possibility that one of the branches of C has more than one intersection with $y = x^n$. Suppose $y(x)$ were such a branch.

Let $f(x) = y(x) - x^n$. Then $f * (x) = y'(x) - ny(x)/x$ is a false derivative for $f(x)$ in the interval (α_1, α_2). Roots of $f(x)$ are separated by roots of $f * (x)$. To find roots of $f * (x)$, we need to eliminate y between $f * (x)$ and $p(x, y)$.

Let $r_2(x) = resultant(p(x, y), xp_x(x, y) + 100yp_y(x, y), y)$. The degree of $r_2(x)$ is 21.

$$r_2(x) = 3764473000x^{21} + ... + 729000000$$

The real roots of $r_2(x)$ are [-1.21285..., -0.8375..., 0.8249..., 1.212927..., 5.6387...]. We are interested to see what happens just above 1, so the only important value here is 1.212927.... Call this γ_2 and let $\gamma_1 = 1$. It is clear that γ_2^{100} is larger than any possible solution of $p(\gamma_2, y) = 0$. Thus $(\gamma_1, \gamma_2, \alpha_2, \infty)$ is a complete decomposition.

The roots of $p(1, y)$ are [-0.1231..., 2.0, 8.123...] and the roots of $p(\gamma_2, y)$ are [0.0057..., 2.0498..., 15.78897...]. So there are two crossings between γ_1 and

γ_2, and thus $p(x, x^n)$ has two roots above in $[1, 1.212927...]$. After successive bisections of this interval it is found that 1.02 separates the two roots.

The resultants and the real roots were found using the computer algebra system Axiom. It took a few minutes interactively.

Axiom was also used directly to solve the original problem, and after about an hour the roots $[1.00694..., 1.02197...]$ were found.

Decomposition of Real Line

In this section we find a complete decomposition for $p(x, x^n)$.

Assume $p(x, x^n)$ is irreducible. Factorization can be done in a time polynomial in the size.

First find $(\alpha_1, ..., \alpha_j)$ with $1 = \alpha_1 < ... < \alpha_j = \infty$ and $\{\alpha_2, ..., \alpha_{j-1}\}$ is the set of roots of $r_1(x) = resultant(p, p_y, y)$ for $x > 1$. In each subinterval between these points the zero set of $p(x, y)$ is a stack of non intersecting implicitly defined graphs of functions. Each of the functions is called a branch of $\mathcal{C}$.

Let $y(x)$ be a branch of $\mathcal{C}$ over (α_i, α_{i+1}). Let

$$f(x, r) = y(x) - rx^n$$

$$f * (x) = y'(x) - ny(x)/x$$

Now, for each fixed value of the parameter r, extend both $f(x, r)$ and $f*(x)$, by continuity, to the closed interval $[\alpha_i, \alpha_{i+1}]$. It may happen that the limiting values at the endpoints are plus or minus infinity.

Theorem.   If $\alpha_i \le u < v \le \alpha_{i+1}$, and $f(u, 1) = f(v, 1) = 0$, then there is a point w in (u, v) so that $f * (w) = 0$.

Proof.

Assume that there is no root of $f(x, 1)$ in (u, v).

$$f'(x, r) = y'(x) - nrx^{n-1}$$

and for each fixed value of r, $f * (x)$ is a false derivative for $f(x, r)$ in (u, v).

For x in the interval $[u, v]$, define $\delta(x)$ to be such that $f(x, \delta(x)) \equiv 0$. We can define $\delta(x)$ explicitly as

$$\delta(x) = y(x)/x^n$$

The function $\delta(x)$ is continuous and is equal to one at the endpoints of the interval $[u, v]$. Therefore $\delta(x)$ has a maximum or minimum inside the interval. Let w be a point inside the interval at which $\delta(x)$ has a maximum or minimum. Let $r = \delta(w)$.

Suppose $f'(x, r) \ne 0$ when $x = w$. Then, by the implicit function theorem, r could be both increased and decreased by making an appropriate change in x. But this is impossible, because w is an extremal point.

Thus $f'(x, r) = 0$ when $x = w$. Since $f * (x)$ is a false derivative for all the functions in the family $f(x, r)$, it must be that $f * (w) = 0$.

That completes the proof of the theorem.

We want to subdivide the intervals (α_i, α_{i+1}) so that no branch of C crosses the graph of $y = x^n$ more than once. According to the theorem, it is sufficient to add all the roots of all the false derivatives $f * (x)$ as described above.

Let $r_2(x) = resultant(p(x, y), xp_x + nyp_y, y)$

If $r_2(x)$ were identically zero, then $p(x, y)$, which is irreducible, must divide $xp_x + nyp_y$. This means that the branches of the zero set of $p(x, y)$ satisfy the same differential equation, namely, $xy' + ny = 0$, as x^n. Therefore there are no intersections, unless $p(x, x^n) \equiv 0$, which we said was not the case.

Therefore we can suppose that $r_2(x)$ is not identically zero. We add the roots of $r_2(x)$ to $\{\alpha_1, ..., \alpha_j\}$ to get the decomposition $(\gamma_1, ..., \gamma_k)$.

Evaluating the polynomial at the points

We are now in the following situation, after a time polynomial in the size:

We have a complete decomposition $= \{\gamma_1 < ... < \gamma_k\}$ of algebraic numbers, with defining equations with degrees bounded by polynomials in d, and with coefficients bounded by polynomials in C and n.

For each number γ in the decomposition, we find an ordered set Y_γ, consisting of points (x, y) such that $p(\gamma, y) = 0$. We can also decide how the adjacent Y_γ sets are connected to each other by the curves [Roy 1987]. This can be done in a time polynomial in the size. Suppose now that we could place (γ, γ^n) in correct order in the ordered set Y_γ. We could then discover how many solutions there were in each interval.

Suppose then that we are given real algebraic numbers γ and β, and a number n, and we wish to decide whether or not $\gamma^n > \beta$. We suppose $1 < \gamma < \beta$.

We have defining polynomials with integral coefficients for γ and β, with a bound on the degrees, which is some power of d, and a bound on the coefficients, M, which is a polynomial in d, C, n.

We have $\gamma^n > \beta$ if and only if $nlog(\gamma) > log(\beta)$. Therefore we can decide these questions by computing the logarithmic linear form

$$\Lambda = nlog(\gamma) - log(\beta)$$

A good deal is known about such forms. see Theorem 3.1 in Baker [1975], also Theorem 2 in Baker[1980].

The specific case we are interested in is dealt with in Mignotte and Waldschmidt [1989].

If α is an algebraic number of degree d, with conjugates $\alpha_1, ..., \alpha_d$ and minimal polynomial

$$c_0 x^d + \dots + c_d = c_0 \prod_{i=1}^{d} (x - \alpha_i)$$

define

$$h(\alpha) = d^{-1}(log(c_0) + \Sigma_{i=1}^{d} log(max(1, \mid \alpha_i \mid)))$$

Theorem (Mignotte and Waldschmidt) *If γ and β are multiplicatively independent,*

$$\mid \Lambda \mid > e^{-f}$$

where f is

$$500(D^4 A_1 A_2 (7.5 + log(n)))^2$$

where D is the degree of the number field $Q(\gamma, \beta)$, and A_1 and A_2 are some numbers satisfying

$$A_1 \geq 1$$

$$A_2 \geq 1$$

$$A_1 \geq h(\gamma) + log(2)$$

$$A_2 \geq h(\beta) + log(2)$$

$$A_1 \geq (2e/D) \mid log(\gamma) \mid$$

$$A_2 \geq (2e/D) \mid log(\beta) \mid$$

Both A_1 and A_2 are bounded by $2log(M) + log(D) + 2$ Thus f is bounded by a polynomial in the size of the original problem.

For our purposes, we need the following.

Theorem *There is a polynomial f in the size of the original problem so that*

$$\mid \Lambda \mid < e^{-f} \Rightarrow \Lambda = 0$$

proof.

The theorem in this form can be found in [Baker 1980], using a polynomial f which is probably too large for practical purposes. To get a smaller bound, take f to be bigger than the Mignotte and Waldschmidt function given above.

Suppose that the absolute value of Λ is below this bound. We know then, from the Mignotte and Waldschmidt Theorem, that γ and β are multiplicatively dependent. (In fact Mignotte and Waldschmidt say that the hypothesis of multiplicative independence is easily removed, and that they intend to do this in the near future. I imagine that they meant something more elegant than the following.)

We know $\gamma^a = \beta^b$ for some integers a and b which are relatively prime.

There is an algebraic number η so that $\eta^b = \gamma$ and $\eta^a = \beta$. Since a and b are relatively prime, η is actually in $Q(\gamma, \beta)$, and is a primitive element. The multiplicative dependence which holds at η also holds at all the conjugates of η. We don't yet know η.

However, $h(\eta) \leq h(\eta^b) = h(\gamma)$, so we have an upper bound, M, on the coefficients in the defining polynomial, $g(x)$, for η,

$$M \leq 2^D e^{Dh(\gamma)}$$

If x is in the interval $(1, 1 + 1/D)$, then $g'(x) < DeM$, since $(1 + 1/n)^n$ tends upwards to e. We can suppose $g(x)$ is irreducible, so $g(1) \neq 0$, but $g(1)$ is an integer. Thus $\eta - 1 > 1/(DeM)$. Thus $log(\eta) > 1/(DeM) > e^{-J}$.

Also $\mid nlog(\gamma) - log(\beta) \mid \geq log(\eta)$ unless $\gamma^n = \beta$. Thus $\gamma^n = \beta$.

That completes the proof of the theorem.

The important feature of this situation is that the polynomial depends on $log(M)$ and $log(n)$, not on n or M. The number of decimal places to which we must compute Λ in the worst case is bounded by a polynomial in the size of the problem.

R. P. Brent has shown that $log(x)$ can be evaluated to k decimal places in $O(M(k)log(k))$ steps, where $M(k)$ is the number of steps needed to multiply two k decimal numbers. So we can get k decimal places of $log(x)$ from $O(k)$ decimal places for x and $O(klog(k)^2log(log(k)))$ steps.

We can also approximate real algebraic numbers to k decimal places, using Newton's method, in a time which is linear in k and some polynomial in the size of the defining polynomials.

Therefore, the time needed in the worst case to confirm or deny that $\mid \Lambda \mid > e^{-J}$ is polynomial in the size of the original problem. If we find that the absolute value of Λ is above the bound we are finished.

Evidently, in order to determine the sign of Λ, we only need to find $log(\gamma)$ and $log(\beta)$ to sufficient precision so that the integer part of $log(\beta)/log(\gamma)$ can be determined. So the average behavior may be somewhat better than indicated above.

Remarks

If we know the number of distinct real roots, we can approximate to them by bisection of intervals. However, I do not know a useful lower bound on the distance between distinct roots of $p(x, x^n)$. In particular, if near γ the zero set of $p(x, y)$ contains several curves which go off to infinity, it seems possible that the roots of $p(x, x^n)$ might be extremely close.

Since we can now find the number of roots of polynomials of the form $p(x, x^n)$ this raises the possibility of solving some related problems.

0) The technique used above will also find the number of distinct roots of $p(x, x^{m/n})$ for $x > 1$. Thus the number of roots of $p(x^n, x^m)$ for $x > 1$ can also be found by making a change of variable $x \to x^{1/n}$. (This was suggested by the referee.)

1) We can find the sign of $p(x, x^n)$ at a given algebraic number γ, with small minimal polynomial $r(x)$. This can be done as follows. Take the resultant, $R(y)$, with respect to x of $p(x, y)$ and $r(x)$. Find roots β of $R(y)$, and compare γ^n to β as above.

2) We can find the sign of $p(x, x^n)$ at an algebraic number γ given as a root of $q(x, x^n)$. This can be done by taking the resultant of $p(x, x^n)$ and $q(x, x^n)$ with respect to x^n, and then using 1).

3) We can find a cylindrical decomposition of the xy plane adapted to $p(x, x^n, y)$. This is done as usual, using 1) and 2).

4) We can find real roots of $p(x, x^n, x^m)$ where n and m are large, and $p(x, y, z)$ is a relatively small polynomial. The technique can be applied as before. A problem occurs at the evaluation stage, however. Here we need to compare γ^m and β where γ and β are roots of polynomials of the form $r(x, x^n)$. The Mignotte and Waldschmidt result does not help in this case.

Thus given a sparse polynomial $s(x)$ of high degree, we can find the roots by finding a small polynomial $p(x, y, z)$ and an n and m so that $s(x) = p(x, x^n, x^m)$. But some more work needs to be done to estimate the efficiency of this method.

References

A. Baker, Transcendental Number Theory, Cambridge University Press, 1975

A. Baker, Acta Arithmetica 37 (1980) pp 257-283

R. P. Brent, JACM 23, pp 242-251, 1976

G. E. Collins and R. G. K. Loos, Real Zeros of Polynomials, Computing Supplementum 4 (ed B. Buchberger, G. E. Collins and R. G. K. Loos) Springer-Verlag, Wien-New York, 1982, pp 83-94

F. Cucker, L. G. Vega, F. Rossello, On Algorithms for real algebraic plane curves, in *Effective Methods in Algebraic Geometry*, Edited by Teo Mora, Carlo Traverso, Birkhauser, 1991

J. H. Davenport, Computer algebra for cylindrical algebraic decomposition, Bath computer science technical report 88-10

M. Mignotte and M. Waldschmidt, Linear forms in two logarithms, Acta Arith 53, 1989, pp 251-287

M. Pohst and H. Zassenhaus, Algorithmic Algebraic Number Theory, Cambridge University Press, 1989

D. Richardson, Finding roots of equations involving functions defined by first order algebraic differential equations, in Effective Methods in Algebraic Geometry, edited by Teo Mora, Carlo Traverso, Birkhauser, Boston, Basel, Berlin 1991

D. Richardson, Computing (in a bounded part of the plane) the topology of a real curve defined by solutions of algebraic differential equations, submitted to JSC

M. F. Roy, Computation of the topology of a real curve, proceeding of the conference on computational geometry and topology, Sevilla 1987

R. Zipple, Probabalistic algorithms for sparse polynomials, Proceedings EUROSAM 79 (Marseille) Springer LNCS 72 pp 216-226

Daniel Richardson
Department of Mathematics
University of Bath
Bath BA2, England
email: dsr@uk.ac.bath.maths

Locally effective objects and algebraic topology

J. Rubio   F. Sergeraert

1 Algebraic topology.

Algebraic topology consists of associating *invariants* to topological spaces; these invariants are of an algebraic nature, describing certain topological properties. For example, since Poincaré, it is known how to associate the group $\pi_1(X, x_0)$ to a topological space X and to one of its points x_0; this group is called the *Poincaré group* or the *first homotopy group* of the space X based on x_0. This group is null if and only if the space X is *simply connected at x_0*; in another case, the group measures the lack of simple connectivity. Many other groups can be associated to a topological space, evaluating certain properties of this space: homology groups, homotopy groups, K-theory groups, etc.

The *results* of algebraic topology are combinatorial objects, while topological spaces are not. Determining invariants of algebraic topology using the computer needs a combinatorial intermediary, of a type suitable for a computer. Several methods are possible; one of the most popular and most efficient of these is that of *simplicial sets*, a short description of which will be given later. Any topologist who wishes to work using a computer, should then proceed in two stages:

1) giving a combinatorial version, as a simplicial set for example, of the topological space on which the topologist wishes to work;

2) being able to use effective methods which allow automatic calculation, using a theoretical or real computer, of the invariant which is sought.

The first stage is purely theoretical and by its nature cannot admit an effective treatment. The second stage is the subject of this paper.

The reasonable objects which can be studied in algebraic topology are of a finite nature. In terms of simplicial sets, we mean *finite* simplicial sets. It is then legitimate to demand that the invariants associated to finite simplicial sets are also of a finite nature. But to show that these invariants are of finite type is often very difficult. It is thus, for example, for homotopy groups: Serre proved at the beginning of the fifties [11] that if a finite simplicial set X is simply connected, then the homotopy groups of X are groups of finite type. This is not at all evident since the definition itself of these groups uses functional spaces which are non-finite for any reasonable definition of finiteness. Sophisticated algebraic methods (spectral sequences) allowed Serre to prove that, in spite of these highly non-finite intermediaries, the results at the end of the calculation were, however, finite.

This surprising result obtained by Serre poses, therefore, the problem of calculating these invariants using a computer. The spectral sequence method is not effective, and needs therefore essential improvements to be implemented in a computer. The improvement found recently (see [12] and [7]) consists of using a strictly richer notion than that of spectral sequence. This notion is called the *perturbation lemma* (it would be better to call it the *fundamental theorem of homological algebra*). The perturbation lemma is apparently effective, but it frequently requires the use of infinite spaces! In this paper, we are trying to explain that this new obstacle is not a real obstacle, whenever these spaces are considered from a new point of view. We will show that these spaces can be, in spite of their non-finite nature, coded in a suitable form (*locally effective* coding), in such a way that *all* information necessary for the calculations (information which is always of a *local* nature) is accessible. The true nature of these spaces is thus shown: they are to be considered as databases ready to answer any type of question. The number of possible questions is infinite, but, as the processes for finding answers are of an algorithmic nature, there is nothing to stop the implementation of these processes on a computer. The adjective *functional* which is used for these spaces acquires in this way a new scope: instead of functions it would be better to speak of algorithms, and, in this spirit, these spaces should be called *algorithmic spaces*.

2   Simplicial sets.

For a detailed study of simplicial sets, see [6] and [9].

Definition 1 — A *simplicial set* X is a sequence $(X_n)_{n \geq 0}$ of sets together with, for every $n > 0$, *face operators* $\partial_i : X_n \to X_{n-1}$, $0 \leq i \leq n$, and, for every $n \geq 0$, *degeneracy operators* $s_i : X_n \to X_{n+1}$, $0 \leq i \leq n$, which satisfy the following identities:

$$\partial_i \partial_j = \partial_{j-1} \partial_i , \; i < j$$
$$s_i s_j = s_{j+1} s_i , \; i \leq j$$
$$\partial_i s_j = \begin{cases} s_{j-1}\partial_i & i < j \\ id_{X_n} & i = j, j+1 \\ s_j \partial_{i-1} & i > j+1 \end{cases}$$

The relation between simplicial sets and topological spaces is neither simple nor is it the objective of this paper. Let us say only that there exists a *realization operator* which associates a topological space $|X|$ to a simplicial set X. A n-*simplex* of X, that is an element of X_n, brings to $|X|$ a n-simplex of R^n (a convex hull of $n+1$ affinely independent points in R^n) whenever the n-simplex of X is not *degenerate*; a simplex is degenerate if it is the image through a degeneracy operator of a simplex of lesser dimension.

We will give only two examples; we ask the reader to check the technical details.

The first example, denoted by Δ^2, is defined as follows: Δ_n^2 is the set of increasing maps (in the broad sense) of $\underline{n} = \{0, 1, \ldots, n-1, n\}$ towards $\underline{2} = \{0, 1, 2\}$. If $\sigma : \underline{n} \to \underline{2}$ is a n-simplex of Δ^2 its i-face $\partial_i\sigma = \sigma'$ is the map defined by $\sigma'(j) = \sigma(j)$ if $j < i$, $\sigma'(j) = \sigma(j+1)$ if $j \geq i$. In the same way, its i-degeneracy $s_i\sigma = \sigma'$ is the map $\sigma'(j) = \sigma(j)$ if $j \leq i$, $\sigma'(j) = \sigma(j-1)$ if $j > i$. It is easy to verify that the compatibility relations required between face operators and degeneracy operators are satisfied. The non-degenerate simplices are the injective maps $\underline{n} \to \underline{2}$; this requires $n \leq 2$ and so there exists one and only one non-degenerate simplex in dimension 2 (a triangle), three non-degenerate simplices in dimension 0 (its three sides) and three simplices in dimension 0 (its three vertices). A 0-simplex is necessarily non degenerate. In this way, we can obtain the presentation of a triangle as a simplicial set. More generally, the geometric p-simplex can be presented as a simplicial set in a similar form, simply by replacing 2 by p.

A simplicial set is *finite* if it only has a finite number of non-degenerate simplices. This restriction in the definition is indispensable since a simplex always generates by successive degeneracies an infinity of degenerate simplices. We will show that algebraic topology naturally leads to the introduction and use of non finite simplicial sets, even if the intention is to calculate objects of a finite nature (homology groups or homotopy groups of finite type). The realization of a simplicial set is compact if and only if this simplicial set is finite.

The second example serves to illustrate the role of degenerate simplices. The simplicial set S^2 is defined as follows. The set of n-simplices S_n^2 is the disjoint union of the set of ordered surjective maps from $\{0, 1, \ldots, n\} = \underline{n}$ to $\{0, 1, 2\} = \underline{2}$ (the set of these surjective maps is denoted by Σ_n) with the set with only one element $\{*_n\}$: $S_n^2 = \Sigma_n \coprod \{*_n\}$. A face of $*_n$ is always $*_{n-1}$, and a degeneracy of $*_n$ is always $*_{n+1}$. If σ is an element of Σ_n and if $0 \leq i \leq n$, then $s_i(\sigma)$ is the surjective map σ' defined by $\sigma'(j) = \sigma(j)$ if $j \leq i$, $\sigma(j-1)$ if $j > i$; notice that if σ is a surjective map towards $\underline{2}$, then σ' is also a surjective map. Similary, $\partial_i(\sigma)$ is defined as the surjective map σ' defined by $\sigma'(j) = \sigma(j)$ if $j < i$, $\sigma(j+1)$ if $j \geq i$; but this map σ' may not be a surjective map (this occurs if and only if i has only one antecedent for σ) in which case, one must then decide that $\partial_i(\sigma) = *_{n-1}$. The reader will verify that the axioms for simplicial sets are satisfied.

The set of non-degenerate simplices of S^2 has only two elements, $*_0$ and $id_{\underline{2}}$, the first in dimension 0, the second in dimension 2. To prove this, if $n > 0$, then $*_n = s_0(*_{n-1})$, and if σ' is a surjective map $\sigma' : \underline{n} \to \underline{2}$ with $n > 2$, then for the first integer i which has two antecedents for σ', the identity $\sigma' = s_i(\partial_i(\sigma'))$ holds. The result of this is that the realization $|S^2|$ of S^2 can be constructed with a point $*$ (which proceeds from the unique 0-simplex $*_0$) and the 2-simplex Δ^2 (which proceeds from the only non-degenerate simplex in dimension 2); the faces of this last simplex are "swallowed" by $*$, and finally $|S^2| = \Delta^2/\partial\Delta^2$, the quotient of the standard 2-simplex by its boundary; it is well known that this topological space is homeomorphic to the 2-sphere. We have thus obtained a combinatorial model for the 2-sphere which only contains two non-degenerate simplices, while the minimal triangulation of the 2-sphere as a simplicial complex requires 4 vertices, 6 edges and 4 triangles. A similar model for the p-sphere as a simplicial set can be obtained by replacing 2 by p; the model obtained will only have two non-degenerate simplices, one of which is in dimension 0, and the other in dimension p.

If a simplex is degenerate, it is, in a certain way, a degeneracy of a unique non degenerate simplex. This property is explained in the following proposition.

Proposition 2 — *Let X be a simplicial set and σ' be a n-simplex of X, that is an element of X_n. Then there exists one and only one system $(m, \sigma, (i_{n-m}, \ldots, i_1))$ verifying:*

- σ *is a non-degenerate m-simplex;*

- *the sequence of integers $(i_{n-m}, \ldots, i_1)$ is strictly decreasing;*

- $\sigma' = s_{i_{n-m}} \cdots s_{i_2} s_{i_1} \sigma$

The proof, although elementary, is not so easy; it is left as an "exercice". This proposition allows us to work only with non degenerate simplices to describe simplicial sets. To achieve the above, we can proceed as follows. A degenerate simplex σ' is going to be represented by a pair $(\sigma, (i_p, \ldots, i_1))$ where σ is a non degenerate simplex, $(i_p, \ldots, i_1)$ is a strictly decreasing sequence of integers and $\sigma' = s_{i_p} \ldots s_{i_1} \sigma$; in addition, if σ is a non degenerate simplex, one can write $\sigma = (\sigma, ())$, where the list of degeneracy indices is empty.

3   A natural example of the construction of an infinite simplicial set.

In this section, we treat the *loop space* construction. The topological version of this construction is defined as follows. Let X be a topological space and x_0 a point in X. The loop space of X based on x_0 is the space $\Omega(X, x_0)$ of the continuous maps from the unit segment $I = [0, 1]$ to X applying the ends 0 and 1 of the segment I to the base point x_0:

$$\Omega(X, x_0) = \mathcal{C}(I, 0, 1; X, x_0, x_0).$$

Similarly, let us consider the *path space* of X, consisting of paths whose origin is x_0:

$$\mathcal{P}(X, x_0) = \mathcal{C}(I, 0; X, x_0).$$

The three spaces $\Omega(X, x_0)$, $\mathcal{P}(X, x_0)$ and X are organized in a sequence of two continuous maps:

$$\Omega(X, x_0) \xrightarrow{i} \mathcal{P}(X, x_0) \xrightarrow{p} X$$

where the map i is the canonical inclusion (a loop is also a path) and the map p is the canonical projection which associates to every path $\gamma : I \to X$ the point $\gamma(1)$. The use of the name *projection* for p is justified by the fact that the path space $\mathcal{P}(X, x_0)$ admits a structure of *twisted product* (*fibred* structure) of the *base space* X and the *fiber* $\Omega(X, x_0)$. The path space is contractible; in other words, from the point of view of algebraic topology, $\mathcal{P}(X, x_0)$ plays a role of *unit* space, which gives to the loop space a role of (twisted) *inverse* of the space X; it can be proved that such a non twisted inverse does not exist, and the torsion, that is to say the fibred structure, is essentially unique. It is in this way that Serre in 1950 introduced in a natural way these spaces which nowadays, 40 years later, play a capital role in numerous questions in algebraic topology and also in differential topology and geometry.

A loop space is a space of functions from the unit interval $[0, 1]$, and therefore a *functional* space, of infinite dimension for every reasonable definition of dimension. It cannot therefore be represented as a simplicial set. But combinatorial versions of this construction exist. To find such a combinatorial version one must, given a simplicial set X and a base point x_0, vertex of X ($x_0 \in X_0$), define a simplicial set $\Omega(X, x_0)$ which verifies properties in the simplicial context similar to those of Serre's loop space. Several solutions are possible; here we are only interested in one of the most important, due to Kan [5]. Let us assume that X is a *reduced* simplicial set, that is to say, with an unique vertex, which therefore will be the only possible base point for X: $X_0 = \{x_0\}$. To construct $\Omega = \Omega(X, x_0)$, we begin by defining $X'_n = X_n - s_0 X_{n-1}$ if $n > 0$. Then we decide that the set of n-simplices Ω_n of the simplicial set to be constructed is the (non-abelian) free group generated by X'_{n+1}. When X'_{n+1} is non empty, this set Ω_n is therefore infinite. It is now necessary to define the face and degeneracy operators of Ω; these operators, whose domain and codomain are *groups* of simplices, are going to be group homomorphisms. Since the domain group is a free group, it is enough to define the operators on the generators of these groups; to avoid any confusion, we denote $\gamma(x)$ the generator of Ω_n which comes from the element x of X_{n+1}. The face and degeneracy operators of Ω are then defined as follows:

$$\begin{aligned}
\partial_0 \gamma(x) &= \gamma(\partial_1 x)\gamma(\partial_0 x)^{-1} \ ; \\
\partial_i \gamma(x) &= \gamma(\partial_{i+1} x), \ i > 0 \ ; \\
s_i \gamma(x) &= \gamma(s_{i+1} x), \ i \geq 0.
\end{aligned}$$

We can see that in the first of these formulae the structure of free group of the codomain is used to define the image of a generator of the domain group. The compatibility relations between face and degeneracy operators of X imply the same relations between the operators of Ω and we have from now on, a simplicial set $\Omega = \Omega(X, x_0)$.

By a similar procedure but more complex, which are not going to explain here, a path space $\mathcal{P} = \mathcal{P}(X, x_0)$ can be defined. These spaces are organised in a *simplicial fibration* (voir [5]):

$$\Omega(X, x_0) \xrightarrow{i} \mathcal{P}(X, x_0) \xrightarrow{p} X$$

exactly the same as in the topological setting.

If $\Omega(X, x_0)$ is in its turn reduced, which is equivalent to requiring that the simplicial set X has no non-degenerate 1-simplex, the process can be iterated, allowing the definition of a second loop space $\Omega^2(X, x_0)$, and so on. Notice that not only Ω^2_n as going to be always infinite, but also the set of its generators as a free group is also going to be infinite! The problem of calculating the homology groups of iterated loop spaces (Adams' problem) is one of the most important in algebraic topology. This problem was posed by Adams in 1956. Adams have solved the above problem for the *first* loop space [1], but it was not until 1980 that Baues [2] solved it for the *second* loop space. Four solutions for the general problem were found independently at the end of the 80's ([4] [8] [10] [13]), but the solutions were of a very different nature. The solution mentioned here, due to the first author (announced in [8], and developed in [7]), uses the methods of *functional programming* to solve the difficulties caused by the presence, essential, of infinite simplicial sets. This simple method is also efficient. It is the only one which up to now has been implemented on the computer, giving very interesting results [9].

4 Effective coding of finite simplicial sets.

The coding of *finite* simplicial sets does not produce any special problems. Let X be a simplicial set, and, for each positive or null integer n, let X_n be the set of its n-simplices, and $X_n^N = X_n - \cup_{0 \leq i \leq n-1} s_i X_{n-1}$ the set of its non degenerate n-simplices. The simplicial set X is finite if and only if the set $X^N = \cup_{n \geq 0} X_n^N$ of all the non degenerate simplicial is finite. It is enough therefore to represent such a simplicial set by giving the list of its non degenerate simplices, and, for each of these, its dimension and the description of its faces. For example, the simplicial set structure of the 2-sphere S^2 can be described with the help of the list:

```
((* 0 ())
 (s2 2 ((* (0)) (* (0)) (* (0)))))
```

In this code the only non-degenerate simplices of S^2 are represented by the symbols * (for the unique 0-simplex of S^2) and s2 (for the unique non-degenerate 2-simplex). So the code is a list of two elements: the first element contains the information about * and the second about s2. Each of these elements is in its turn a list of three elements: the first element is the symbol which represents the non-degenererate simplex; the second element is the dimension of this simplex and finally the third element is a list which contains its faces. For example, in the list (* 0 ()), which contains the information about the only 0-simplex of S^2, the third element is the empty list, because a 0-simplex has no face.

In the general case, each face is coded in the way explained in section 2: it is a pair whose first element is a non-degenerate simplex and the second is a sequence of integers (the degeneracy indices of the simplex). For example, in the list (s2 2 ((* (0)) (* (0)) (* (0)))) above, one can read that the three faces of s2 are all the degeneracy through the operator s_0 of the 0-simplex * (each face is represented by the list (* (0)).

In the same way the 3-sphere S^3 will be described with the help of the list:

```
((* 0 ())
 (s3 3 ((* (1 0)) (* (1 0)) (* (1 0)) (* (1 0)))))
```

This time each face of the maximal simplex s3 is the degeneracy through $s_1 s_0$ of the base point *.

We leave the reader with the job of verifying that, with the same conventions, the torus T^2 can be represented with the help of the list:

```
((* 0 ())
 (east 1 ((* ()) (* ())))
 (north-east 1 ((* ()) (* ())))
 (north 1 ((* ()) (* ())))
 (triangle-1 2 ((north ()) (north-east ()) (east ())))
 (triangle-2 2 ((east ()) (north-east ()) (north ()))))
```

The manipulation on the computer of such simplicial sets does not pose any problem. In particular it is elementary to write programs computing the *homology groups* of such simplicial sets. These groups are defined as follows. Let X be a simplicial set. For every integer n we define the *group of n-chains* $C_n X$ as the free **Z**-module generated by the n-simplices of X:

$$C_n X = Z^{(X_n)}.$$

We can then define for each n a boundary operator $d_n : C_n X \to C_{n-1} X$ defining its image for each generator

$$d_n \sigma = \sum_{i=0}^{n} (-1)^i \partial_i \sigma,$$

and extending to $C_n X$ by **Z**-linearity. The diverse relations imposed on the face operators then *imply* that a composite $d_{n-1} \circ d_n$ is necessarily null. This allows a definition of the group of n-cycles $Z_n X = \ker d_n$ and the group of n-boundaries $B_n X = \operatorname{im} d_{n+1}$, and finally the n-dimensional homology group $H_n X = Z_n X / B_n X$. For example, for the sphere S^2 described previously, the groups $H_n S^2$ are null except $H_0 S^2 \simeq Z$ and $H_2 S^2 \simeq Z$. The collection $\{C_n X, d_n\}_n$ of chain groups and boundary operators is called the *chain complex* $C_* X$ of the simplicial set X.

We can also define the *normalized chain complex* $C_*^N X$ of X in the same way except that $C_n^N X$ is generated by the non degenerate simplices of X. To be more precise, it can be shown that the set of degenerate simplices of X generates a subcomplex $C_*^D X$ of the complex $C_* X$, because the boundary of a degenerate simplex is always a combination of degenerate simplices. This allows the definition of the quotient complex $C_*^N X = C_* X / C_*^D X$. It can be proved that the homology groups of the latter complex are the same as those of the initial complex: $H_n^N X \simeq H_n X$. The normalized complex is much smaller than the initial complex, which greatly eases the calculation of homology groups. For example, it is trivial, using the normalized complex, to verify that the homology groups of S^2 are those previously mentioned: they inmediately result from the fact that S^2 has only two non degenerate simplices, one of which is in dimension 0 and the other in dimension 2.

Notice that in no part in the previous definitions has the simplicial set X required to be *finite*, so that these algebraic manipulations make sense even if the finiteness hypothesis is not satisfied. If the simplicial set X is finite, its homology groups are necessarily **Z**-modules of finite type, and it is elementary to write programs calculating these groups. But the remarkable results obtained by Serre, mentioned in the introduction, affirm that in numerous natural situations, even if the simplicial set X is not finite, the homology groups of X are of finite type. Here we state a typical example.

Theorem 3 (Serre) — *Let $n \geq 2$ be an integer and X be a finite simplicial set with a unique vertex and X has no non degenerate i-simplex, $0 < i < n$. Then if $p < n$, every homology group of the simplicial set $\Omega^p X$ is a group of finite type.*

For example, the sphere S^4 is a finite simplicial set without non degenerate simplex in dimension 1, 2 and 3. The simplicial set $\Omega^3 S^4$ is highly infinite: each set of n-simplices is a non-abelian free group having as generators the elements of a non-abelian free group whose generators are the elements of a non-abelian free group having a non-empty finite set of generators! But, according to Serre's theorem, the homology groups of $\Omega^3 S^4$ are, however, of finite type. For example, $H_5 \Omega^3 S^4 \simeq Z \oplus Z/6Z$. This then poses the problem of calculating these homology groups. The second author introduced a general method, called *effective homology* (see [12] and [7]), allowing the solution of this problem and many others. This method requires the ability to code in the computer these non-finite simplicial sets. This question is treated in the following section.

5 Locally effective coding of infinite simplicial sets.

If X is a simplicial set where, for example, the set X_1^N of non degenerate 1-simplices is infinite, it is then evident that this simplicial set will not be able to be represented or, rather, *coded* as was explained in the previous section.

A mathematician who wishes to work with abelian groups in a computer, can, in the same spirit as in the previous section, choose for coding such a group, the list of its elements. He will only be able to work with *finite* groups. The group $\mathbf{Z}$ of integers is not finite; but this, however, will not prevent work with integes on a computer! Several solutions are possible. The first one consists in deciding that we will work in a "general" way with integers of limited sized, for example of absolute value smaller than an integer **maxint**. Better still, with *multiprecision* algorithms, we can manipulate integers of arbitrary size, the only limitation being the finite size of memory in the computer; this memory is, in principle, extensible "on demand". In this situation, "the list of the elements of the group", which in fact is infinite, is not accessible to the programmer. Only the addition, the multiplication and the change of sign operators are accessible to the programmer; the unit elements for addition and multiplication are also accessible.

We can proceed in the same way with our infinite simplicial sets. If X is an infinite simplicial set, we can code in a certain way the simplices of X (similar to the way the integers of $\mathbf{Z}$ are "coded", but $\mathbf{Z}$ is not coded), and to supply the face and degeneracy operators on these simplices. In Scratchpad language, we can define the category of simplicial sets and then a particular simplicial set will be a domain of this category. A simplicial set then appears as an object together with an algebraic structure, where the composition laws are external laws with integer coefficients. So, to apply a face operator, we must know not only a n-simplex but also a face index i between 0 and n. So the face operator is an external composition law (not defined at every pair) $Z \times X \rightarrow X$ where we have denote $X = \coprod_n X_n$, the set of all the simplices of X. We can proceed in the same way for degeneracy operators.

Let us consider, for example, the loop space construction. Let X be a simplicial set, where we also denote X the set of its simplices. We want to construct a computer version of the loop space $\Omega X = \Omega(X, x_0)$; we suppose that x_0 is the unique vertex (0-simplex) of X. Let us also suppose that we know how to code the simplices of X on the computer. Let us denote $X'_n = X_n - s_0 X_{n-1}$ and $(\Omega X)_n$ the non-abelian free group generated by X'_{n+1}. The set $(\Omega X)_n$ is the set (in general infinite) of the n-simplices of ΩX. The representation of the simplices of X then easily allow the representation of the simplices of ΩX, for example as lists where each element of the list is a pair $(\sigma \; \mathbf{k})$: the first component σ represents an element of X'_{n+1} and the integer $\mathbf{k}$ is an exponent in the writing of an element of a free group. So the computer object:

$$((\sigma_1 k_1)(\sigma_2 k_2) \ldots (\sigma_p k_p))$$

codes the simplex $\sigma_1^{k_1} \sigma_2^{k_2} \ldots \sigma_p^{k_p}$ of ΩX, if each k_i is a computer integer and each σ_i is the computer code of a simplex of X. All this is only coherent if the σ_i are all of the same dimension $n+1$ and if none of these is 0-degenerate.

It is then necessary to supply the user with face and degeneracy operators. These operators have been defined in section 3 and are described using the operators of the initial simplicial set X. If the above operators are available, it is easy to write the corresponding operators of ΩX as little programs using, among other things, the operators of X. In Scratchpad language, the domain ΩX of the simplicial sets category can be defined using, among other things, the operators of the domain X of the same category. Better still, a builder of domains can be written taking, as input, the domain X and returning, as output, the domain ΩX. The procedure can be iterated to construct the domains $\Omega^2 X$, $\Omega^3 X$, etc. We can see that the situation is very similar to the usual situation in commutative algebra, where from a ring R, one can construct the polynomial ring with coefficients in R, and then, by iteration, the polynomial rings with any number of variables.

In contrast to the type of coding described in the previous section, such a coding of a simplicial set is called a *locally effective* coding, to explain that the only information available in such a situation is of a *local* nature. For example, we can, given two simplices of such a set, ask if their intersection is non-empty. On the contrary, in general, the information of a global nature will be inaccessible. So it is impossible to ask if such a simplicial set is empty or not, or if it is simply connected, or to ask for its homology groups, which is information of a global nature.

The *effective homology* methods [12], use both types of coding, effective and locally effective, allowing an overall gain: the locally effective coding of "infinite" objects allows to solve apparent difficulties of non-finiteness, while the effective coding of appropriated objects allows to obtain homological information of a global nature. The two types of coding are related by hybrid objects, essentially, homotopy equivalences between effective chain complexes and locally effective chain complexes. An example of such a hybrid object is the homotopy equivalence between the chain complex of a path space $\mathcal{P}(X, x_0)$ and the chain complex whose only no-null group of chains is that of dimension 0, which is isomorphic to $\mathbf{Z}$. Since the set of generators of each group $C_n(\mathcal{P}(X, x_0))$ is, in general, infinite, we cannot code $C_*(\mathcal{P}(X, x_0))$ in an *effective* way. But the information contained in a locally effective code for this chain complex is rich enough to define on the computer the homotopy equivalence between $C_*(\mathcal{P}(X, x_0))$ and the trivial complex $\mathbf{Z}$. The length of this paper does not allow more explanations; refer to [12] for more information.

6   A report about a computer calculation.

The methods sketched in the previous section allowed the proof of computability results sought for a long time, specially the solution to Adams' problem about the calculation of the homology of iterated loop spaces, posed since 1956. Three other solution were found independently by other researchers ([4] [10] [13]), but these solutions up to now have not been implemented on the computer.

As an example of use, let us consider the calculation of the homology group $H_5\Omega^3 Moore(Z_2, 4)$. The space $Moore(Z_2, 4)$ is a topological space well defined up to homotopy whose only non-trivial homology group is Z_2 in dimension 4. This space is the third suspension of the real projective plane. This calculation is interesting, since the theorists use a sophisticated spectral sequence to calculate this group. This is the *Bockstein spectral sequence* which uses the particular properties of suspensions. These theorists thought that the group $H_5\Omega^3 Moore(Z_2, 4)$ was isomorphic to $Z_2^4 \oplus Z_4$.

To make this calculation with our software, it is enough to use four of its functions. We give a brief specification of these functions:

moore *function:*

- **Input:** two positive integer numbers n and m;

- **Output:** the code for the Moore space whose only non-trivial homology group is Z_n in dimension m;

- **Comments:** the output space is a *finite* simplicial set and it is coded in an *effective* way (as explained in section 4, for example).

ess-sseh *function:*

- **Input:** the *effective* code for a *finite* simplicial set;

- **Output:** the same simplicial set but coded as a simplicial set with effective homology;

- **Comments:** ess-sseh stands for "effective simplicial set to simplicial set with effective homology".

loop-space-eh *function:*

- **Input:** a positive integer p and a simplicial set X with effective homology;

- **Output:** the simplicial set with effective homology $\Omega^p X$;

- **Comments:** the simplicial set X must be reduced and should have no non-degenerate i-simplex, $0 < i \leq p$.

homology *function:*

- **Input:** a simplicial set X with effective homology and an integer $n \geq 0$;

- **Output:** the homology group $H_n(X)$;

- **Comments:** the output group is necessarily a group of finite type, because X is a simplicial set *with effective homology.*

To compute the group $H_5\Omega^3 Moore(Z_2, 4)$ using these functions, you must proceed in two steps. First you run the Lisp instruction

```
(setf ooom4 (loop-space-eh 3 (ess-sseh (moore 2 4))))
```

which takes a little less than two CPU seconds on a Sun-4-490 under Lucid-Common-Lisp. The function **moore** constructs the Moore space ad hoc. The function **ess-sseh** transforms this space into a *simplicial set with effective homology.* This is quite a big set of algorithms ready to answer a multitude of questions. Finally, these algorithms will allow the function **loop-space-eh** to be called. This function will construct the espace $\Omega^3 Moore(Z_2, 4)$, assigned to the symbol **ooom4**.

At this stage, a lot of domains in the Scratchpad sense were constructed, about 800. The following were constructed:

- 20 simplicial sets of which 8 are effective and 12 locally effective;

- 137 chain complexes of which 103 are locally effective;

- 9 differential graded coalgebras of which 6 are locally effective;

- 15 differential graded comodules of which 6 are locally effective;

- 487 morphisms of chain complexes;

- 40 reductions between chain complexes;

- 16 homotopy equivalences between chain complexes.

Each of these domains is to be considered as a set of algorithms which can be interrogated if necessary. Very few of them are *effective*, in other words, with a finite underlying set. This does not prevent work with them. We can, for example, (and this is the second step to compute $H_5\Omega^3 Moore(Z_2, 4)$) ask for the fifth homology group of our loop space, in the following way:

```
(homology ooom4 5)
```

and in 2h27mn CPU time, the program answers that this group is Z_2^5 and not $Z_2^4 \oplus Z_4$! The theorists who were mentioned previously have not know, up to now, how to confirm or deny our result. The examination of the trace shows that the calculation is very complex, as one should expect. The calculation makes an intervention of the tens of thousands of generators of chain complexes.

The situation briefly explained in this paper seems to us to be a field of study which is ideal for symbolic computation, due to the wealth and variety of the structures involved, which cannot be compared to what is normally done in commutative algebra. And also, because of the large number of construction processes, which reflect the great complexity of topology as regards algebra. In particular, test examples of great interest can be found in this field to compare the respective merits of different computer algebra systems. All our software was written directly in Common Lisp, but the translation for example into Scratchpad is, in principle, only a simple exercise. It would be interesting to compare the results then obtained (facility of development, maintenance, time and space efficiency) with ours.

References

[1] J. Frank Adams. *On the Cobar construction.* Proceedings of the National Academy of Sciences of the U.S.A., 1956, vol. 42, pp 409-412.

[2] H. J. Baues. *Geometry of loop spaces and the cobar construction.* Memoirs of the American Mathematical Society, 1980, vol. 230.

[3] Clemens Berger. *Une version effective du théorème de Hurewicz.* Thèse Institut Fourier, Grenoble, 1991.

[4] Billera, Kapranov, Sturmfels. *Convex geometry and a conjecture of Baues in the theory of loop spaces.* Preprint.

[5] Daniel M. Kan. *A combinatorial definition of homotopy groups.* Commentarii Mathematici Helvetici, vol. 67, pp 282-312.

[6] J. Peter May. *Simplicial objects in algebraic topology.* Van Nostrand, 1967.

[7] Julio Rubio, Francis Sergeraert. *An algorithm for the automatic calculation of the homology groups of iterated loop spaces.* To appear.

[8] Julio Rubio, Francis Sergeraert. *Suites spectrales d'Eilenberg-Moore et homologie effective.* Comptes-Rendus hebdomadaires des séances de l'Académie des Sciences, Paris, Série A, 1988, vol. 306, pp 723-726.

[9] Julio Rubio, Francis Sergeraert. *A program computing the homology groups of loop spaces.* SIGSAM Bulletin ACM, 1991, vol. 25, pp. 20-24.

[10] Rolf Schön. *Effective algebraic topology*. Memoirs of the American Mathematical Society, 1991, vol. 451.

[11] Jean-Pierre Serre. *Groupes d'homotopie et classes de groupes abéliens*. Annals of Mathematics, 1953, vol. 58, pp. 258-294.

[12] Francis Sergeraert. *The computability problem in algebraic topology*. To appear in Advances in Mathematics.

[13] V.A. Smirnov. *On the chain complex of an iterated loop space*. Mathematics of the USSR, Izvestiya,1990, vol. 35, pp 445-455.

Appendix.

In this appendix, we give a list of groups which have been calculated by our program. A time reference such as 6 mns or 5 hrs means the time taken by the program from the beginning of the calculation (before H_0). These calculations were done on a Sun-4-490 under Lucid-Common-Lisp.

$\Omega^2 S^3$.

$H_1 = Z$
$H_2 = Z/2Z$
$H_3 = Z/2Z$
$H_4 = Z/2Z \oplus Z/3Z$
$H_5 = Z/2Z \oplus Z/3Z$ (1 mn)
$H_6 = (Z/2Z)^2$ (5 hrs)
$H_7 = (Z/2Z)^2$ (7 hrs)

$\Omega^3 S^4$.

$H_1 = Z$
$H_2 = Z/2Z$
$H_3 = Z/2Z$
$H_4 = Z \oplus Z/2Z \oplus Z/3Z$
$H_5 = Z \oplus Z/2Z \oplus Z/3Z$ (20 mns)

$\Omega^4 S^5$.

$H_1 = Z$
$H_2 = Z/2Z$
$H_3 = Z/2Z$
$H_4 = (Z/2Z)^2 \oplus Z/3Z$
$H_5 = (Z/2Z)^2 \oplus Z/3Z$ (13 mns)

$\Omega^3 Moore(Z/2Z, 4)$.

$H_1 = Z/2Z$
$H_2 = Z/2Z$
$H_3 = Z/4Z \oplus Z/2Z$
$H_4 = Z/4Z \oplus (Z/2Z)^2$
$H_5 = (Z/2Z)^5$ (2 hrs 27 mns)

$\Omega^3 (Moore(Z/2Z, 4) \vee S^5)$.

$H_1 = Z/2Z$
$H_2 = Z \oplus Z/2Z$
$H_3 = Z/4Z \oplus (Z/2Z)^2$
$H_4 = Z \oplus Z/4Z \oplus (Z/2Z)^3$ (52 mns)

$\Omega(\Omega S^3 \bigcup_f D^3)$.

A 3-disk is attached to ΩS^3 through a map $f : S^2 \to \Omega S^3$ of degree 2 so that $\pi_2(\Omega S^3 \bigcup_f D^3) = Z/2Z$.

$H_1 = Z/2Z$
$H_2 = Z/2Z$
$H_3 = Z \oplus (Z/2Z)^2$
$H_4 = (Z/2Z)^4$
$H_5 = Z \oplus (Z/2Z)^6$
$H_6 = (Z/2Z)^{13} \oplus Z/3Z$
$H_7 = (Z/2Z)^{20}$ (49 mns)

$\Omega(\Omega^2 S^4 \bigcup_f D^3)$.

A 3-disk is attached to $\Omega^2 S^4$ through a map $f : S^2 \to \Omega^2 S^4$ of degree 2 so that $\pi_2(\Omega^2 S^4 \bigcup_f D^3) = Z/2Z$.

$H_1 = Z/2Z$
$H_2 = Z/2Z$
$H_3 = Z \oplus (Z/2Z)^2$
$H_4 = Z \oplus (Z/2Z)^4$ (1 mn)
$H_5 = Z \oplus (Z/2Z)^6$ (1 hr 27 mns)

$\Omega^2(\Omega S^4 \bigcup_f D^4)$.

A 4-disk is attached to ΩS^4 through a map $f : S^3 \to \Omega S^4$ of degree 2 so that $\pi_3(\Omega S^4 \bigcup_f D^3) = Z/2Z$.

$H_1 = Z/2Z$
$H_2 = Z/2Z$
$H_3 = Z/4Z \oplus Z/2Z$
$H_4 = Z \oplus (Z/2Z)^3$ (1 mn)
$H_5 = (Z/2Z)^5$ (5 hrs)

$\Omega^2(\Omega^2 S^5 \bigcup_f D^4)$.

A 4-disk is attached to $\Omega^2 S^5$ through a map $f : S^3 \to \Omega^2 S^5$ of degree 2 so that $\pi_3(\Omega^2 S^5 \bigcup_f D^4) = Z/2Z$.

$H_1 = Z/2Z$
$H_2 = Z/2Z$
$H_3 = Z/4Z \oplus Z/2Z$
$H_4 = (Z/2Z)^4$
$H_5 = (Z/2Z)^5$ (34 mns)

Julio Rubio
Dpto. Informática, Zaragoza (SPAIN)
email: rubio@cc.unizar.es

Francis Sergeraert
Institut Fourier – L.M.C.,
B.P. 74
38402 Saint Martin d'Hères (FRANCE)
email: sergerar@imag.fr

Decision of Algebra Isomorphisms
Using Gröbner Bases

K. Shirayanagi

Abstract. Given two finite-dimensional non-commutative finitely presented algebras A and B over a field k, this paper presents two algorithms for deciding whether or not they are isomorphic as k-algebras. We reduce this problem to a radical ideal membership problem in a commutative ring, by Hilbert's zero point theorem. That is, A and B are isomorphic if and only if the determinant f of a k-linear mapping: $A \to B$ does not lie in the radical of an ideal I derived from the relations of A and B. Moreover an efficient technique for computing f is provided.

We propose two Gröbner basis methods for solving our problem. One is to compute the radical of I directly and then to rewrite the above determinant f modulo this radical. The other method is to judge solvability of a certain system of algebraic equations: the union of I and a new polynomial. As a result, it is shown that the isomorphism problem for finitely presented algebras is decidable.

1 Introduction

Finitely presented algebras are universal algebras defined by a finite number of generators and a finite number of relations. The word and isomorphism problems for such algebras are fundamental and hence have been profoundly studied in mathematics and computer science. For instance, it is a classical result that both problems are undecidable for finitely presented semigroups and groups. A survey on the word and isomorphism problems for various types of universal algebras can be found in [6]. Glass ([6]) concluded with a sentence *"the isomorphism problem for commutative rings should be considered since this variety has uniformly soluble word problem."* We investigate the isomorphism problem for non-commutative finitely presented algebras over a field k. This class includes commutative polynomial rings, since we can incorporate the commutative relations among all generators in the defining relations. As to the word problem for commutative polynomial rings over a field, it is well-known that Buchberger ([2]) has already provided an affirmative solution by the Gröbner basis method, since it is equivalent to an ideal membership problem. In this paper, we consider *finite-dimensional* finitely presented algebras as k-vector spaces. In particular, it is shown that the isomorphism problem is decidable (uniformly soluble) when k is algebraically closed.

In the paper [12], monomial algebras (defined by the form "monomial $= 0$" only), the simplest class of finitely presented algebras, have been already shown to have a very effective solution, based on the presentation uniqueness theorem. That is, every finite-dimensional monomial algebra has a unique irredundant presentation up to a permutation of generators. In this case, surprisingly, it turns out that the answer is independent of the ground field k. Moreover, as to non-degenerate binomial algebras (defined by the form "monomial $= 0$" or "binomial $= 0$ (*non-degenerate relations*) "), the paper [13] has provided some necessary conditions for an existence of isomorphism, which enable us to construct an effective procedure for solving the problem. These conditions were described in terms of a partially ordered set, which was defined by the set of k-linear bases endowed with a left-prefix ordering. In addition, this poset was also used to prove the aforementioned unique presentation theorem for monomial algebras.

This paper deals with the problem for general finitely presented algebras: given two finitely presented algebras A and B of the same dimension over a field k, decide whether they are isomorphic as k-algebras. We can no longer define the posets as above from A and B. In Section 2, we provide a general criterion for deciding an algebra isomorphism. When k is algebraically closed, it is shown in Section 3 that this problem is reduced to a radical membership problem in a commutative ring, by Hilbert's zero point theorem. That is, let f be the determinant of a generic k-linear mapping from A to B and I an ideal derived from the defining relations of A and B. Then A is isomorphic to B if and only if f is not in $\sqrt{I}$, the radical of I. Moreover, in Section 4, an efficient technique for computing f is presented. This technique is based on the fact that the result of f does not include any entry of the first column vector. In Section 5, we propose two Gröbner basis methods for solving our problem: rewriting method and adjoining method. The first is to compute the Gröbner basis of $\sqrt{I}$ and then to rewrite f modulo $\sqrt{I}$. An appropriate ordering of variables for the basis is also proposed. The second method is a well-known technique: to adjoin a new indeterminate t to the commutative ring and to judge solvability of a certain system of algebraic equations; the union of I and $tf - 1$. The rewriting method is useful mainly when k is an algebraically closed field or when we demonstrate that two algebras are *not* isomorphic. The adjoining method can provide a more precise solution when k is not necessarily algebraically closed. If two algebras are isomorphic over an algebraic extension k' of k, then this method can find k'. We present the respective algorithms based on the two methods.

Throughout this paper, $k\langle X|R\rangle$ denotes the non-commutative associative algebra generated by variables in X having relations R.

2   Isomorphism Criterion

Let k be a field of arbitrary characteristic and A and B be finite-dimensional finitely presented algebras over k, defined by $k\langle X_A | R_A \rangle$ and $k\langle X_B | R_B \rangle$ respectively. Take $S_A = \{t_i\}$ and $S_B = \{s_j\}$ as k-linear bases of A and B so that $t_1 = 1_A$ and $s_1 = 1_B$. By computing *finite* non-commutative Gröbner bases of R_A and R_B, we construct S_A and S_B using the bases, by a method similar to that in the commutative case ([2], METHOD 6.6). See [10], [9], and [7] for the details of non-commutative Gröbner bases. Mora's algorithm ([10]) can always find a *finite* non-commutative Gröbner basis in the finite-dimensional case. In addition, using the finite Gröbner basis, we can determine whether a given finitely presented algebra is finite-dimensional, by its Ufnarovskij graph (see [14] and [3]). However, note that the main purpose of this paper is not to use *non-commutative* Gröbner bases here, but to use *commutative* Gröbner bases as discussed in Section 5.

Now assume that $\sharp S_A = \sharp S_B = n$ (i.e. $\dim A = \dim B$). Let us define a k-linear mapping $\varphi : A \to B$ by image of each generator of A as follows:

$$\varphi(x_\lambda) = \sum_{j=1}^{n} a_{\lambda j} s_j, \quad a_{\lambda j} \in k \tag{1}$$

where $X_A = \{x_\lambda\}$. If φ acts on each element t_i of S_A as ring homomorphism, then $\varphi(t_i)$ is formally determined under R_B in the following manner:

$$\varphi(t_i) = \sum_{j=1}^{n} b_{ij} s_j, \quad b_{ij} \in k$$

Thus we have a square matrix $M_\varphi = (b_{ij})_{1 \le i,j \le n}$ of degree n. Here we assume that $\varphi(1_A) = 1_B$. That is,

$$b_{1j} = \begin{cases} 1 & \text{if } j = 1 \\ 0 & \text{otherwise} \end{cases}$$

Let $\varphi(R_A)$ be the set of all relations applied to the respective relations in R_A by φ as ring homomorphism. Then we have the following isomorphism criterion.

Theorem 1 (Isomorphism Criterion) *A and B are isomorphic as k-algebras* $\Longleftrightarrow$ *for the above φ, both $\varphi(R_A)$ under R_B and $\det M_\varphi \neq 0$ can be satisfied in the field k.*

Proof. ($\Longrightarrow$) We apply the mapping φ to an isomorphism between A and B. Then $\varphi(R_A)$ under R_B and $\det M_\varphi \neq 0$ can be simultaneously satisfied in k

because φ is a homomorphism from A to B and a bijective k-linear mapping. ($\Longleftarrow$) We can naturally view φ as a homomorphism from $k\langle X_A\rangle$ to B, where $k\langle X_A\rangle$ is the free associative algebra generated by X_A over k. By assumption, φ is a surjective homomorphism with $R_A \subseteq$ Ker φ. Thus there exists a surjective homomorphism $\bar{\varphi}$: $k\langle X_A\rangle/\langle R_A\rangle \to B$ such that $\varphi = \bar{\varphi} \circ c$, where $\langle R_A\rangle$ is the two-sided ideal generated by R_A and c is the canonical mapping from $k\langle X_A\rangle$ onto $k\langle X_A\rangle/\langle R_A\rangle$. However, the assumption that $\dim A = \dim B$ implies the result since we can identify A as $k\langle X_A\rangle/\langle R_A\rangle$. ■

Example Decide whether $A = k\langle x|x^2 = x\rangle$ and $B = k\langle x|x^2 = 1\rangle$ are isomorphic. (In this case, both are *degenerate binomial algebras*, which were left out of consideration in [13].)

Take $S_A = \{1, x\}$ and $S_B = \{1, x\}$ (using the same symbol x, if there is no confusion possible). Now, for a k-linear mapping φ from A to B, set

$$\varphi(x) = a + bx, \quad a, b \in k$$

Thus we have

$$M_\varphi = \begin{pmatrix} 1 & 0 \\ a & b \end{pmatrix}$$

$\varphi(R_A)$ is $\{\varphi(x^2) = \varphi(x)\}$. Since "left hand side" $= \varphi(x)^2 = a^2 + 2abx + b^2x^2$, "right hand side" $= a + bx$, and $x^2 = 1$ in B, we have

$$\begin{cases} 2ab & = & b \\ a^2 + b^2 & = & a \end{cases}$$

Since $\det M_\varphi = b$, when char $k \neq 2$, $\det M_\varphi \neq 0$ can also be satisfied by setting $a = 1/2$ and $b = \pm 1/2$. Consequently,

$$\begin{cases} A \simeq B & \text{if char } k \neq 2 \\ A \not\simeq B & \text{if char } k = 2 \end{cases}$$

3   Radical Membership Problem

All relations in $\varphi(R_A)$ can be expressed in the form of equalities in commutative ring (since k is a commutative ring). Hence we can reduce our problem into a radical membership problem.

Let each $a_{\lambda j}$ in the expression (1) correspond one to one with an indeterminate $x_{\lambda j}$ over k.

$$a_{\lambda j} \longrightarrow x_{\lambda j}$$

Of course, each $x_{\lambda j}$ is not in X_A or X_B. We define a commutative ring $R = k[x_{\lambda j}]$ (generated by $x_{\lambda j}$'s over k). φ is naturally extended to the k-linear mapping $\tilde{\varphi}: R\langle X_A|R_A\rangle \to R\langle X_B|R_B\rangle$ (coefficient ring extension). Thus

$$\tilde{\varphi}(x_\lambda) = \sum_{j=1}^{n} x_{\lambda j} s_j. \tag{2}$$

Let $M_{\tilde{\varphi}} = (y_{ij})$ denote the formal image of $M_\varphi = (b_{ij})$ by the above correspondence as ring homomorphism. Obviously, we have $det M_{\tilde{\varphi}} \in R$. Put $f(x_{\lambda j}) = det M_{\tilde{\varphi}}$.

Similarly, $\tilde{\varphi}(R_A)$ is the set of all relations applied to the respective relations in R_A by $\tilde{\varphi}$ as ring homomorphism. Now let the relations among $x_{\lambda j}$'s derived from $\tilde{\varphi}(R_A)$ be as follows:

$$\begin{cases} f_1(x_{\lambda j}) &= 0 \\ &\vdots \\ f_r(x_{\lambda j}) &= 0 \end{cases}$$

We have the polynomials $f_1(x_{\lambda j}), \cdots, f_r(x_{\lambda j})$ in R. Now put $I = (f_1, \cdots, f_r)$ (the ideal in R generated by these polynomials).

We say that $A \simeq B$ for an extension (k') if $k'\langle X_A|R_A\rangle \simeq k'\langle X_B|R_B\rangle$ for some algebraic extension k' of k. Then we have the following isomorphism test.

Theorem 2 (Isomorphism Test) $A \simeq B$ for an extension $\Longleftrightarrow f \notin \sqrt{I}$, where $\sqrt{I} = \{x \in R \mid \exists n \text{ s.t. } x^n \in I\}$.

Proof. $A \simeq B$ for an extension k'

$\Longleftrightarrow$ both $\varphi(R_A)$ under R_B and $f \neq 0$ can be satisfied in k' (by Theorem 1)

$\Longleftrightarrow$ there exist solutions of I satisfying $f \neq 0$

$\Longleftrightarrow$ ¬"every solution of I is a solution of $f = 0$"

$\Longleftrightarrow$ ¬"$f \in \sqrt{I}$" (by Hilbert's zero point theorem ([16]))

$\Longleftrightarrow$ $f \notin \sqrt{I}$ ∎

Corollary 1 *When k is algebraically closed, $A \simeq B \Longleftrightarrow f \notin \sqrt{I}$.*

4   Constant Elimination Technique

Before considering Gröbner basis methods, we show an efficient technique for computing the determinant $det M_{\tilde{\varphi}}$. First, the following theorem shows that it does not include any entry of the first column vector.

Theorem 3 (Determinant Property) $det M_{\tilde{\varphi}} \in k[x_{\lambda j}; j \geq 2]$.

Proof. First, from $\tilde{\varphi}(x_\lambda) = \sum_{j=1}^{n} x_{\lambda j} s_j$, the matrix $M_{\tilde{\varphi}}$ contains the row vectors $(x_{\lambda 1}, \cdots, x_{\lambda n})$ for all λ's. Every element t_i of S_A is a k-linear combination of monomials of x_λ's in R_A. Since the image by $\tilde{\varphi}$ of a monomial of x_λ's contains products between $x_{\mu 1}$ and $\sum_{j=1}^{n} x_{\lambda j} s_j$ for some μ's and λ's, the image of t_i makes a row vector containing products between $x_{\mu 1}$ and $(x_{\lambda 1}, \cdots, x_{\lambda n})$ for some μ's and λ's. Conversely, all $x_{\nu 1}$'s appearing except in the first column vector derive from these products. Thus, by a property of determinant, we can eliminate all these products and in the new matrix, no $x_{\nu 1}$'s appear except in the first column vector. Finally, since the first row vector is $(1, 0, \cdots, 0)$, the determinant can be calculated without using any entry of the first column vector. ∎

By this theorem, we can propose a technique for calculating $det M_{\tilde{\varphi}}$ that we eliminate constant terms in the initial definition ((2), Section 3) of $\tilde{\varphi}$. That is, **Constant Elimination Technique:** We assume that $\tilde{\varphi}$ is settled in the following manner:

$$\tilde{\varphi}(x_\lambda) = \sum_{j=2}^{n} x_{\lambda j} s_j. \tag{3}$$

We want to stress, however, that this assumption is *only valid for computing* $det M_{\tilde{\varphi}}$ and that we have to come back to (2) for developing $\tilde{\varphi}(R_A)$. In fact, there is not always an isomorphism such as (3) (see Example in Section 2).

The elimination of constant terms makes it easier to compute $\tilde{\varphi}(t_i)$ and $det M_{\tilde{\varphi}}$.

5 Gröbner Basis Methods

In Section 3, we have reduced our problem into a radical ideal membership problem. So we can consider two Gröbner basis methods: rewriting method and adjoining method. Hereafter $GB(J)$ denotes the (reduced) Gröbner basis of an ideal J calculated by Buchberger Algorithm ([2], ALGORITHM 6.3).

5.1 Rewriting Method

It is immediate from Theorem 2 that the following method is correct.

Rewriting Method: $A \simeq B$ for an extension $\Longleftrightarrow f \not\rightarrow^* 0$ modulo $GB(\sqrt{I})$ in R.
When k is algebraically closed, $A \simeq B \Longleftrightarrow f \not\rightarrow^* 0$ modulo $GB(\sqrt{I})$ in R.

Moreover, by Theorem 3, an ordering of variables such that $x_{\lambda 1}$'s *are the smallest* is the most appropriate, since the determinant will be rewritten.

We can use several algorithms to construct the Gröbner basis of the radical $\sqrt{I}$ ([1], [8], [15] etc). However, we do not go to details here. When I is zero-

dimensional and k is perfect, we can employ a simple method based on Lemma 92 of Seidenberg ([11]). The Seidenberg method is as follows ([5], Section 9):
(1) Find univariate polynomials f_i in each variable x_i by the lexicographic Gröbner basis w.r.t. a variable ordering with x_i the smallest.
(2) Put $g_i = f_i/gcd(f_i, f_i')$, where f_i' is the derivative of f_i taken w.r.t. x_i.
(3) Then $\sqrt{I} = (I, g_1, \cdots, g_n)$.

In general, the Seidenberg method is very inefficient. A more efficient algorithm is proposed in [4]. In the example (Example 2) given later, however, we will find the Seidenberg method to be sufficient.

In addition, since $I \subseteq \sqrt{I}$, we have
$1 \in GB(I)$ or $f \to^* 0$ modulo $GB(I) \Longrightarrow A \not\simeq B$ for any extension.

Consequently, together with Constant Elimination Technique in Section 4, we can construct an algorithm for solving our problem.

Algorithm ITR $(k; X_A, R_A; X_B, R_B)$: **Isomorphism Test by Rewriting**

> **Input:** Field k; two finite presentations $X_A, R_A; X_B, R_B$.
> **Assumption:** $dimA = dimB < \infty$, where $A = k\langle X_A | R_A \rangle$ and $B = k\langle X_B | R_B \rangle$.
> **Output:** TRUE if $A \simeq B$ for an extension, otherwise FALSE.
> **Step 1:** Find linear bases $S_A = \{t_i\}$ and $S_B = \{s_j\}$.
> **Step 2:** Define a k-linear mapping: $\tilde{\varphi}(x_\lambda) = \sum_{j=1}^n x_{\lambda j} s_j$, where $X_A = \{x_\lambda\}$ and each $x_{\lambda j}$ is an indeterminate over k. Put $R = k[x_{\lambda j}]$.
> **Step 3:** Define a k-linear mapping: $\tilde{\varphi}^*(x_\lambda) = \sum_{j=2}^n x_{\lambda j} s_j$ and make the matrix expression $M_{\tilde{\varphi}^*}$ of $\tilde{\varphi}^*$ under S_A and S_B. % Constant Elimination Technique
> **Step 4:** Compute $f = det M_{\tilde{\varphi}^*}$. % $det M_{\tilde{\varphi}^*} = det M_{\tilde{\varphi}}$ by Theorem 3
> **Step 5:** Construct the set of polynomials $\{f_1, \cdots, f_r\}$ which consists of all coefficients of $\tilde{\varphi}(R_A)$ (under the assumption that $\tilde{\varphi}$ behaves as a ring homomorphism) w.r.t. each basis in S_B, using the relation R_B. Put $I = (f_1, \cdots, f_r)$ (the ideal in R).
> **Step 6:** Compute $G_I = GB(I)$ of I w.r.t. a variable ordering such that $x_{\lambda 1}$'s are the smallest.
> **Step 7:** If $1 \in G_I$, then return FALSE (not even homomorphic); otherwise if $f \to^* 0$ modulo G_I then return FALSE.
> **Step 8:** Compute the Gröbner basis of $G_{\sqrt{I}}$ of $\sqrt{I}$.
> **Step 9:** If $f \to^* 0$ modulo $G_{\sqrt{I}}$ then return FALSE otherwise TRUE.

Let us give three examples of running **ITR**, assuming that k is an algebraically closed field, say $\mathbf{C}$, the field of complex numbers. We used REDUCE 3.3 on NEC PC-98 (BUG, Sapporo Japan) for calculating Gröbner bases.

Example 1 First let us observe monomial algebras, although this case is immediate from the result of [12]. Are $A = k\langle x, y | xy = yx = y^2 = x^3 = 0 \rangle$ and

$B = k\langle x, y | x^2 = yx = y^2 = 0\rangle$ isomorphic?

Input: C; $\{x, y\}, \{xy = yx = y^2 = x^3 = 0\}$; $\{x, y\}, \{x^2 = yx = y^2 = 0\}$.

Output: FALSE.

[Trace of ITR]

Step 1: $S_A = \{1, x, y, x^2\}$ and $S_B = \{1, x, y, xy\}$.

Step 2: Define $\tilde{\varphi}$ as follows:

$$\begin{cases} \tilde{\varphi}(x) & = & a + bx + cy + dxy \\ \tilde{\varphi}(y) & = & a' + b'x + c'y + d'xy \end{cases}$$

Here $a, b, c, d, a', b', c', d'$ are indeterminates over k. $R = k[a, b, c, d, a', b', c', d']$.

Step 3: Define $\tilde{\varphi}^*$ as follows:

$$\begin{cases} \tilde{\varphi}^*(x) & = & bx + cy + dxy \\ \tilde{\varphi}^*(y) & = & b'x + c'y + d'xy \end{cases}$$

We have

$$M_{\tilde{\varphi}^*} = \begin{pmatrix} 1 & 0 & 0 & 0 \\ 0 & b & c & d \\ 0 & b' & c' & d' \\ 0 & 0 & 0 & bc \end{pmatrix}$$

Step 4: $f = det M_{\tilde{\varphi}^*} = bc(bc' - b'c)$.

Step 5: From $\tilde{\varphi}(xy) = \tilde{\varphi}(yx) = \tilde{\varphi}(y^2) = \tilde{\varphi}(x^3) = 0$, we have $I = (aa', ab' + a'b, ac' + a'c, ad' + a'd + bc', ad' + a'd + b'c, a'^2, a'b', a'c', 2a'd' + b'c', a^3, a^2b, a^2c, a^2d + abc)$.

Step 6: With respect to the purely lexicographical monomial ordering with $b > b' > c > c' > d > d' > a > a'$, we have $G_I = GB(I) = \{a(da + bc), a'd + c'b + d'a, a^2b, a'b + b'a, b'^2a, a'd + b'c + d'a, 2a'd' + b'c', b'd'a, b'a^2, a'b', a^2c, a'c + c'a, c'^2a, c'd'a, c'a^2, a'c', d'^2a^2, a^3, a'a, a'^2\}$.

Step 7: $1 \notin G_I$. But since $f = bc(bc' - b'c) \rightarrow 0$ modulo G_I, FALSE is returned.

Example 2 Decide whether $A = k\langle x | x^3 = x^2 - 1\rangle$ and $B = k\langle x, y | x^2 = 4x, xy = yx = 0, y^2 = 3y\rangle$ are isomorphic.

Input: C; $\{x\}, \{x^3 = x^2 - 1\}$; $\{x, y\}, \{x^2 = 4x, xy = yx = 0, y^2 = 3y\}$.

Output: TRUE.

[Trace of ITR]

Step 1: $S_A = \{1, x, x^2\}$ and $S_B = \{1, x, y\}$.

Step 2: $\tilde{\varphi}(x) = a + bx + cy$, where a, b, c are indeterminates over k. $R = k[a, b, c]$.

Step 3: $\tilde{\varphi}^*(x) = bx + cy$, and we have $M_{\tilde{\varphi}^*} = \begin{pmatrix} 1 & 0 & 0 \\ 0 & b & c \\ 0 & 4b^2 & 3c^2 \end{pmatrix}$.

Step 4: $f = det M_{\tilde{\varphi}^*} = bc(3c - 4b)$.

Step 5: From $\tilde{\varphi}(x^3) = \tilde{\varphi}(x^2 - 1)$, $I = (a^3 - a^2 + 1, 12ab^2 + 3a^2b + 16b^3 - 4b^2 - 2ab, 9ac^2 + 3a^2c + 9c^3 - 3c^2 - 2ca)$.

Step 6: $G_I = GB(I) = \{16b^3 + 12b^2a - 4b^2 + 3ba^2 - 2ba, 9c^3 + 9c^2a - 3c^2 + 3ca^2 - 2ca, a^3 - a^2 + 1\}$, w.r.t. the purely lexicographical ordering with $b > c > a$.

Step 7: $1 \notin G_I$. And $f \not\to 0$ modulo G_I.

Step 8: I is zero-dimensional by the form of G_I ([2], METHOD 6.9). Hence by the Seidenberg method, we have $G_{\sqrt{I}} = G_I$ since $f_1(a) = a^3 - a^2 + 1$, $f_2(b) = b^7 - 1/8b^5 + 1/256b^3 + 23/4096b$, $f_3(c) = c^7 - 2/9c^5 + 1/81c^3 + 23/729c$, and so $gcd(f_i, f_i') = 1$ for all i.

Step 9: Because $f \not\to 0$ modulo $G_{\sqrt{I}} \ (= G_I)$, TRUE is returned.

Example 3 (non-commutative case) What about $A = k\langle x, y | xy = x, yx = 2y \rangle$ and $B = k\langle x, y | xy = x, yx = 0, y^2 = 1 \rangle$?

Input: C; $\{x, y\}, \{xy = x, yx = 2y\}$; $\{x, y\}, \{xy = x, yx = 0, y^2 = 1\}$.

Output: FALSE.

[Trace of ITR]

Step 1: $S_A = \{1, x, y\}$ and $S_B = \{1, x, y\}$.

Step 2: $\tilde{\varphi}(x) = a + bx + cy$, $\tilde{\varphi}(y) = a' + b'x + c'y$. $R = k[a, b, c, a', b', c']$.

Step 3: $\tilde{\varphi}^*(x) = bx + cy$, $\tilde{\varphi}^*(y) = b'x + c'y$. And $M_{\tilde{\varphi}^*} = \begin{pmatrix} 1 & 0 & 0 \\ 0 & b & c \\ 0 & b' & c' \end{pmatrix}$.

Step 4: $f = det M_{\tilde{\varphi}^*} = bc' - b'c$.

Step 5: $I = (aa' + cc' - a, ab' + ba' + bc' - b, ac' + ca' - c, aa' + cc' - 2a', ab' + ba' + cb' - 2b', ac' + ca' - 2c')$.

Step 6: $G_I = GB(I) = \{b, b', c - 2c', c'^2 + a'^2 - a', c'a' - 1/2c', a - 2a', a'^3 - 3/2a'^2 + 1/2a'\}$, w.r.t. the purely lexicographical ordering with $b > b' > c > c' > a > a'$.

Step 7: $1 \notin G_I$. But $f \to 0$ modulo G_I. Hence FALSE is returned.

5.2 Adjoining Method

We propose another method for our problem as follows:

Adjoining Method: Introduce a new indeterminate t. Then, $A \simeq B$ for an extension $\iff 1 \notin GB((I, tf - 1))$ in $R[t]$.

When k is algebraically closed, $A \simeq B \iff 1 \notin GB((I, tf - 1))$ in $R[t]$.

The adjoining method is a well-known technique for the radical membership, based on an alternative description of Hilbert's zero point theorem. In fact, the correctness of this method is shown by that "I and $f \neq 0$ in R have a solution" $\iff$ "I and $tf - 1$ in $R[t]$ have a solution".

We can construct an algorithm based on the adjoining method. Steps 1 to 5 are exactly the same as **ITR**.

Algorithm ITA $(k; X_A, R_A; X_B, R_B)$: **Isomorphism Test by Adjoining**

   Input: Field k; two finite presentations $X_A, R_A; X_B, R_B$.

Assumption: $dimA = dimB < \infty$, where $A = k\langle X_A|R_A\rangle$ and $B = k\langle X_B|R_B\rangle$.

Output: TRUE if $A \simeq B$ for an extension, otherwise FALSE.

Step 1: Find linear bases $S_A = \{t_i\}$ and $S_B = \{s_j\}$.

Step 2: Define a k-linear mapping: $\tilde\varphi(x_\lambda) = \sum_{j=1}^n x_{\lambda j}s_j$, where $X_A = \{x_\lambda\}$ and each $x_{\lambda j}$ is an indeterminate over k. Put $R = k[x_{\lambda j}]$.

Step 3: Define a k-linear mapping: $\tilde\varphi^*(x_\lambda) = \sum_{j=2}^n x_{\lambda j}s_j$ and make the matrix expression $M_{\tilde\varphi^*}$ of $\tilde\varphi^*$ under S_A and S_B.

Step 4: Compute $f = det M_{\tilde\varphi^*}$.

Step 5: Construct the set of polynomials $\{f_1, \cdots, f_r\}$ which consists of all coefficients of $\tilde\varphi(R_A)$ (under the assumption that $\tilde\varphi$ behaves as a ring homomorphism) w.r.t. each basis in S_B, using the relation R_B. Put $I = (f_1, \cdots, f_r)$ (the ideal in R).

Step 6: Compute $G_{\tilde I} = GB(\tilde I)$ for the ideal $\tilde I = (I, tf - 1)$ in $R[t]$, w.r.t. the purely lexicographical monomial ordering with a variable ordering such that t is the smallest.

Step 7: If $1 \in G_{\tilde I}$ then return FALSE otherwise return TRUE.

Remark The purely lexicographical ordering of monomials is appropriate for finding an extension field such that A and B are isomorphic over it. This is based on the well-known property that the lexicographical Gröbner basis contains a univariate polynomial in the variable with the smallest ordering. The variable ordering such that t is the smallest may be just heuristic for obtaining a univariate polynomial with minimal degree. An example will be given later.

Let us give examples of running **ITA** for the same problems as Section 5.1. Steps 1 to 5 are omitted in each trace.

Example 1
Input: C; $\{x,y\}, \{xy = yx = y^2 = x^3 = 0\}$; $\{x,y\}, \{x^2 = yx = y^2 = 0\}$.
Output: FALSE.
[Trace of ITA]
Step 6: $G_{\tilde I} = GB((I, tbc(bc' - b'c) - 1)) = \{1\}$ in $k[a,b,c,d,a',b',c',d',t]$, w.r.t. the purely lexicographical ordering with $a > a' > b > b' > c > c' > d > d' > t$.
Step 7: FALSE since $1 \in G_{\tilde I}$.

Example 2
Input: C; $\{x\}, \{x^3 = x^2 - 1\}$; $\{x,y\}, \{x^2 = 4x, xy = yx = 0, y^2 = 3y\}$.
Output: TRUE.
[Trace of ITA]
Step 6: $G_{\tilde I} = GB((I, tf - 1)) = \{a + 9/25c^2 + 69/200ct + 3/2c - 9/25, b - 27/100c^2 - 207/800ct - 3/8c + 1/50, c^3 - 1/9c + 23/324t, t^2 + 144/23\}$ in $k[a,b,c,t]$, w.r.t. the purely lexicographical ordering with $a > b > c > t$.
Step 7: TRUE since $1 \notin G_{\tilde I}$.

Example 3

Input: $\mathbf{C}$; $\{x, y\}, \{xy = x, yx = 2y\}$; $\{x, y\}, \{xy = x, yx = 0, y^2 = 1\}$.
Output: FALSE.
[Trace of ITA]
Step 6: $G_{\bar{I}} = GB((I, tf - 1)) = \{1\}$ in $k[a, b, c, a', b', c't]$, w.r.t. the purely lexicographical ordering with $a > a' > b > b' > c > c' > t$.
Step 7: FALSE since $1 \in G_{\bar{I}}$.

Furthermore, when k is not algebraically closed, if two algebras are isomorphic over an algebraic extension k' of k, then we can find k' by simply inspecting $G_{\bar{I}}$ (in **Step 6** of **ITA**). In Example 2, when k is $\mathbf{Q}$, the field of rational numbers, $t^2 + 144/23$ has no solution in k. This implies that $A \not\simeq B$ for $\mathbf{Q}$, since if a, b, c (a solution of I) $\in \mathbf{Q}$, then $f \in \mathbf{Q}$ and hence $t = f^{-1} \in \mathbf{Q}$. From $G_{\bar{I}}$, we can obtain the result that:

$$\mathbf{Q}(\sqrt{-23}, \alpha)\langle x | x^3 = x^2 - 1 \rangle \simeq \mathbf{Q}(\sqrt{-23}, \alpha)\langle x, y | x^2 = 4x, xy = yx = 0, y^2 = 3y \rangle,$$

where α is a root of $x^3 - 1/9x + \sqrt{-23}/27 = 0$.

Finally, let us note the decidability of the isomorphism problem.

Theorem 4 (Decidability) *It is decidable whether two given finite dimensional finitely presented algebras are isomorphic for an extension.*

Proof. As mentioned in Section 2, we can find finite (non-commutative) Gröbner bases for two given relations since the given algebras are finite-dimensional. Thus, the dimensions of the underlying linear spaces are computable by a method similar to the Gröbner basis method ([2], METHOD 6.6). If the dimensions are not equal, return FALSE. Otherwise employ **ITA** (or **ITR**). All we need to check therein is **Step 1**, i.e., to find the linear bases, but it is also possible by the same method. ∎

6   Conclusion

We have proposed two Gröbner basis methods for solving the isomorphism problem for finitely presented algebras. As a result, we have obtained the decidability of this problem.

The membership test in the radical $\sqrt{I}$ of the ideal I defined by two presentations is an essential part for our problem. This means that computing the whole set of $\sqrt{I}$ itself is not necessary. Thus, naturally, the adjoining method is more tractable than the rewriting method in principle. However, in **ITA**, we have to introduce a new variable (t) and a polynomial $(tf - 1)$ of high degree, which may require a lot of time and memory for the Gröbner basis computation. On the other hand, in **ITR**, the answer is sometimes obtained in the stage of **Step 7**, without needing to compute the radical. Moreover, it should be noted that the capacities of **ITR** and **ITA** are both dependent on the power limit of the computation of symbolic determinants.

Acknowledgement

The author is grateful to Professor T. Mora for his remarks on non commutative Gröbner basis. The author also would like to thank anonymous referees for their invaluable comments including the latest results on the radical computation.

References

[1] Alonso, M. E., Mora, T. and Raimondo, M., Local decomposition algorithms, *Proc. AAECC 8, L. N. Comp. Sci.* **508** (1991), 208-221.

[2] Buchberger, B., Gröbner Bases: An Algorithmic Method in Polynomial Ideal Theory, *Chapter 6 in Multidimensional Systems Theory (N. K. Bose ed.), D. Reidel Publishing Company* (1985), 184-232.

[3] Gateva-Ivanova, T. and Latyshev, V., On recognisable properties of associative algebras, *J. Symbolic Computation* **6** (1988), 371-388.

[4] Gianni, P. and Mora, T., Algebraic solution of systems of polynomial equations using Groebner bases, *Proc. AAECC 5, L. N. Comp. Sci.* **356** (1989), 247-257.

[5] Gianni, P., Trager B., and Zacharias G., Gröbner Bases and Primary Decomposition of Polynomial Ideals, *J. Symbolic Computation* **6** (1988), 149-168.

[6] Glass, A. M. W., The Word and Isomorphism Problems in Universal Algebra, *Proc. Universal Algebra and Lattice Theory, L. N. Math.* **1149** (1984), 123-128.

[7] Kandri-Rody, A. and Weispfenning, V., Non-commutative Gröbner Bases in Algebras of Solvable Type, *J. Symbolic Computation* **9** (1990), 1-26.

[8] Krick, T. and Logar, A., An algorithm for the computation of the radical of an ideal in the ring of polynomials, *Proc. AAECC 9, L. N. Comp. Sci.* **539** (1991), 195-205.

[9] Le Chenadec, P., Canonical Forms in Finitely Presented Algebras, *Research Notes in Theoretical Computer Science, Pitman*, (1986).

[10] Mora, T., Groebner bases for non-commutative polynomial rings, *Proc. AAECC 3, L. N. Comp. Sci.* **229** (1986), 353-362.

[11] Seidenberg, A., Constructions in algebra, *Trans. Am. Math. Soc.* **197** (1974), 273-313.

[12] Shirayanagi, K., A Classification of Finite-dimensional Monomial Algebras, *Effective Methods in Algebraic Geometry (T. Mora and C. Traverso eds.), Progress in Mathematics 94, Birkhäuser* (1991), 469-482.

[13] Shirayanagi, K., On the Isomorphism Problem for Finite-dimensional Binomial Algebras, *Proc. International Symposium on Symbolic and Algebraic Computation (ISSAC-90)* (1990), 106-111.

[14] Ufnarovskij, V., A growth criterion for graphs and algebras defined by words, *Mat. Zametki* **31**, 465-472 (in Russian); English transl., *Math. Notes* **31** (1982), 238-241.

[15] Vasconcelos, W., Jacobian matrices and constructions in algebra, *Proc. AAECC 9, L. N. Comp. Sci.* **539** (1991), 48-64.

[16] Zariski, O. and Samuel, P., Commutative Algebra Vol. II, *Van Nostrand,* (1960).

Kiyoshi Shirayanagi
NTT Communication Science Laboratories
Hikaridai, Seika-cho, Soraku-gun, Kyoto 619-02 Japan
email: shirayan@progn.kecl.ntt.jp

COMPLEXITY OF BEZOUT'S THEOREM II
VOLUMES AND PROBABILITIES

M. Shub[*] S. Smale[**]

Abstract. In this paper we study volume estimates in the space of systems of n homegeneous polynomial equations of fixed degrees d_i with respect to a natural Hermitian structure on the space of such systems invariant under the action of the unitary group. We show that the average number of real roots of real systems is $\mathcal{D}^{1/2}$ where $\mathcal{D} = \prod d_i$ is the Bezout number. We estimate the volume of the subspace of badly conditioned problems and show that volume is bounded by a small degree polynomial in n, N and $\mathcal{D}$ times the reciprocal of the condition number to the fourth power. Here N is the dimension of the space of systems.

Section 1. Introduction.

This paper can be read independently of **Shub-Smale** hereafter referred to as [I], but is closely related to it. Here we confine ourselves to homogeneous polynomials and projective spaces although some extensions to the affine case may be dealt with as in [I].

The paper [I] can be read for background and more references.

First consider a real polynomial system $f : \mathbb{R}^{n+1} \to \mathbb{R}^n$ so that $f(z) = (f_1(z_0, \ldots, z_n), \ldots, f_n(z_0, \ldots, z_n))$ and each f_i is a homogeneous polynomial of degree $d_i > 0$. Let $\mathcal{H}^{\mathbb{R}}_{(d)}$ be the linear space of all such f where $d = (d_1, \ldots, d_n)$ (permitting $f_i \equiv 0$).

There is a natural inner product on $\mathcal{H}^{\mathbb{R}}_{(d)}$ invariant under the induced action of the orthogonal group $O(n+1)$ acting on $\mathbb{R}^{n+1}$ (so that $\langle f \circ O, g \circ O \rangle = \langle f, g \rangle$ for $O \in O(n+1)$). See Section 2 for this. This inner product defines a Riemannian structure and volume element on the corresponding projective space $P(\mathcal{H}^{\mathbb{R}}_{(d)})$ of lines (Fubini-Study). In turn this volume element defines a probability measure so that the following makes sense.

---

[*]Some of this work was carried out when Shub was visiting the Berkeley Math Department for 2 months   in 1992,

[**]Supported partially by NSF funds

Theorem A. *The average number of real zeros in $P_n(\mathbb{R})$ of $f \in P(\mathcal{H}_{(d)}^{\mathbb{R}})$ is $\mathcal{D}^{1/2}$ where $\mathcal{D} = \prod_{i=1}^{n} d_i$.*

According to Bezout's theorem, the (average) number of zeros over $\mathbb{C}$ is $\mathcal{D}$.

Eric **Kostlan-1991** proved this result earlier in the case that all of the d_i's are the same. In this paper Kostlan has a similar result for under-determined systems, or gives average volumes of real varieties. Moreover, **Kostlan-1987** suggested using an orthogonally (or unitarily over $\mathbb{C}$) invariant metric in these complexity matters. Such a metric was already used in the theory of group representations and harmonic analysis (see e.g. **Stein-Weiss**).

In 1943, M. **Kac** had found in one variable, the expected number of real roots to be asymptotic to $\frac{2}{\pi} \log d$ using the traditional measure.

Next consider the complex case, $f : \mathbb{C}^{n+1} \to \mathbb{C}^n$, with each coordinate function f_i a homogeneous polynomial of degree d_i. Let $\mathcal{H}_{(d)}$ be the linear space of such f, and $P(\mathcal{H}_{(d)})$ the corresponding complex projective space. Define

$$V_{(d)} = \{(f, \zeta) \in P(\mathcal{H}_{(d)}) \times P(\mathbb{C}^{n+1}) \mid f(\zeta) = 0\}.$$

This complex non-singular subvariety of codimension n plays a central role in our work.

Inspired by **Wilkinson** we define the condition number $\mu : V_{(d)} \to \mathbb{R}^+ \cup \infty$ by

$$\mu(f, \zeta) = \|Df(\zeta)|_{N_\zeta}^{-1} \Delta(d_i^{1/2} \|\zeta\|^{d_i - 1}) \| \, \|f\|.$$

Here $\Delta(y_i)$ is the diagonal matrix with y_i as the (i, i) entry and

$$N_\zeta = \{w \in \mathbb{P}(\mathbb{C}^{n+1}) \mid \langle w, \zeta \rangle = 0\}.$$

The careful reader will have noted our customary practice of identifying objects in linear spaces and their quotient projective spaces. But appropriate homogenization gives sense to our definitions as $\mu(f, \zeta)$ above. If $Df(\zeta)|_{N_\zeta}$ is singular then $\mu(f, \zeta) = \infty$. In [I] μ was defined and called μ_{proj}.

Example. Let $e_0 = (1, 0, \ldots, 0)$ and $f_i(z) = z_0^{d_i - 1} z_i$, $i = 1, \ldots, n$. We claim that $\mu(f, e_0) = \left(\sum_{i=1}^{n} \frac{1}{d_i}\right)^{1/2} D^{1/2}$, $D = \max_i d_i$, and thus (f, e_0) is a very well conditioned set of pairs varying over d.

Observe $N_{e_0} = \{(0, v_1, \ldots, v_n), v_i \in \mathbb{C}\}$ and that $Df(e_0)|_{N_{e_0}}$ is represented by the identity matrix. Moreover it is checked that $\|f\| = \left(\sum_{i=1}^{n} \frac{1}{d_i}\right)^{1/2}$. This yields our statement.

Typical numerical algorithms follow paths in $V_{(d)}$ and loss of precision due to roundoff errors can be controlled by upper bounds on the corresponding condition number μ. Moreover μ plays a primary role in the (exact arithmetic) complexity analysis of such algorithms (see [I]). Thus it is important to understand the probability distribution of μ. We will show:

Theorem B. *If $n > 1$, and $D \geq 1$, then the probability that $\mu(f,\zeta) > \mu_0$ is less than $K\frac{nN}{\mu_0^2}$. More precisely*

$$\frac{\text{Vol}\{(f,\zeta) \in V_{(d)} \mid \mu(f,\zeta) \geq \mu_0\}}{\text{Vol } V_{(d)}} \leq K\frac{nN}{\mu_0^2}.$$

Here N is the dimension of $\mathcal{H}_{(d)}$.

The constant K is a universal constant less than 25. The case $n = 1$ will be dealt with in Theorem D.

Note that as a consequence of Theorem B, we see that most (f,ζ) in $V_{(d)}$ are well conditioned. The bound is independent of $\mathcal{D}$ and a low polynomial in n and N.

Numerical analysis has a useful tradition of relating the condition number to the distance ρ of the nearest ill-posed problem (see e.g. **Eckart-Young, Demmel**). In our setting the ill-conditioned pairs $\Sigma' \subset V_{(d)}$ are described by

$$\Sigma' = \{(f,\zeta) \in V_{(d)} \mid Df(\zeta)|_{N_\zeta} \text{ is singular}\}.$$

It can be shown that Σ' is a non-singular hypersurface in $V_{(d)}$ by a transversality argument. For $\zeta \in \mathbb{C}^{n+1}$ let $\hat{V}_\zeta = \{f \in \mathcal{H}_{(d)} \mid f(\zeta) = 0\}$ and $V_\zeta = P(\hat{V}_\zeta)$. Then V_ζ can be naturally identified with $\pi_2^{-1}(\zeta) \subset V_{(d)}$ where $\pi_2 : V_{(d)} \to P_n$ is the restriction of the projection $P(\mathcal{H}_{(d)}) \times P_n \to P_n$, and $P_n = P(\mathbb{C}^{n+1})$.

Define $\rho(f,\zeta)$ to be the distance in V_ζ of (f,ζ) to $\Sigma' \cap V_\zeta$. This projective space distance is taken for convenience to be $d_P = \sin d_R$ where d_R is the Riemannian (or Fubini-Study) distance. The diameter of projective space is one.

A main result of [I] is:

Condition Number Theorem.

$$\mu(f,\zeta) = \frac{1}{\rho(f,\zeta)}.$$

A sketch of the proof is given in Section 2.

Thus Theorem B may be interpreted as giving distribution estimates of pairs (f,ζ) close to ill-conditioned ones.

We now pass to the more subtle situation corresponding to all roots of a given $f \in P(\mathcal{H}_{(d)})$. Define the condition number $\mu(f)$ of $f \in P(\mathcal{H}_{(d)})$, $\mu : P(\mathcal{H}_{(d)}) \to \mathbb{R}^+ \cup \infty$ by

$$\mu(f) = \max_{\substack{\zeta \\ f(\zeta)=0}} \mu(f,\zeta).$$

Example. $n = 1$, $f_d(z) = z_0^d - z_1^d$. The zeros of f_d are $(1, q)$, whre q is a dth root of unity. Take typically $q = 1$. Then $\|\zeta\| = \sqrt{2}$, $\zeta = (1, 1)$ and $\|f_d\| = \sqrt{2}$. Also $N_\zeta = \{(v, -v), v \in \mathbb{C}\}$, $\|Df_d(\zeta)|_{N_\zeta}^{-1}\| = \frac{1}{2d}$ and it follows that $\mu(f_d) = \frac{2^{d/2}}{4d^{1/2}}$.

Note that for this series f_d the condition number grows exponentially in d, and so is extremely ill-conditioned. Yet this f_d and its variations with exponential behavior are used in numerical analysis as a model starting polynomial whose roots are known. We will prove the existence of well-conditioned sequences $\{g_d\}$ (even for general $n \geq 1$), yet are unable to exhibit them even for the 1-variable case. In a further account we hope to develop this subject which in the case $n = 1$, is intimately connected to elliptic capacity and transfinite diameter (see **Tsuji**).

Defining $\rho(f) = \min_{\zeta, f(\zeta)=0} \rho(f, \zeta)$ we have:

Corollary of the Condition Number Theorem.

$$\mu(f) = \frac{1}{\rho(f)}.$$

Theorem C. Let $N_\rho = \{f \in P(\mathcal{H}_{(d)}) \mid \rho(f) < \rho\}$. Then for $\rho < \frac{1}{\sqrt{n}}$, $n > 1$, $D \geq 1$

$$\frac{\mathrm{Vol}\, N_\rho}{\mathrm{Vol}\, P(\mathcal{H}_{(d)})} \leq \frac{\rho^4 n^2 (n+1)(N-1)(N-2)D}{4}$$

where $N = \dim \mathcal{H}_{(d)}$.

The closest previous results in the direction of Theorem 3 are due to **Renegar**.

Remark. The exponent 4 in Theorem C is somewhat surprising. Since $\Sigma \subset P(\mathcal{H}_{(d)})$ is a hypersurface, dimensional considerations say that the volume of the tube of radius ρ around Σ should vary like ρ^2.

Problem. Is $d(f, \Sigma) = O(\rho_{(f)}^2)$?
The theorem says that on the average it is. There is one exceptional case where $d = (1, \ldots, 1)$ in which case $\Sigma \subset P(\mathcal{H}_{(1,1,\ldots,1)})$ is of complex codimension 2. In this exceptional case the ρ^4 could be expected.

If we consider $x^2 - 2\varepsilon xy = f(x, y)$ $f \in \mathcal{H}_{(2)}$, $x^2 - 2\varepsilon xy + \varepsilon^2 y^2 \in \Sigma$ but $\rho(f) = O(\varepsilon)$.

Corollary. For $n > 1$ and each $d = (d_1, \ldots, d_n)$ there is $f_{(d)} \in P(\mathcal{H}_{(d)})$ such that $\mu(f_{(d)}) \leq \left(\frac{n^2(n+1)(N-1)(N-2)D}{4}\right)^{1/4}$.

This is a consequence of Theorem 3 and the Corollary of the Condition Number Theorem.

As we have suggested, it is an interesting open problem to exhibit such $f_{(d)}$. These could serve as better starting points of numerical algorithms than those currently used.

The probabilistic estimates given in Theorems B and C, together with the results of [I] have implications on the efficiency of algorithms for solving non-linear systems. We hope to develop this point in a future paper.

In Theorem D we give the results of Theorem B and C for the case $n = 1$.

Theorem D. *For* $n = 1$ *and* $d > 1$

i)
$$\frac{\text{Vol}\{(f,\xi) \in V_{(d)} | \mu(f,\xi) \geq \mu_0 \geq 1\}}{\text{Vol } V_{(d)}} =$$

$$1 - \frac{(d+1)}{2}\left(1 - \frac{1}{\mu_0^2}\right)^{d-1} + \frac{(d-1)}{2}\left(1 - \frac{1}{\mu_0^2}\right)^d$$

ii) $\dfrac{\text{Vol } N_\rho}{\text{Vol } P(\mathcal{H}_{(d)})} \leq d(1 - (1 - \rho^2)^{d-1}(1 + (d-1)\rho^2))$ for $0 \leq \rho \leq 1$.

Section 2. Background.

The following diagram motivates the more abstract treatment of the next section.

$$P(\mathcal{H}_{(d)}) \times P_n$$
$$\cup$$
$$\pi_1 \qquad V_{(d)} \qquad \pi_2$$
$$\swarrow \qquad\qquad\qquad \searrow$$
$$P(\mathcal{H}_{(d)}) \qquad\qquad\qquad P_n$$

The map $\pi_1 : V_{(d)} \rightarrow P(\mathcal{H}_{(d)})$ is a branched covering, branched along Σ' so that on $V_{(d)} - \pi_1^{-1}(\Sigma)$, where $\Sigma = \pi_1(\Sigma')$, π_1 is a covering map. The fiber $\pi_1^{-1}(f)$, $f \in P(\mathcal{H}_{(d)})$, $f \notin \Sigma$ consists of $\mathcal{D}$ points, corresponding to the zeros of f.

Moreover, $\pi_2 : V_{(d)} \rightarrow P_n$ is a fiber map with fiber $\pi_2^{-1}(\zeta) = V_\zeta$ over $\zeta \in P_n$.

The Unitary Group $U(n + 1)$ is the group of linear automorphisms of $\mathbb{C}^{n+1}$ preserving the standard Hermitian inner product. This induces an action $x \rightarrow ux$, $u \in U(n + 1)$ on P_n as well as the action $f \rightarrow fu^{-1}$ on $P(\mathcal{H}_{(d)})$. Moreover $U(n + 1)$ acts on $P(\mathcal{H}_{(d)}) \times P_n$ by $(f, x) \rightarrow (fu^{-1}, ux)$ leaving $V_{(d)}$ invariant. The fiber map $\pi_2 : V_{(d)} \rightarrow P_n$ is also invariant under these actions.

The invariant norm on $\mathcal{H}_{(d)}$ is given by $\|f\|^2 = \sum_i \sum_\alpha |a_\alpha^i|^2 \begin{pmatrix} d_i \\ \alpha_1, \ldots, \alpha_n \end{pmatrix}^{-1}$

where $f_\alpha^i(z) = \sum_\alpha a_\alpha^i z^\alpha$, $\alpha = (\alpha_1, \ldots, \alpha_n)$ is a multi-index and $\begin{pmatrix} d_i \\ \alpha_1, \ldots, \alpha_n \end{pmatrix}$

is the multi-nomial coefficient, $\frac{d_i!}{\alpha_1! \ldots \alpha_n!}$.

Here is a sketch of the proof of the condition number theorem. The condition number $\mu : V_{(d)} \to \mathbb{R}$ is invariant under $U(n+1)$ using the chain rule (as in Lemma 1 of Section III-1 of [I]).

The variety $\Sigma' \subset V_{(d)}$ is unitarily invariant as well as the distance in V_ζ. This implies that $\rho : V_{(d)} \to \mathbb{R}$ is invariant under $U(n+1)$. Given $(f, x) \in V_{(d)}$ pick $u \in U(n+1)$ with $ux = e_0 = (1, 0, \ldots, 0)$. Then

$$\rho(f, x) = \rho(fu^{-1}, e_0), \text{ and } \frac{1}{\mu(fu^{-1}, e_0)} = \frac{1}{\mu(f, x)}.$$

Thus it is sufficient to prove the condition number theorm for $x = e_0$. For $f \in \mathcal{H}_{(d)}$, write $f = (f_1, \ldots, f_n)$

$$(*) \qquad f_i(z) = a_i z_0^{d_i} + z_0^{d_i - 1} \Sigma a_{ij} z_j + \ldots$$

If $f \in \hat{V}_{e_0}$ then the a_i are all zero, and note thus that $K = \{f \in \mathcal{H}_{(d)} \mid f_i(z) = a_i z_0^{d_i}\}$ is the orthogonal space to $\hat{V}_{e_0}$. Let $L(d) = \{f \in \mathcal{H}_{(d)} \mid f_i(z) = z_0^{d_i-1} \Sigma a_{ij} z_j\}$, $J(d) = $ the orthogonal complement of $L(d)$ in $\hat{V}_{e_0}$, and $\pi : V_{e_0} \to L(d)$ the projection.

Note that $\text{Null}(e_0) = \{u \in \mathbb{C}^{n+1} \mid \langle u, e_0 \rangle = 0\}$ is $\{(0, u_1, \ldots, u_n) \in \mathbb{C}^{n+1}\}$ and can be identified with $\mathbb{C}^n$. In this way, $Df(e_0)$ may be regarded as a linear map from $\mathbb{C}^n \to \mathbb{C}^n$ and as a matrix. This is the matrix (a_{ij}) when f has the form of $(*)$.

Let $\mathcal{M}(n)$ be the space of $n \times n$ matrices endowed with the Frobenius norm $\|M\|^2 = \Sigma |m_{ij}|^2$. Recall that $\Delta(y_i)$ denotes the diagonal matrix with entries $(y_1, \ldots, y_n)$.

Lemma. *The map $L(d) \to \mathcal{M}(n)$ sending $f \to \Delta(d_i^{-1/2}) Df(e_0)$ is a norm preserving linear isomorphism.*

The proof follows from the definition of the norm on $\mathcal{H}_{(d)} \supset L(d)$.

Now suppose $\|f\| = 1$. Then $\mu(f, e_0) = \|Df(e_0)^{-1} \Delta d_i^{1/2}\|$ using the operator norm. By the theorem of **Eckardt-Young** (see [I]), $\mu(f, e_0) = d(\Delta(d_i^{-1/2}) Df(e_0), S)^{-1}$ in $\mathcal{M}(n)$ where S is the set of singular matrices. It follows using the previous lemma that $\mu(f, e_0)$ is $\frac{1}{d(f, \Sigma' \cap \hat{V}_{e_0})}$ in $\mathcal{H}_{(d)}$. In the projective space this translates into our $\frac{1}{\rho(f, e_0)}$ completing the sketch of our proof.

Remark 1. In [I] it was part of the definition that $\mu(f, \zeta) \geq 1$. In fact it is a consequence of the original definition that $\mu(f, \zeta) \geq \sqrt{n}$. One uses the condition number theorem and the fact that $\rho \leq \sqrt{n}$ for any any matrix on the unit sphere in $\mathcal{M}(n)$.

Remark 2. There is a real version of this section where the unitary group is replaced by the orthogonal group, $O(n+1)$, acting on $\mathbb{R}^{n+1}$, etc.

We end this section by stating some standard facts that we use in our proofs. The volume $\mathrm{Vol}(S^{n-1})$ of the unit $(n-1)$-sphere S^{n-1} is $\frac{2\pi^{n/2}}{\Gamma(n/2)}$ using the gamma function. Let $P_n(\mathbb{R})$ be real projective space of dimension n and P_n complex projective space of (complex) dimension n. Then

$$\mathrm{Vol}\, P_n(\mathbb{R}) = \frac{1}{2} \mathrm{Vol}(S^n)$$

$$\mathrm{Vol}\, P_n = \frac{1}{2\pi} \mathrm{Vol}(S^{2n+1}).$$

We also use the following integral formula:

$$\int_0^1 t^p (1-t^2)^q \, dt = \frac{1}{2} \frac{\Gamma\left(\frac{p+1}{2}\right) \Gamma(q+1)}{\Gamma\left(\frac{p+1}{2} + q + 1\right)}.$$

Section 3. Some General Integral Formulae.

Let M, N be (real) compact Riemannian manifolds and V a compact submanifold of the product $M \times N$ with $\dim V = \dim M$. Suppose that the restriction $\pi_2 : V \to N$ of the projection $M \times N \to N$ is a locally trivial fibration. Let $V_y = \pi_2^{-1}(y)$. Let x be a regular value of $\pi_1 : V \to M$, the restriction of the projection $M \times N \to M$. Define $A(x,y) : T_y(N) \to T_x(M)$ to be the linear map whose graph is the orthogonal complement to $TV_y(x,y)$ in $TV(x,y)$. Let U be an open subset of V and $\#(x)$ be the number of points in $\pi_1^{-1}(x) \cap U$.

Theorem 1.

$$\int_{x \in \pi_1 U} \#(x) dM = \int_N \int_{V_y \cap U} \det(A^*(x,y) A(x,y))^{1/2} dV_y dN.$$

Proof.

$$\int_{\pi_1 U} \#(x) dM = \int_U |\det(D\pi_1)| dV = \int_N \int_{V_y \cap U} |\det(D\pi_1)| \frac{1}{N_J \pi_2} dV_y dN.$$

Here $N_J \pi_2 = |\det(D\pi_2 \mid T(V_y)^\perp)|$ is the normal Jacobian and $T(V_y)^\perp$ is the orthogonal complement of TV_y in $TV(x,y)$. Here the first equality is a version of the usual change of variable formula for integrals. The second equality is the coarea formula (**Morgan**).

By Sard's Theorem it is sufficient to show that

$$(*) \qquad |\det(D\pi_1)| \frac{1}{N_J \pi_2} = (\det A^* A)^{1/2}$$

where both sides are evaluated at $(x,y) \in M \times N$ and x is a regular value of $\pi_1 : V \to M$.

Let H_1 and H_2 be finite dimensional real vector spaces (complex vector spaces) with inner product (Hermitian product). Let $A : H_1 \to H_2$ be linear and define the graph of A as $\Gamma(A) = \{(x, A(x)) \mid x \in H_1\}$. Let $\pi : \Gamma(A) \to H_1$ be the restriction of the projection. Let $\Gamma(A)$ inherit the inner product structure of the product.

Lemma 1.

$$|\det \pi| = \frac{1}{\det(I + A^*A)^{1/2}}$$

where A^* is the adjoint of A.

Proof. Let $\begin{pmatrix} I \\ A \end{pmatrix} : H_1 \to H_1 \times H_2$ be the map $x \to (x, Ax)$. There is an orthogonal (unitary in the complex case) automorphism) of H_1 such that

$$O\left(\left(\begin{pmatrix} I \\ A \end{pmatrix}^* \begin{pmatrix} I \\ A \end{pmatrix}\right)^{1/2}\right) = \begin{pmatrix} I \\ A \end{pmatrix}$$

and hence $\left|\det \begin{pmatrix} I \\ A \end{pmatrix}\right| = \det(I + A^*A)^{1/2}$. Also $\operatorname{Det} \pi = \dfrac{1}{\operatorname{Det}\begin{pmatrix} I \\ A \end{pmatrix}}$ since

$\pi \circ \begin{pmatrix} I \\ A \end{pmatrix} = I$. These two equalities give the proof.

Lemma 2. *Let $B(x, y) : TM(x) \to TN(y)$ be the linear map whose graph $\Gamma(B(x, y))$ is $TV(x, y)$ for a regular value x of π_1. Then*

$$\det(I + B^*(x, y)B(x, y)) = \det(I + A^{-1*}(x, y)A^{-1}(x, y)).$$

Proof. $TV_y(x, y)$ is contained in the $TM(x)$ factor of $TM(x) \times TN(y)$. As $TV(x, y)$ is the orthogonal direct sum of $TV_y(x, y)$ and $\Gamma(A(x, y))$, it follows that $TV(x, y)$ is the graph of the linear map $B : TM(x) \to TN(y)$ which is zero on $TV_y(x, y)$ and $A^{-1}(x, y)$ on $(D\pi_1(x, y))(\Gamma(A(x, y)))$ which in turn is orthogonal to $TV_y(x, y)$ in $TM(x)$. Consequently $\det(I + B^*(x, y)B(x, y)) = \det(I + A^{-1*}(x, y)A^{-1}(x, y))$.

Now we return to the proof of (∗) and the theorem. By Lemmas 1 and 2, using $A = A(x, y)$,

$$|\det(D\pi_1)(x, y)| = \det(I + A^{-1*}A^{-1})^{-1/2}.$$

By Lemma 1 and the definition of A,

$$\frac{1}{N_J \pi_2(x, y)} = \det(I + A^*A)^{1/2}.$$

Thus

$$|\det(D\pi_1(x, y))|\frac{1}{N_J \pi_2(x, y)} = \frac{\det(I + A^*A)^{1/2}}{\det(I + A^{-1*}A^{-1})^{1/2}}.$$

The next lemma finishes the proof.

Lemma 3. *Let $A : V \to W$ be a linear isomorphism of finite dimensional vector spaces with inner product, then*

$$\frac{\det(I + A^*A)}{\det(I + A^{-1*}A^{-1})} = \det(A^*A).$$

Proof. Multiply numerator and denominator by $\det(AA^*)$

$$\frac{\det(I + A^*A)}{\det(I + A^{*-1}A^{-1})} = \frac{\det(AA^*)\det(I + A^*A)}{\det(AA^*)\det(I + A^{*-1}A^{-1})}$$
$$= \frac{\det(AA^*)\det(I + A^*A)}{\det(AA^* + I)}.$$

But then $\det(AA^* + I) = \det(I + A^*A)$ since AA^* and A^*A have the same eigenvalues.

Let U, V, M, π_1, π_2 etc. be as in Theorem 2.

Theorem 2.

$$\operatorname{Vol} U = \int_N \int_{V_y \cap U} \det(I + A^*A)^{1/2} dV_y dN.$$

Proof.

$$\operatorname{Vol} U = \int_U 1 dV = \int_N \int_{V_y \cap U} \frac{1}{N_J \pi_2} dV_y dN$$
$$= \int_N \int_{V_y \cap U} \det(I + A^*A)^{1/2} DV_y dN$$

by Lemma 1 and the definition of A.

Remark. The complex versions of Theorems 1 and 2 are true with the same proof. Then the exponents of $\frac{1}{2}$ on $\det(I + A^*A)^{1/2}$, $(\det A^*A)^{1/2}$ musts be removed as we pass to the real determinant. That is |real determinant| = |determinant|^2 for a complex matrix.

Section 4. Integration Formulae in $V_{(d)}$.

We use the notations introduced in Section 1, 2 and specialize the complex versions of Theorems 1 and 2 of Section 3 (cf. the remark at the end of Section 3) with $M = P(\mathcal{H}_{(d)})$ and $N = P_n$. Section 2 provides the background for the following.

Theorem 1. *Let U be an open set in $V_{(d)}$ which is unitarily invariant, $e_0 = (1,0,\ldots,0) \in P_n$ and $\#(f)$ be the number of points in $\pi_1^{-1}(f) \cap U$ (i.e. the number of zeros of f in U). Then*

(a) $\operatorname{Vol} U = \operatorname{Vol} P(n) \int_{V_{e_0} \cap U} \det(I + Df(e_0^*)Df(e_0))$

(b) $\int_{\pi_1 U} \#(f) = \operatorname{Vol} P(n) \int_{V_{e_0} \cap U} \det(Df(e_0)^* Df(e_0))$.

We will use:

Proposition. *$\det A^*(f,x)A(f,x)$ is invariant under the unitary group acting on $V_{(d)}$, and*

$$(*)\qquad\qquad A^*(f,e_0)A(f,e_0) = Df(e_0)^* Df(e_0).$$

Postponing the proof of the proposition for the moment, we will prove Theorem 1(b). Use Theorem 1 of the previous section and both parts of this proposition to obtain

$$\int_{\pi_1 U} \#(f) = \int_{P_n} \int_{V_{e_0} \cap U} \det(Df(e_0)^* Df(e_0))$$

$$= \operatorname{Vol} P_n \int_{V_{e_0} \cap U} \det(Df(e_0)^* Df(e_0)).$$

The proof of Theorem 1(a) is similar.

Since $\pi_2 : V_{(d)} \to P_n$ is unitarily invariant, so is the orthogonal complement of $TV_x(f,x)$ in $TV_{(d)}(f,x)$ where $V_x = \pi_2^{-1}(x)$. Then by definition $A(f,x)$ transforms by unitary compositions and thus $\det A^*(f,x)A(f,x)$ is unitarily invariant. This proves the first part of the proposition.

Working in the corresponding vector spaces, write as in Section 2, $\mathcal{H}_{(d)} = K + L(d) + J(d)$. Also

$$T(V_{(d)})(f,e_0) = \{(h,w) \mid Df(e_0)w = h(e_0)\}$$

so that $A : \mathbb{C}^n \to K \subset T_f(P(\mathcal{H}_{(d)}))$ is characterized by $Aw = h \in K$ and $h(e_0) = Df(e_0)(w)$. Here we are using the notation of Section 2 and the definition of $A = A(f,e_0)$. Then

$$A(w)_i = h_i = \sum_j a_{ij} w_j z_0^{d_i}$$

recalling $Df(e_0)(w) = \sum_j a_{ij} w_j$ and the Hermitian structure on K. Then $A^* A = Df(e_0)^* Df(e_0)$.

Section 5. Proof of Theorem A.

We start with:

Proposition 1. *Let* $\pi : S_1^l \to D^k$ *be orthogonal projection of the unit sphere* $S_1^l \subset \mathbb{R}^{l+1}$ *on the unit disk* $D^k \subset \mathbb{R}^k$ *a subspace of* $\mathbb{R}^l$, $0 < k < l$. *Let* $\phi : D^k \to \mathbb{R}$ *be continuous and* $U \subset D^k$ *be open then*

$$\int_{S^l} (\phi \circ \pi) \mathcal{X}(\pi^{-1}(U)) = \int_{D^k} \mathcal{X}(U)(1 - \|x\|^2)^{\frac{l-k-1}{2}} \phi(x) \operatorname{Vol} S_1^{l-k}.$$

For the proof we use the following lemma.

Lemma. *The normal Jacobian of* $\pi : S^l \to D^k$ *at* $x \in S^l$ *is* $(1-\|\pi(x)\|^2)^{1/2}$.

Proof. Let $\|\pi(x)\| = r$. Let S_r^{k-1} be the sphere of radius r about 0 in $\mathbb{R}^k$. Then $\pi^{-1}(S_r^{k-1}) = S_r^{k-1} \times S_{\sqrt{1-r^2}}^{l-k}$ which leaves only one normal direction to $\pi^{-1}(\pi(x))$, namely the one which maps to the day in D^k. Thus the problem reduces to $S^1 \subset \mathbb{R}^2$ and the norm of the projection is easily seen to be $\sqrt{1 - r^2}$.

Proof of Proposition 1.

$$\int_{S^N} (\phi \circ \pi) \mathcal{X}(\pi^{-1}(U)) = \int_{\pi^{-1}(U)} \phi \circ \pi$$

$$= \int_U \int_{\pi^{-1}(x)} (\phi \circ \pi)(\pi^{-1}(x)) \frac{1}{NJ(x)}$$

$$= \int_U \int_{S_1^{l-k}} \phi(x)(1 - \|x\|^2)^{\frac{l-k}{2}} \frac{1}{(1 - \|x\|^2)^{1/2}}$$

$$= \operatorname{Vol} S_1^{l-k} \int_U (1 - \|x\|^2)^{\frac{l-k-1}{2}} \phi(x)$$

$$= \operatorname{Vol} S_1^{l-k} \int_{D^k} \mathcal{X}(U)(1 - \|x\|^2)^{\frac{l-k-1}{2}} \phi(x).$$

Let $d = (d_1, \ldots, d_n)$, $\mathcal{D} = \pi_1^n d_i$. Then

$$A_d = \frac{\int_{P(\mathcal{H}_{(d)}^{\mathbb{R}})} \#(f)}{\operatorname{Vol} P(\mathcal{H}_{(d)}^{\mathbb{R}})}$$

is by definition the average number of real roots. We will prove that $A_d = \mathcal{D}^{1/2}$.

To prove this we will show

Lemma 1. $A_d = \mathcal{D}^{1/2} G(n)$ where G is a function of n.

Apply the lemma to the case $d = (d_1, \ldots, d_n) = (1, \ldots, 1)$. Then clearly $A_d = 1$ and $\mathcal{D} = 1$. Therefore $G(n)$ is identically 1, and $A_d = \mathcal{D}^{1/2}$.

Thus it is sufficient to prove Lemma 1.

For this we use the real analogues of results of the previous sections, with orthogonal invariance replacing unitary invariance. Thus the real version of Theorem 1(b) of Section 4 gives (with $U = P(\mathcal{H}_{(d)}^{\mathbb{R}})$):

Lemma 2.

$$A_d = \frac{1}{2} \frac{\text{Vol } P_n(\mathbb{R})}{\text{Vol } P(\mathcal{H}_{(d)}^{\mathbb{R}})} \int_{S_{e_0}} (\det Df(e_0)^* Df(e_0))^{1/2}.$$

Here $e_0 = (1, 0, \ldots, 0)$, $Df(e_0) : \mathbb{R}^n \to \mathbb{R}^n$ is the derivative and $Df(e_0)^*$ its adjoint. Moreover S_{e_0} is the unit sphere in $\hat{V}_{e_0}^{\mathbb{R}} = \{f \in \mathcal{H}_{(d)}^{\mathbb{R}} \mid f(e_0) = 0\}$ and $V_{e_0} = P(\hat{V}_{e_0}^{\mathbb{R}})$ has $\frac{1}{2}$ the volume of S_{e_0}. Let $N = \dim \mathcal{H}_{(d)}^{\mathbb{R}}$ so that $\dim S_{e_0} = N - n - 1$.

Lemma 3.

$$\int_{S_{e_0}} (\det Df(e_0)^* Df(e_0))^{1/2} = \mathcal{D}^{1/2} \frac{\pi^{N/2}}{\Gamma\left(\frac{N}{2}\right)} H(n)$$

where $H(n)$ depends only on n.

Since $\text{Vol } P(\mathcal{H}_{(d)}^{\mathbb{R}}) = \frac{1}{2} \text{Vol}(S^{N-1}) = \frac{\pi^{N/2}}{\Gamma(N/2)}$, Lemmas 2 and 3 yield that $A_d = \mathcal{D}^{1/2} \text{Vol } P_n(\mathbb{R}) H(n)$ proving Lemma 1. Thus it remains to prove only Lemma 3.

As in Section 2 let $L(d)$ be the linear subspace of $f \in \hat{V}_{e_0}^{\mathbb{R}}$ of the form $f_i = x_0^{d_i - 1} \sum_{j=1}^{n} a_{ij} x_j$ and $\pi : \hat{V}_{l_0}^{\mathbb{R}} \to L(d)$ the natural projection. Then $J(d)$ consists of polynomial systems with no terms of the form $z_0^{d_i - 1}$, $z_0^{d_i - 1} \Sigma a_{ij} z_j$ in the ith coordinate. Next we may identify $L(d)$ with the space of $n \times n$ matrices $A = (a_{ij})$ as above. Here $L(d)$ comes endowed with the Hermitian invariant norm from $\hat{V}_{e_0}^{\mathbb{R}}$.

Let $\mathcal{M}_{\mathbb{R}}(n)$ denote the space of $n \times n$ real matrices endowed with the Frobenius norm.

By Proposition 1,

$$\int_{\substack{f \in V_{e_0} \\ \|f\|=1}} (\text{Det } Df(e_0)^* Df(e_0))^{1/2} =$$

$$\int_{\substack{A \in L(n) \\ \|A\| \leq 1}} (\text{Det } A^* A)^{1/2} (1 - \|A\|^2)^{\frac{N - n^2 - n - 2}{2}} \text{Vol}$$

where Vol is the volume of S^{N-n^2-n-1} or $\dfrac{2\pi^{\frac{N-n^2-n}{2}}}{\Gamma\left(\frac{N-n^2-n}{2}\right)}$.

Now use the change of variables $\Delta^{-1}A = M \in \mathcal{M}_{\mathbb{R}}(n)$, $\Delta = \Delta(d_i^{1/2})$ noting $\|A\|_{L(n)} = \|\Delta^{-1}A\|_{\mathcal{M}_{\mathbb{R}}(n)}$. Thus this last integral is:

$$\text{Vol} \int_{\substack{M \in \mathcal{M}_{\mathbb{R}}(n) \\ \|M\| \le 1}} (\text{Det}((\Delta M)^* \Delta M))^{1/2} (1 - \|M\|^2)^{\frac{N-n^2-n-2}{2}}$$

or yet since $\det \Delta^2 = \mathcal{D}$

$$\text{Vol}\, \mathcal{D}^{1/2} \int_{\substack{M \in \mathcal{M}_{\mathbb{R}}(n) \\ \|M\| \le 1}} (\text{Det}(M^*M))^{1/2} (1 - \|M\|^2)^{\frac{N-n^2-n-2}{2}}.$$

Use polar coordinates to obtain that this last is

$$\mathcal{D}^{1/2} \frac{\pi^{\frac{N-n^2-n}{2}}}{\Gamma\left(\frac{N-n^2-n}{2}\right)} \int_0^1 (1 - r^2)^{\frac{N-n^2-n}{2}-1} dr \int_{\substack{\|M\|=r \\ M \in \mathcal{M}(n)}} (\det M^*M)^{1/2}.$$

But $\int_{\|M\|=r} (\det M^*M)^{1/2} = r^{n^2+n-1} \int_{\|M\|=1} (\det M^*M)^{1/2}$ and where r^{n^2-1} scales the volume element and r^n the $(\det M^*M)^{1/2}$.

$$\int_0^1 r^{n^2+n-1}(1 - r^2)^{\frac{N-n^2-n-1}{2}} dr = \frac{1}{2} \frac{\Gamma\left(\frac{N-n^2-n}{2}\right) \Gamma\left(\frac{n^2+n}{2}\right)}{\Gamma\left(\frac{N}{2}\right)}.$$

The last follows from the identity

$$\int_0^1 s^{k-1}(1 - s)^{l-1} ds = \frac{\Gamma(k)\Gamma(l)}{\Gamma(k + l)}$$

and the substitution $r^2 = s$ (compare Section 2). Putting these together yields Lemma 3 and finishes the proof of Theorem A.

Section 6. Proof of Theorem C.

Our proof of Theorem C depends heavily on the following simple corollary of a result of **Edelman-1992a**

Theorem. For $n \ge 2$, $0 < \rho < \frac{1}{\sqrt{n}}$,

$$\int_{\substack{A \in \mathcal{M}(n) \\ \|A\|=1 \\ d(A,S) \le \rho}} \det(A^*A) \le \frac{\rho^4 n^2 \Gamma(n^2)\Gamma(n + 2)}{4\Gamma(n^2 + n - 2)} \text{Vol}(S^{2n^2-1}).$$

Here $\mathcal{M}(n)$ is the space of $n \times n$ complex matrices with the Frobenius norm, S is the set of singular matrices in $\mathcal{M}(n)$ and d is the distance in $\mathcal{M}(n)$.

We use Proposition 1 of Section 5, just as in the last section, but now we are working over $\mathbb{C}$ so that the real dimension of $\mathcal{M}(n)$ is $2n^2$ etc. The argument of the last section yields (where $N = $ complex dimension $\mathcal{H}_{(d)}$)

$$\int_{f \in N_\rho \cap S_{e_0}} \det(Df(e_0)^* Df(e_0)) =$$

$$\mathcal{D} \int_{\substack{M \in \mathcal{M}(n) \\ d(M,S) < \rho \\ \|M\| \leq 1}} \det(M^* M)(1 - \|M\|^2)^{N - n^2 - n - 1} \text{Vol}$$

where $\text{Vol} = \text{Vol}\, S^{2(N - n^2 - n) - 1}$ and S is the set of singular matrices. Use polar coordinates to evaluate the integral on the right as (following Section 5)

$$\int_{\substack{M \in \mathcal{M}(n) \\ \|M\| \leq 1 \\ d(M,S) \leq \rho}} \det(M^* M)(1 - \|M\|^2)^{N - n^2 - n - 1} =$$

$$\int_0^1 r^{2n} r^{2n^2 - 1}(1 - r^2)^{N - n^2 - n - 1} \int_{\substack{M \in \mathcal{M}(n) \\ \|M\| = 1 \\ d(M,S) \leq \frac{\rho}{r}}} \det M^* M.$$

Now apply Edelman's Theorem to obtain the upper bound for both sides as (for $n > 1$):

$$\text{Vol}(S^{2n^2 - 1}) \frac{\rho^4}{4} \int_0^1 r^{2n^2 + 2n - 1}(1 - r^2)^{N - n^2 - n - 1} r^{-4} dr \frac{n^2 \Gamma(n^2) \Gamma(n + 2)}{\Gamma(n^2 + n - 2)}$$

$$= \frac{\pi^{n^2}}{\Gamma(n^2)} \frac{\rho^4}{4} \frac{n^2 \Gamma(n^2) \Gamma(n + 2)}{\Gamma(n^2 + n - 2)} \frac{\Gamma(n^2 + n - 2) \Gamma(N - n^2 - n)}{\Gamma(N - 2)}.$$

Now we put this information together to obtain:

$$\frac{\text{Vol}\, N_\rho}{\text{Vol}\, P(\mathcal{H}_{(d)})} \leq \frac{\int_{\pi_1 N_\rho \subset P(\mathcal{H}_{(d)})} \#(f)}{\text{Vol}\, P(\mathcal{H}_{(d)})} \text{ since } \#(f) \geq 1$$

$$= \frac{\text{Vol}\, P_n}{\text{Vol}\, P(\mathcal{H}_{(d)})} \frac{1}{2\pi} \int_{f \in N_\rho \cap S_{e_0}} \det(Df(e_0)^* Df(e_0))$$

by Section 4, Theorem 1(b). Continuing, this is less than (by the above calculation)

$$\frac{\pi^{n^2}}{\Gamma(n^2)} \frac{n^2 \rho^4 \mathcal{D}}{4} \frac{\text{Vol}\, S^{2n+1}}{\text{Vol}\, S^{2N-1}} \frac{1}{2\pi} \text{Vol}\, S^{2(N - n^2 - n) - 1} \frac{\Gamma(N - n^2 - n) \Gamma(n^2) \Gamma(n + 2)}{\Gamma(N - 2)}$$

which by a further easy calculation turns out to be

$$\frac{\rho^4}{4}Dn^2(n+1)(N-1)(N-2).$$

Remark. Theorem C and Theorem B might be improved especially for reasonable ranges of ρ and μ_0. For example, Edelman's result which we have used above actually says that

$$\int_{\substack{A\in\mathcal{M}(n)\\ \|A\|=1\\ d(A,S)\leq\rho}} \det(A^*A) =$$

$$\int_0^{\rho^2} \sum_{r=0}^{n-1} \lambda^{n-r}(1-n\lambda)^{n^2+r-2} \frac{\Gamma(n^2)\Gamma(n+1)\Gamma(n+2)}{\Gamma(r+1)\Gamma(n-r)\Gamma(n^2+r-1)\Gamma(n+2-r)}\, \text{Vol}$$

where $\text{Vol} = \text{Vol}(S^{2n^2-1})$ and we have somewhat crudely estimated by disregarding the $(1-n\lambda)$ factors and then maximize the terms of the sum at $r = n-1$ and finally multiplying by n.

Section 7. Proof of Theorem B.

For the proof we use several lemmas and a theorem of Alan Edelman.

Lemma 1. *For $y \in [0,1]$, $(1-y)^k \geq 1 - ky$.*

Proof. It is true for $y = 0$. The rest follows from a comparison of the derivatives.

Lemma 2. *For $x \geq 0$ and $n \in \mathbb{N}$*

$$1 - (1 - \min(1, nx))^{n^2-1} \leq n(n^2 - 1)x.$$

Proof.

$$1 - (1 - \min(1, nx))^{n^2-1}$$
$$\leq 1 - (1 - (n^2 - 1)\min(1, nx)) \text{ by Lemma 1}$$
$$\leq n(n^2 - 1)x.$$

Lemma 3. *For $0 \leq x \leq 1$ $D, n \in \mathbb{N}$*

$$\left(1 + \frac{Dx}{n}\right)^n (1-x)^D \leq 1.$$

Proof.

$$\left(\frac{1}{1-x}\right)^{D/n} = (1 + x + x^2 + \ldots)^{D/n} \geq 1 + \frac{D}{n}x.$$

Lemma 4. *Let M be an $n \times n$ matrix with eigenvalues $\lambda_1, \ldots, \lambda_n$ then $\text{Det}(I + M) = \prod(1 + \lambda_i)$.*

Lemma 5. *Let A be an $n \times n$ matrix which is a weak dilation, i.e.*

$$\|A(v)\| \geq \|v\| \qquad \forall\, v \neq 0 \qquad v \in \mathbb{R}^n.$$

Let M be a matrix and $0 \leq \lambda_1 \leq \cdots \leq \lambda_n$ the eigenvalues of M^*M $0 \leq \mu_1 \leq \cdots \leq \mu_n$ the eigenvalues of M^*A^*AM. Then $\mu_i \geq \lambda_i \; \forall\, i$.

Proof. The eigenvalues of N^*N for any $n \times n$ matrix N are the principal axes of the elipse hich is the image of the unit sphere by N. The image of the unit sphere by AM is outside the image of the unit sphere by M since A is a weak dilation.

Lemma 6. *Let M be an $n \times n$ matrix then $\text{Det}(I + M^*\Delta(d_i)M) \leq \left(1 + \frac{D\|M\|^2}{n}\right)^n$ where $D = \max d_i$.*

Proof. $\Delta(D^{1/2})\Delta(d_i^{-1/2})$ is a weak dilation, so by Lemmas 4 and 5

$$\text{Det}(I + M^*\Delta(d_i)M) \leq \text{Det}(I + M^*\Delta(D)M) = \text{Det}(I + DM^*M).$$

Let $0 \leq \lambda_1 \leq \lambda_{1n} \leq \cdots \leq \lambda_n \leq P_n$ be the eigenvalues of M^*M. Then $\text{Det}(I + \Delta M^*M) = \prod(1 + D\lambda_i) \leq \left(1 + \frac{D\Sigma\lambda_i}{n}\right)^n$ by the inequality of the geometric and arithmetic means, and the last equals $\left(1 + \frac{D\|M\|^2}{n}\right)^n$ by defintion of the Frobenius norm.

Theorem (Edelman). *The volume of $N_\zeta \cap S_1^{2n^2-1}$ is*

$$\left(1 - (1 - \min(1, n\zeta^2))^{n^2-1}\right) \text{Vol}\, S_1^{2n^2-1}$$

where $S_1^{2n^2-1}$ is the unit sphere in $\mathcal{M}(n)$ with the Frobenius norm.

Proof. Actually this follows immediately from Corollary 3.2—**Edelman-1992** and the fact that $\lambda_{\min} = (\min_{\|v\|=1} \|A(v)\|)^2 = \zeta^2$.

Now we pass to the proof of Theorem B. First assume $D > 1$. By taking $N_\rho = \mathcal{U}$ and applying Theorem 1(a) of Section 4, this amounts to estimating:

$$Q(\rho, d) = \frac{\int_{N_\rho \cap V_{e_0}} \det(I + Df(e_0)^* Df(e_0))}{\int_{V_{e_0}} \det(I + Df(e_0)^* Df(e_0))}.$$

Here $Q(\rho, d) = \frac{\text{Vol}\, N_\rho(\Sigma')}{\text{Vol}\, V_{(4)}}$.

Just as in Section 5, we obtain:

$$Q(\rho, d) = \frac{\int_{M \in \mathcal{M}(n), \|M\| \leq 1, d(M,S) \leq \rho} \det(I + M^*\Delta(d_i)M)(1 - \|M\|^2)^{N-n^2-n-1}}{\int_{M \in \mathcal{M}(n), \|M\| \leq 1} \det(I + M^*(\Delta d_i)M)(1 - \|M\|^2)^{N-n^2-n-1}}.$$

Apply Lemmas 3 and 6 to obtain an upper bound for the numerator. Use the fact that $\det(I + M^*\Delta(d_i)M) \geq 1$ to bound the denominator from below. This leads to

$$Q(\rho, d) \leq \frac{\int_{\|M\| \leq 1, d(M,S) \leq \rho}(1 - \|M\|^2)^{N-n^2-n-1-D}}{\int_{\|M\| \leq 1}(1 - \|M\|^2)^{N-n^2-n-1}}.$$

Using polar coordinates

$$Q(\rho, d) \leq \frac{\int_0^1(1 - r^2)^{N-n^2-n-1-D}r^{2n^2-1}\, \mathrm{Vol}(N_{\rho/r} \cap S^{2n^2-1})}{\int_0^1(1 - r^2)^{N-n^2-n-1}r^{2n^2-1}\, \mathrm{Vol}(S^{2n^2-1})}.$$

Apply Edelman's Theorem and Lemma 2

$$Q(\rho, d) \leq \frac{\rho^2 n(n^2 - 1)\int_0^1(1 - r^2)^{N-n^2-n-1-D}r^{2n^2-3}\,dr}{\int_0^1(1 - r^2)^{N-n^2-n-1}r^{2n^2-1}\,dr}.$$

$$Q(\rho, d) \leq \rho^2 n(n^2 - 1)\frac{\Gamma(N - n^2 - n - D)\Gamma(n^2 - 1)\Gamma(N - n)}{\Gamma(N - n - D - 1)\Gamma(N - n^2 - n)\Gamma(n^2)}$$

$$\leq \rho^2 n\frac{\Gamma(N - n)}{\Gamma(N - n - D - 1)} \cdot \frac{\Gamma(N - n^2 - n - D)}{\Gamma(N - n^2 - n)}$$

$$\leq \rho^2 n\left(\frac{N - n - D}{N - n^2 - n - D}\right)^D(N - n - D - 1).$$

We conclude the proof with

Lemma 7. Let $n > 1$, $D > 1$. Then $\left(\frac{N-n-D}{N-n^2-n-D}\right)^D \leq K$.

Proof. Let $P = N - n^2$. Then the estimate is

$$\left(\frac{P + n^2 - n - D}{P - n - D}\right)^D \leq K$$

or

$$\left(1 + \frac{n^2}{P - n - D}\right)^D \leq K$$

$$\left(1 + \frac{n^2}{N - n^2 - n - D}\right)^D \leq K.$$

The worst case is seen to be $d = (1, 2)$ and $N - n^2 - n - D = 1$.

Finally, the condition number theorem allows to pass from the above estimates to Theorem B. The case where $D = 1$ is simpler

$$Q(\rho, d) = \frac{\int_{N_\rho \cap S_1^{2n^2-1}} \det(1 + M^*M)}{\int_{S_1^{2n^2-1}} \det(1 + M^*M)}$$

where $S_1^{2n^2-1}$ is the unit sphere in $\mathcal{M}(n)$. Now $1 \leq \det(1 + M^*M) \leq (1 + \frac{1}{n})^n \leq \rho$ so

$$Q(\rho, d) \leq e \frac{\mathrm{Vol}(S_1^{2n^2-1} \cap N_\rho)}{\mathrm{Vol}(S_1^{2n^2-1})} \leq e\rho^2 n(n^2 - 1)$$

by Edelman's theorem and lemma 2.

Section 8. Proof of Theorem D.

For the proof of Theorem D we will use a simple lemma on indefinite integrals which can be checked by differentiating.

Lemma. i) $\int (1 + dr^2)(1 - r^2)^{d-2} r\, dr = \frac{1}{2}((1 - r^2)^d - \left(\frac{d+1}{d-1}\right)(1 - r^2)^{d-1})$

ii) $\int dr^2(1 - r^2)^{d-2} r\, dr = \frac{1}{2}((1 - r^2)^d - \frac{d}{d-1}(1 - r^2)^{d-1})$

Now we turn to the proof of Theorem D i). Identifying $\mathcal{M}(1)$ with $\mathbb{C}$ we have as in Theorem B that

$$\frac{\mathrm{Vol}\, N_\rho(\Sigma')}{\mathrm{Vol}\, V_{(d)}} = Q(\rho, d) = \frac{\int_{z \in \mathbb{C}, |z| < \rho}(1 + d|z|^2)(1 - |z|^2)^{d-2}}{\int_{z \in \mathbb{C}, |z| < 1}(1 + d|z|^2)(1 - |z|^2)^{d-2}}.$$

Now use polar coordinates and Lemma i).

For the proof of Theorem D ii). Remark that by Theorem 1 (b)

$$\mathcal{D}\, \mathrm{Vol}\, P(\mathcal{H}_{(d)}) = \mathrm{Vol}\, P(n) \int_{V_{e_0}} \det(Df(e_0)^* Df(e_0))$$

so for $n = 1$ as in the proof of Theorem C

$$\frac{\mathrm{Vol}\, N_\rho}{\mathrm{Vol}\, P(\mathcal{H}_{(d)})} \leq \frac{\mathrm{Vol}\, P(1)}{\mathrm{Vol}\, P(\mathcal{H}_{(d)})} \frac{1}{2\pi} \int_{f \in N_\rho \cap S_{e_0}} \det(Df(e_0)^* Df(e_0))$$

$$= \frac{\mathrm{Vol}\, P(1)}{\mathrm{Vol}\, P(\mathcal{H}_{(d)})} \frac{1}{2\pi} \mathrm{Vol}\, S^{2(d-1)-1} \int_{z \in \mathbb{C} \atop |z| < \rho} d|z|^2(1 - |z|^2)^{d-2}$$

$$= \frac{d(d-1)}{\pi} \int_{z \in \mathbb{C} \atop |z| < \rho} d|z|^2(1 - |z|^2)^{d-2}$$

Now use polar coordinates and lemma ii) to finish the proof.

REFERENCES

Demmel, J., *On condition numbers and the distance to the nearest ill-posed problem*, Numerische Math. **51** (1987), 251–289.

Eckart, C. and Young, G., *The approximation of one matrix by another of lower rank*, Psychometrika 1 (1936), 211–218.

Edelman, A., *On the distribution of a scaled condition number*, Math. of Comp. **58** (1992), 185–190.

Edelman, A., *On the moments of the determinant of a uniformly distributed complex matrix*, in preparation (1992a).

Kac, M., *On the average number of real roots of a random algebraic equation*, Bull. Amer. Math. Soc. **49** (1943), 314–320.

Kostlan, Eric, *Random polynomials and the statistical fundamental theorem of algebra*, Preprint, Univ. of Hawaii (1987).

Kostlan, E., *On the distribution of the roots of random polynomials*, to appear in "From Topology to Computation" Proceedings of the Smalefest, Hirsch, M., Marsden, J. and Shub, M. (Eds) (1991).

Morgan, F., *Geometric Measure Theory, A Beginners Guide*, Academic Press, 1988.

Renegar, J., *On the efficiency of Newton's Method in approximating all zeros of systems of complex polynomials*, Math. of Operations Research **12** (1987), 121–148.

Shub, M. and Smale, S., *Complexity of Bezout's Theorem I, Geometric Aspects*, Preprint (referred to as [I]) (1991).

Stein, E. and Weiss, G., *Introduction to Fourier Analysis on Euclidean Spaces*, Princeton Univ. Press, Princeton, NJ, 1971.

Tsuji, M., *Potential Theory in Modern Function Theory*, Maruzen Co. Ltd., Tokyo, 1959.

Wilkinson, J., *Rounding Errors in Algebraic Processes*, Prentice Hall, Englewood Cliffs, NJ.

Michael Shub
IBM, T.J. Watson Research Center,
Yorktown Heights, NY 10598 USA
email: SHUB @ WATSON.IBM.COM

Steve Smale
Math Department, University of California,
Berkeley, CA 94720 USA
email: SMALE @ MATH.BERKELEY.EDU

A parametrized nullstellensatz

F. Smietanski

Introduction

Let k be a field and $f, F_1, \ldots, F_s \in k[X_1, \ldots, X_n]$, with $\deg(F_i) = d_i$ and $\max\{d_1, \ldots, d_s\} \leq d$. When there exist some polynomials $A_1, \ldots, A_s$ and power N such that we have a representation $1 = A_1 F_1 + \ldots + A_s F_s$ or $f^N = A_1 F_1 + \ldots + A_s F_s$, the question arises how to bound optimally $a := \sup\{\deg A_i\}$ and N.

The first effective result, $a \leq 2(2d)^{2^{n-1}}$, was proved by G.Hermann [7] using elimination theory.

The first fundamental step towards the optimal simply exponential bound is $a \leq n\, m\, d^m + md$ where $m = \min\{n, s\}$, established by W.D.Brownawell [4] with analytical representation techniques in zero characteristic.

In any characteristic, L.Caniglia A.Galligo and J.Heintz [6] obtained a less better bound, $d^{n(n+3)/2}$, with completely elementary methods.

The optimal bound, $N \leq d_1 \ldots d_m$, was eventually proved by J.Kollár [8] using cohomologicals tools, but with a technical restriction $d_i \geq 3$.

Replace the field k by a ring R. Then after reduction of the fractions to the same denominator, the problem becomes to study a representation like $r = A_1 F_1 + \ldots + A_s F_s$ or $r \cdot f^N = A_1 F_1 + \ldots + A_s F_s$ with r in R. And the new question is to bound also r and the coefficients of $A_1, \ldots, A_s$, when R has some suitable structure.

The first result in this direction is that of C.Berenstein and A.Yger [3] when R is the ring of integers Z. It states:
$\deg(A_i) \leq 10(n+1)^5(2d+1)^{2n}$ and $\sup(|r|, |A_i|) \leq c(n) \cdot d^{9n+3}(H + d \cdot \log d + \log n)$ with $|P| = \max\{|\text{coefficients of} P|\}$ and $H = \sup\{|Fi|\}$.

The existence of a bound for the denominator r has been also established, in the very general case of rings which only verify some norm-like properties, by P.Philippon [11] at the cost of letting down the bounds for the coefficients of $A_1, \ldots, A_s$.

See [13] and [1, 2] for an extended bibliography and overview in this field.

We want to present in this article a less general intermediary statement, restricted to the case of a polynomial ring $R = k[T_1, \ldots, T_m]$, but which gives also bounds of r and of the coefficients of $A_1, \ldots, A_s$ in addition to those of N and of the degrees of the A_i.

The proof associates two ingredients, the cohomological approach introduced by J.Kollár [8] under the form that W.D.Brownawell [5] adopted for the part dealing with the degrees of the polynomials, to the approach developed by P.Philippon [11], using Chow forms to get, in addition, bounds on the coefficients.

Part I

Let k be a field, $R = k[T_1, \ldots, T_m]$ (denoted by $k[T]$) a base ring, $K = Frac(R)$ its fraction field, $A = R[X_0, \ldots, X_n]$ (denoted by $R[X]$) a graded algebra of polynomials in X_i. Let $J = (f_1, \ldots, f_s)$ an ideal of A, d_i the degree of f_i in $X, \partial i$ the degree of f_i in T, and H the maximum of the ∂i. We suppose that the f_i are reordered such that $d_2 \geq d_3 \geq \ldots \geq d_s \geq d_1$. An ideal I of A will be called **proper** if it verifies $I \cap R = (0)$. We will suppose in all the sequel $d_i \geq 3, \quad i = 1, \ldots, s$.

Theorem. Let $J = (f_1, \ldots, f_s)$ be given, then there exists a family of proper prime ideals $P_1, \ldots, P_r$ each of them containing J, among which appear at least each proper prime isolated ideal of J, a set of integers $e_1, \ldots, e_r$, a family $m_1, \ldots, m_q$ of prime ideals in R with height ≥ 2 and integers $\lambda_1, \ldots, \lambda_q$, and $r(T)$ a non-nul element of R such that

$$m_1^{\lambda_1} \ldots m_q^{\lambda_q} \cdot r(T) \cdot P_1^{e_1} \ldots P_r^{e_r} \subset J, \quad \text{with}$$

$$\sum_1^r e_i \leq 3 \cdot d_1 \ldots d_\mu \quad \text{where} \quad \mu = \min(s, n) \quad \text{and}$$

$$\sum_1^r e_i + \sum_1^q \lambda_i \leq (d_1 + H) \ldots (d_v + H) \quad \text{where} \quad v = \min(s, n+1)$$

$$\text{and} \quad \deg r(T) \leq 3^{v-1} \cdot d_1 \ldots d_v \cdot (\sum_1^v \frac{1}{d_i}) \cdot H$$

Corollaries: we suppose k algebraically closed in corollaries 1 and 2.

Corollary 1. Let $R = k[T_1, T_2, T_3]$ a graded ring, $f_1, \ldots, f_s \in A = R[X_0, \ldots, X_n]$ bihomogeneous polynomials of homogeneous degree d_i in X and homogeneous degree in T less than H. Then there exists a non-nul $s(T)$ and a power N depending only on $I = (f_1, \ldots, f_s)$ and verifying :

$$N \leq 3 \cdot d_1 \ldots d_\mu \quad \text{and}$$

$$\deg(s(T)) \leq (d_1 + H) \ldots (d_v + H) + 3^{v-1} \cdot d_1 \ldots d_v \cdot (\sum_1^v \frac{1}{d_i}) \cdot H$$

such that: for every $f \in A$, i) and ii) are equivalents:

    i) there exists non-nul $a(T) \in R$ and $M \geq 0$ such that $a(T) \cdot f^M \in I$

    ii) there exists $A_1, \ldots, A_s \in R[X]$ such that $s(T) \cdot f^N = \sum_{i=1}^{s} A_i \cdot f_i$ and

$$\deg_T(A_i \cdot f_i) \leq (d_1 + H) \ldots (d_v + H) + 3^{v-1} \cdot d_1 \ldots d_v \cdot (\sum_1^v \frac{1}{d_i}) \cdot H + N \cdot \deg_T(f)$$

$$\deg_x(A_i \cdot f_i) \leq N \cdot \deg_x(f)$$

Corollary 2. Let $R = k[T_1, T_2]$ be a affine ring, $f_1, \ldots, f_s \in A = R[X_1, \ldots, X_n]$ non-homogeneous polynomials of degree d_i in X and degree in T less than H. Then there exists a non-nul $s(T)$ and a power N depending only on $I = (f_1, \ldots, f_s)$ and verifying :

$$N \leq 3 \cdot d_1 \ldots d_\mu \quad \text{and}$$

$$\deg(s(T)) \leq (d_1 + H) \ldots (d_v + H) + 3^{v-1} \cdot d_1 \ldots d_v \cdot (\sum_1^v \frac{1}{d_i}) \cdot H$$

such that:

    a) for every $f \in A$, i) and ii) are equivalents :

      i) there exists non-nul $a(T) \in R$ and $M \geq 0$ such that $a(T) \cdot f^M \in I$

      ii) there exists $A_1, \ldots, A_s \in R[X]$ such that $s(T) \cdot f^N = \sum_{i=1}^{s} A_i \cdot f_i$ and

$$\deg_T(A_i \cdot f_i) \leq (d_1 + H) \ldots (d_v + H) + 3^{v-1} \cdot d_1 \ldots d_v \cdot (\sum_1^v \frac{1}{d_i}) \cdot H + N \cdot \deg_T(f)$$

$$\deg_x(A_i \cdot f_i) \leq N \cdot (\deg_x(f) + 1)$$

    b) i) and ii) are equivalents :

      i) $V(I)$ is empty in L^n where L is the algebraic closure of $k(T_1, T_2)$

      ii) there exists $A_1, \ldots, A_s \in R[X]$ such that $s(T) = \sum_{i=1}^{s} A_i \cdot f_i$ and

$$\deg_T(A_i \cdot f_i) \leq (d_1 + H) \ldots (d_v + H) + 3^{v-1} \cdot d_1 \ldots d_v \cdot (\sum_1^v \frac{1}{d_i}) \cdot H$$

$$\deg_x(A_i \cdot f_i) \leq N$$

Corollary 3. Let $R = k[T_1]$ be a principal affine ring, $f_1, \ldots, f_s \in A = R[X_0, \ldots, X_n]$ homogeneous polynomials of homogeneous degree d_i in X and degree in T less than H. Then there exists a non-nul $s(T)$ and a power N depending only on $I = (f_1, \ldots, f_s)$ and verifying:

$$N \leq 3 \cdot d_1 \ldots d_\mu \quad \text{and} \quad \deg(s(T)) \leq 3^{v-1} \cdot d_1 \ldots d_v \cdot (\sum_1^v \frac{1}{d_i}) \cdot H$$

such that for every $f \in A$, i) and ii) are equivalents :

i) there exists non-nul $a(T) \in R$ and $M \geq 0$ such that $a(T) \cdot f^M \in I$

ii) there exists $A_1, \ldots, A_s \in R[X]$ such that $s(T) \cdot f^N = \sum_{i=1}^s A_i \cdot f_i$ and

$$\deg_x(A_i \cdot f_i) \leq N \cdot (\deg_x(f) + 1)$$

Part II

First step

a) Definition: Let R be a domain and q a primary ideal of $A = R[X_0, \ldots, X_n]$. The ideal q is said to be proper if $q \cap R = (0)$.

Rem 0: We will use in the sequel the following properties of proper ideals: Let $K = frac(R)$, q a primary ideal of $A = R[X_0, \ldots, X_n]$, and $q' = q \cdot K[X_0, \ldots, X_n]$. Then:

i) q is proper iff $q = q' \cap R[X_0, \ldots, X_n]$ and $q \neq (1)$

ii) q is not proper iff $q' = (1)$

It suffices to prove the direct implication of i).

Let us prove that $\overline{q' \cap R[X_0, \ldots, X_n]} \subset q$ under the hypothesis that q is proper:

$$x \in q' \cap R[X_0, \ldots, X_n] \Longrightarrow \exists \ r \in R \backslash \{0\}/r \cdot x \in q$$

but $r \notin p = rad(q)$ for: $p \cap R = rad(q) \cap R = rad(q \cap R) = rad(0) = (0)$ and $r \neq 0$

then, as q is p-primary : $(r \cdot x \in q$ and $r \notin p) \Longrightarrow x \in q$.

b) Let $R = k[T_1, \ldots, T_m]$ an affine ring over the field k, $K = Frac(R)$ its field of fractions, and $A = R[X_0, \ldots, X_n]$ the graded ring over R.

Lemma 1. Let $I = (f_1, \ldots, f_s)$ an ideal of A where $f_1, \ldots, f_s$ are bihomogeneous in T and X. There exists $t \leq v = \min(n + 1, s)$ and a sequence $g_1, \ldots, g_t \in I$ such that $(g_1, \ldots, g_t)$ and I have the same isolated proper prime ideals and for each $i = 1, \ldots, t - 1$:

1) the isolated proper prime ideals of $(g_1, \ldots, g_i)$ which do not contain I do not contain g_{i+1} (the others containing I, already contain g_{i+1}).

2) there is at least one isolated proper prime ideal of $(g_1, \ldots, g_i)$ which does not contain g_{i+1}.

Proof: If the field k is infinite, we define successively:

$$g_1 := f_1, \quad g_i = u^{H-\delta_i} \cdot f_i + \sum_{j=i+1}^{s} \lambda_{i,j} \cdot u^{H-\delta_j} \cdot L^{d_i - d_j} \cdot f_j \quad i \geq 2$$

where u is any linear form of $R = k[T_1, \ldots, T_m]$, L a generic linear form of $k[X_0, \ldots, X_n]$ and the $\lambda_{i,j}$ are generic coefficients in k. Let j be the injection

$j : R[X_0, \ldots, X_n] \rightarrow K[X_0, \ldots, X_n]$. The images $j(g_i)$ verify the properties of [5, lemma 0 p5] or [9]. The pullback in $A = R[X_0, \ldots, X_n]$ gives the above statement as non-proper ideals of A have trivial images (the ideal (1)) by the injection j(see Rem 0).If the field k is finite, we use [5, lemma 0 p7] to bring us back to the former case.

Rem 1: if R is affine and the f_i homogeneous in X only, the proof is easier as there is no more need to homogenize in T with the linear form u.

Second step

a) Now we can rename the generators and replace J by $(g_1, \ldots, g_t) \subset J$, and then suppose that $s = t \leq n + 1$ and that the sequence $(f_1, \ldots, f_s)$ verifies the conditions of lemma 1 above. Indeed the hypothesis with regard to H and the bounds of degrees in X of the first t polynomials remain unchanged. Furthermore J and $(g_1, \ldots, g_t)$ have the same isolated proper prime ideals. As a consequence, it will be sufficient to prove the theorem for the ideal $(g_1, \ldots, g_t) \subset J$, in order to prove it also for J. At last, as in [8] we can divide by the PGCD of the generators to consider only the case $ht(J) > 1$.

b) Let us define the following sequence of primary decomposition for $i = 1, \ldots, s$:

- $I_0 = (0)$ and

- $(I_{i-1}, f_i) = I_i \cap Q_i \cap E_i$ where $Q_i = K_i \cap L_i \cap M_i$

are defined as follows:

- I_i is the intersection of the isolated **proper** primary components whose radical does not contain I (according to condition 1 of lemma 1, they also do not contain f_{i+1}, so that f_{i+1} is regular in A/I_i).

- K_i is the intersection of the remaining isolated **proper** primary components (whose radical does contain I)

- L_i is the intersection of the remaining **non-proper** isolated primary components $L_{i,j}$ such that: $ht_R(L_{i,j} \cap R) = 1$

- M_i is the intersection of the remaining **non-proper** isolated primary components $M_{i,j}$ such that: $ht_R(M_{i,j} \cap R) \geq 2$

- E_i is an intersection of the embeded primary components.

As properties 1) and 2) of lemma 1 stand at each step of the above decomposition, it results that:

- I_i and Q_i are pure equidimensional ideals of height i,

- Each isolated proper prime ideal of height i associated to I also appears among the associated primes of K_i,

- $I_s = (1)$ and $ht(E_i) \geq i + 1$. (Rem: $ht(I) \geq 1$ implies that $K_1 = A$ and $E_1 = E_2 = A$)

Third step

a) The fundamental lemma of J.Kollár [8] reformulated by W.D.Brownawell [5] is used to prove the first essential inclusion (1) below. The demonstration given here does not use exactly either the same decomposition as in [8] or in [5], or that of P.Philippon in [11, part II] but is inspired by the last one with regard to the treatment of the coefficients in R. However, the fundamental lemma stated here remains unchanged because its proof only depends on the three following properties:

- f_{i+1} is regular in the ring A/I_i,

- multiplication by K_i annihilates $I_i/(I_i \cap K_i)$,

- $ht(E_i) \geq i + 1$.

All these properties are kept with respect to the conditions defined above.
b) Fundamental Lemma: [5, Prop 1 p9] after [8]. Using the previous notations:

$$\text{Let:} \quad N_i := Q_1^{3^{i-2}} \cdot Q_2^{3^{i-3}} \ldots Q_{i-1}^{3^0} \quad i = 1, \ldots, s-1$$

$$\text{Then:} \quad N_{i-1} \cdot (I_i \cap Q_i) \subset (I_{i-1}, f_i) \quad i = 2, \ldots, s \tag{1}$$

c) Combining them all, noticing that:

$$I_i \cdot Q_i \subset (I_i \cap Q_i) \quad \text{and} \quad Q_1 \cap I_1 - (f_1)$$

we derive:

$$(Q_1 \cdot N_1) \cdot (Q_2 \cdot N_2) \ldots (Q_{i-1} \cdot N_{i-1}) \cdot (I_i \cap Q_i) \subset (I_1 \cap Q_1, f_2, \ldots, f_i) = (f_1, \ldots, f_i)$$

$$\text{then:} \quad \left(\prod_{j=1}^{i-1} Q_j^{\frac{1+3^{i-j}}{2}} \right) \cdot (I_i \cap Q_i) \subset (f_1, \ldots, f_i) \quad i = 1, \ldots, s$$

$$\text{and:} \quad \left(\prod_{j=1}^{i} Q_j^{\frac{1+3^{i-j}}{2}} \right) \cdot I_i \subset (f_1, \ldots, f_i) \quad i = 1, \ldots, s \tag{2}$$

as $I_s = (1)$, this leads to:

$$\left(\prod_{j=1}^{s} Q_j^{\frac{1+3^{s-j}}{2}} \right) \subset (f_1, \ldots, f_s) \subset I \tag{3}$$

Fourth Step

Notations: If I is an ideal of $R[X] = k[T_1, \ldots, T_m][X_0, \ldots, X_n]$, we define $\deg(I)$ to be the degree of I as an ideal of $k[T_1, \ldots, T_m, X_0, \ldots, X_n]$ (denoted by $k[T, X]$), and $\deg_x(I)$ to be the degree of its extension I' to $K[X_0, \ldots, X_n]$ where $K = Frac(R)$. Thus $\deg_x(I) := \deg(I') = \deg(I \cdot K[X_0, \ldots, X_n])$.

a) The proper primary components of Q_i.
They are the $K_{i,j}$. Their intersection is K_i.

Let $q \subset A = R[X_0, \ldots, X_n]$ such a proper primary component. Its extension q' to $K[X_0, \ldots, X_n]$ has degree $\deg_x(q) = \deg(q')$ and verifies :

$$rad(q')^{\deg(q')} \subset q' \quad \text{(See for ex. [9]or [5, V A])}$$

But $q = q' \cap A$. Consequently $rad(q) = rad(q') \cap A$. Let $d := \deg(q')$, then we have:

$$rad(q)^d = (rad(q') \cap A)^d \subset (rad(q'))^d \cap A \subset q' \cap A = q$$

Therefore the previous inclusion lifts in A giving: $rad(q)^{\deg_x(q)} \subset q$

then: $\quad rad(K_{i,j})^{k_{i,j}} \subset K_{i,j} \quad$ where $\quad k_{i,j} := \deg_x(K_{i,j})$

and: $\quad \displaystyle\prod_j rad(K_{i,j})^{k_{i,j}} \subset K_i \quad i = 1, \ldots, s \qquad (4)$

b) The non-proper primary components of Q_i.
Their intersection is $L_i \cap M_i$.

1) M_i: Let $M_{i,j}$ be its primary components. Their intersection is M_i. Let $\mu_{i,j} := \deg(M_{i,j})$ (degree in $k[T, X]$). Thus:

$$\prod_j rad(M_{i,j})^{\mu_{i,j}} \subset M_i \quad i = 1, \ldots, s \qquad (5)$$

2) L_i: Let $L_{i,j}$ be its primary components. Their intersection is L_i. They all verify $ht_R(L_{i,j} \cap R) = 1$ in R factorial, thus the $L_{i,j} \cap R$ are principal and their intersection $L_i \cap R$ is generated by a single non-null element of R, $r_i(T)$. The Chow forms methods used by P.Philippon [11] will allow us to bound its degree in T.

c) We derive from (4), (5) and b) 2) for $i = 1, \ldots, s$:

$$\left(\prod_j rad(M_{i,j})^{\mu_{i,j}}\right) \cdot r_i(T) \cdot \left(\prod_j rad(K_{i,j})^{k_{i,j}}\right) \subset M_i \cap L_i \cap K_i = Q_i \qquad (6)$$

Multiplying them according to (3) leads to:

$$\left(\prod_{i=1}^s \prod_j rad(M_{i,j})^{\mu_{i,j} \cdot \frac{1+3^{s-i}}{2}}\right) \cdot r(T) \cdot \left(\prod_{i=1}^s \prod_j rad(K_{i,j})^{k_{i,j} \cdot \frac{1+3^{s-i}}{2}}\right) \subset (f_1, \ldots, f_s)$$

$$\text{with:} \quad r(T) := \prod_{i=1}^{s} r_i(T)^{\frac{1+3^{s-i}}{2}} \tag{7}$$

$$\text{we define:} \quad k_i := \sum_j k_{i,j} = \sum_j \deg_x(K_{i,j}) = \deg_x(K_i) \tag{8}$$

$$\text{and:} \quad \mu_i := \sum_j \mu_{i,j} = \sum_j \deg(M_{i,j}) = \deg(M_i) \tag{9}$$

$$N := \sum_{i=1}^{s} \sum_j k_{i,j} \cdot \frac{1+3^{s-i}}{2} = \sum_{i=1}^{s} k_i \cdot \frac{1+3^{s-i}}{2} \tag{10}$$

$$\text{and} \quad L := \sum_{i=1}^{s} \sum_j \mu_{i,j} \cdot \frac{1+3^{s-i}}{2} = \sum_{i=1}^{s} \mu_i \cdot \frac{1+3^{s-i}}{2} \tag{11}$$

d) Some properties of Chow forms.
See [10] or [12] for the proofs of those properties.

Let R be a commutative noetherian ring, $A = R[X_0, \ldots, X_n]$ a graded algebra, $d = (d_1, \ldots, d_{n+1})$ a $n+1$-uplet of integers ≥ 1, M_{d_i} the set of monomials of total degree d_i, U_i the set of indeterminates $U_i := \{u_m^{(i)}, m \in M_{d_i}\}$, and U their union for $i = 1, \ldots, n+1$. We also note S_i the generic homogeneous polynomial of total degree d_i in $A[U]$ defined by $S_i := \sum_{m \in M_{d_i}} u_m^{(i)} \cdot m$.

Definition: Let I be an ideal of A, $M := (X_0, \ldots, X_n)$, we call the U-eliminating ideal of I with index $\Delta = (d_1, \ldots, d_t), t \leq n+1$, the ideal of $R[U]$:

$$F_\Delta(I) := \{(I + (S_1, \ldots, S_t)) : M^*\} \cdot A[U] \cap R[U],$$

where $(a : b^*)$ is the increasing union of the ideals $(a : b^m) m \geq 0$, and $(a : c) := \{x / x \cdot c \subset a\}$.

i) We take $I := I_j$ of height j, $\Delta := (d_1, \ldots, d_s, 1, \ldots, 1)$ the $n+1$-uplet of the degrees of the index, and we note $\Delta_j := (d_{j+1}, \ldots, d_s, 1, \ldots, 1)$ the $(n+1-j)$-uplet of the last components of Δ.

Notation: if $f_i = \sum_{m \in M_{d_i}} f_m^{(i)} \cdot m$ with $f_m^{(i)} \in R$, we denote by $U_i := f_i$ the specialization $\forall m \in M_{d_i}, u_m^{(i)} := f_m^{(i)}$.

By specialization we get :

$$F_{\Delta_{i-1}}(I_{i-1})[U_i := f_i] \subset F_{\Delta_i}(I_{i-1} + (f_i))$$

$$\text{and as:} \quad I_{i-1} + (f_i) = I_i \cap Q_i \cap E_i \subset I_i \cap K_i \cap L_i$$

$$\text{we get ([10] lemma3):} \quad F_{\Delta_{i-1}}(I_{i-1})[U_i := f_i] \subset F_{\Delta_i}(I_i) \cap F_{\Delta_i}(K_i) \cap F_{\Delta_i}(L_i) \tag{12}$$

ii) a few properties.
For every prime ideal P of $A = R[X]$, and J a P-primary ideal, if $K = Frac(R)$, we have :

$$F_{\Delta_i}(P) \text{ is prime, and } F_{\Delta_i}(J) \text{ is } F_{\Delta_i}(P) - \text{primary}, \quad [12, \text{Prop. } 1.3]$$

$$F_{\Delta_i}(J \cdot K[X]) = F_{\Delta_i}(J) \cdot K[U]. \quad [10, \text{Rem.1 p247}]$$

if moreover $P \cdot K[X]$ and $J \cdot K[X]$ have height i in $K[X]$, then:

$$F_{\Delta_i}(P \cdot K[X]) \text{is prime principal in} K[U] \quad [10, \text{lemma5 p246}]$$

and thus $F_{\Delta_i}(J \cdot K[X])$ is primary principal, and a power of $F_{\Delta_i}(P \cdot K[X])$. If P and J are proper, as $J = (J \cdot K[X]) \cap A$, we have in addition :

$$F_{\Delta_i}(J) = F_{\Delta_i}(J \cdot K[X]) \cap R[U] \text{ (and as well for P).}$$

Finally, the principal ideal $F_{\Delta_i}(P)$ can be considered as an irreducible polynomial of $R[U]$, and $F_{\Delta_i}(J)$ as a power of this one with exponent e such that :

$$F_\Delta(J) = F_\Delta(P)^e \quad \text{and} \quad e = exp(J) = inf\{e/P^e \subset J\}$$

Moreover, $F_{\Delta_i}(P)$ characterizes P: if P and P' are two distinct proper prime ideals of height i, then $F_{\Delta_i}(P)$ and $F_{\Delta_i}(P')$ are two distinct irreducible polynomials of $R[U]$ (and then relatively prime and with content equal to 1).

Thus if J is an non-redundant intersection of primary ideals J_α proper with height i then $F_{\Delta_i}(J)$ is the intersection of the $F_{\Delta_i}(J_\alpha)$. As they are polynomials with content equal to 1 and relatively prime, it is also their product. This applies to $F_{\Delta_i}(I_i)$ and $F_{\Delta_i}(K_i)$. Let $P_{I,i}(T, U)$ and $P_{K,i}(T, U)$ be the elements of $R[U]$ respectively generating $F_{\Delta_i}(I_i)$ and $F_{\Delta_i}(K_i)$. They are relatively prime and of content equal to 1.

e) Application to the expression of $F_{\Delta_i}(I_i) \cap F_{\Delta_i}(K_i)$:

$$\begin{aligned} F_{\Delta_i}(I_i) \cap F_{\Delta_i}(K_i) &= (P_{I,i}(T, U)) \cap (P_{K,i}(T, U)) \\ &= (P_{I,i}(T, U) \cdot P_{K,i}(T, U)) \end{aligned} \quad (13)$$

f) Expression of $F_{\Delta_i}(L_i)$.

Lemma 2. Let R be a factorial ring, L a primary ideal of $R[X_0, \ldots, X_n]$ with height $\leq i$ such that $L \cap R$ is **principal** and non-null $(ht_R(L \cap R) = 1)$, then:

$$F_{\Delta_i}(L) = L \cap R$$

Proof: See [10, lemma 6] where the property is stated with the restrictive hypothesis that R is a principal ring, but the demonstration of J.V.Nesterenko uses only the fact that R is a factorial ring and that $L \cap R$ is principal in R.

Application: Each primary component $L_{i,j}$ of L_i verifies the above conditions and $ht_R(L_{i,j} \cap R) = 1$. Thus $L_i \cap R$ is also principal; let $r_i(T)$ be one of its generators, with lemma 2:

$$F_{\Delta_i}(L_i) = \cap_j F_{\Delta_i}(L_{i,j}) = \cap_j (L_{i,j} \cap R) = L_i \cap R = (r_i(T)) \quad (14)$$

g) We deduce from the lines (12),(13) and (14) the inclusion:

$$\begin{aligned} F_{\Delta_{i-1}}(I_{i-1})[U_i := f_i] &\subset (P_{I,i}(T, U) \cdot P_{K,i}(T, U)) \cap (r_i(T)) \\ &= (P_{I,i}(T, U) \cdot P_{K,i}(T, U) \cdot r_i(T)) \end{aligned}$$

as $P_{I,i}$ and $P_{K,i}$ have content 1 in $R[T,U]$ and then also in $(R[T])[U]$.

thus: $P_{I,i}(T,U) \cdot P_{K,i}(T,U) \cdot r_i(T)$ divides $P_{I,i-1}(T,U)[Ui := f_i]$ (15)

Fifth step

a) Let y_i, q_i, r_i be the degrees in T of $P_{I,i}(T,U), P_{K,i}(T,U)$ and $r_i(T)$. We derive from (15) the following inequality:

$$
\begin{aligned}
y_i + q_i + r_i \ & \leq \ \deg_T(P_{I,i-1}(T,U)[U_i := f_i]) \\
& = \ y_{i-1} + \deg_T(f_i) \cdot \deg_{U_i}(P_{I,i-1}(T,U))
\end{aligned}
\tag{16}
$$

Let us define: $\begin{aligned} D_i \ & := \ \deg_{U_i}(P_{I,i-1}(T,U)) \quad (\ [12]\ \text{lemma } 1.8) \\ & \leq \ \deg_x(I_{i-1}) \cdot d_{i+1} \ldots d_s \\ & \leq \ d_1 \ldots d_{i-1} \cdot d_{i+1} \ldots d_s \quad (\text{Bezout inequality})(17) \end{aligned}$

and as $\deg_T(f_i) = \delta_i \leq H$, it comes from (16)(with $y_0 = 0$):

$$
(y_i - y_{i-1}) + q_i + r_i \leq H \cdot D_i \quad i = 1, \ldots, s
\tag{18}
$$

Sixth step

a) The bound of $\deg(r(T))$:

$$
\text{from (7): } \deg(r(T)) \leq \sum_{i=1}^{s} r_i \cdot \frac{1 + 3^{s-i}}{2}
$$

from (18) we deduce: $\displaystyle \sum_{i-1}^{s}((y_i - y_{i-1}) + q_i + r_i) \cdot \frac{1 + 3^{s-i}}{2} \leq H \cdot \sum_{i-1}^{s} D_i \cdot \frac{1 + 3^{s-i}}{2}$

defining: $A_i := \dfrac{1 + 3^{s-i}}{2}$ $A_i - A_{i+1} := 3^{s-(i+1)}$

we get: $\displaystyle \sum_{i=1}^{s}(y_i - y_{i-1}) \cdot A_i \ = \ y_s \cdot A_s - y_0 \cdot A_1 + \sum_{i=1}^{s-1} y_i(A_i - A_{i+1})$

$$
= \ y_s \cdot A_s + \sum_{i=1}^{s-1} y_i \cdot 3^{s-(i+1)} \geq 0
$$

thus: $\displaystyle \sum_{i=1}^{s} r_i \cdot \frac{1 + 3^{s-i}}{2} \ \leq \ H \cdot \sum_{i=1}^{s} D_i \cdot \frac{1 + 3^{s-i}}{2}$

$$
\leq \ 3^{s-1} \cdot H \cdot d_1 \ldots d_s \cdot \sum_{i=1}^{s} \frac{1}{d_i}
$$

finaly: $\quad \deg_T(r(T)) \leq H \cdot 3^{s-1} \cdot d_1 \ldots d_s \cdot \sum_{i=1}^{s} \frac{1}{d_i}$ $\qquad$ (19)

b) Induction relations on the degrees of the components of the K_i:
1) Repeated application of the Bezout theorem applied to:

$$(I_{i-1}, f_i) \cdot K[X] = (I_i \cap K_i \cap E_i) \cdot K[X]$$

leads to: $\quad \deg_x(I_{i-1}) \cdot d_i \geq \deg_x(I_i) + \deg_x(K_i) \quad i = 1, \ldots, s$

$((L_i \cap M_i) \cdot K[X] = (1)$ for all their components are non-proper, and E_i does not play any role because its components all have height greater than $i+1$)

and then: $\deg_x(I_i) + \sum_{j=1}^{i} \deg_x(K_j) \cdot d_{j+1} \ldots d_i \leq \prod_{j=1}^{i} d_j \quad i = 1, \ldots, s$ $\quad$ (20)

2) A special case when $s = n+1$ and $i = s$:
A theorem of Macaulay adapted by J.Kollár (see [8] or [5, lemma 4]) shows that:

$$(X_0, \ldots, X_n)^{(\deg_x(I_n) + d_{n+1} - 1)} \subset (I_n, f_{n+1}) \cdot K[X] \subset K_{n+1} \cdot K[X]$$

but: $\quad rad(K_{n+1} \cdot K[X]) = (X_0, \ldots, X_n) \quad$ because $\quad ht(K_{n+1}) = n+1$

thus: $\quad k_{n+1} \leq exp(K_{n+1} \cdot K[X]) \quad \leq \quad \deg_x(I_n) + d_{n+1} - 1$
$$\leq \quad \deg_x(I_n) + d_n - 1 \qquad (21)$$

3) We always have:

$$\frac{1 + 3^{s-i}}{2} \leq 3^{s-i} \leq d_{i+1} \ldots d_s \quad i = 1, \ldots, s \qquad (22)$$

4) Bound of N:
Case 1) $s \leq n$:
from (8),(22) and (20):

$$N \quad := \quad \sum_{i=1}^{s} k_i \cdot \frac{1 + 3^{s-i}}{2}$$

$$\leq \quad \sum_{i=1}^{s} \deg_x(K_i) \cdot d_{i+1} \ldots d_s = \prod_{i=1}^{s} d_i - \deg(I_s) \leq \prod_{i=1}^{s} d_i$$

Case 2) $s = n+1$:
We derived from (21) that

$$N \quad := \quad k_s + \sum_{i=1}^{s-1} k_i \cdot \frac{1 + 3^{s-i}}{2} \leq (\deg_x(I_n) + d_n - 1) + \sum_{i=1}^{s-1} k_i \cdot \frac{1 + 3^{s-i}}{2}$$

$$= (\deg_x(I_n) + \sum_{i=1}^{s-1} k_i \cdot \frac{1+3^{s-i}}{2}) + (d_n - 1) + (\sum_{i=1}^{s-1} k_i \cdot 3^{s-i-1})$$

$$\leq \prod_{i=1}^{s-1=n} d_i + \prod_{i=1}^{s-1=n} d_i + \prod_{i=1}^{s-1=n} d_i \quad ((8)\ (22)\ \text{and}\ (20))$$

$$\leq 3 \cdot \prod_{i=1}^{s-1=n} d_i$$

c) Dealing with the non-proper components of M_i:

1) Repeated application of the Bezout theorem applied to:

$$(I_{i-1}, f_i) \cdot R[X] = (I_i \cap K_i \cap L_i \cap M_i \cap E_i) \cdot R[X]$$

leads to (with the notation $\quad \underline{d_i} := d_i + H \quad i = 1, \ldots, s$)

$$\deg(I_i) + \deg(K_i) + \deg(L_i) + \deg(M_i) \leq \deg(I_{i-1}) \cdot (d_i + H) \quad i = 1, \ldots, s$$

thus: $$\deg(I_i) + \sum_{j=1}^{i}(\deg(K_j) + \deg(L_i) + \deg(M_i)) \cdot \underline{d}_{j+1} \cdots \underline{d}_i \leq \prod_{j=1}^{i} \underline{d}_j \quad (23)$$

as: $$k_i \cdot \frac{1+3^{s-i}}{2} \leq \deg_x(K_i) \cdot 3^{s-i} \leq \deg_x(K_i) \cdot d_{i+1} \ldots d_s \leq \deg(K_i) \cdot \underline{d}_{i+1} \ldots \underline{d}_s$$

and: $$\mu_i \cdot \frac{1+3^{s-i}}{2} \leq \deg(M_i) \cdot 3^{s-i} \leq \deg(M_i) \cdot \underline{d}_{i+1} \ldots \underline{d}_s$$

thus the terms of the sums:
$N = \sum_{i=1}^{s} k_i \cdot \frac{1+3^{s-i}}{2}$ (9) and $L = \sum_{i=1}^{s} \mu_i \cdot \frac{1+3^{s-i}}{2}$ (10), are bounded by a sub-sum extracted of the expression (23) when $i = s$,

thus we derive: $$L + N \leq \prod_{i=1}^{s}(d_i + H) \quad (24)$$

2) Let $m_{i,j} := (M_{i,j} \cap R)$
as $rad(m_{i,j}) \subset rad(M_{i,j})$, we get from (6):

$$(\prod_{i=1}^{s}\prod_{j} rad(m_{i,j})^{\mu_{i,j} \cdot \frac{1+3^{s-i}}{2}}) \cdot r(T) \cdot (\prod_{i=1}^{s}\prod_{j} rad(K_{i,j})^{k_{i,j} \cdot \frac{1+3^{s-i}}{2}}) \subset (f_1, \ldots, f_s)$$

Let rename the family $\{rad(K_{i,j})\}$ of proper prime ideals in $\{P_i, \quad i = 1, \ldots, r\}$, the exponents $k_{i,j} \cdot \frac{1+3^{s-i}}{2}$ in $\{e_i, i = 1, \ldots, r\}$, and as well for the family $\{rad(m_{i,j})\}$ in the $\{m_i, \quad i = 1, \ldots, q\}$, and the $\mu_{i,j} \cdot \frac{1+3^{s-i}}{2}$ in $\{\lambda_i, i = 1, \ldots, q\}$.

finally: $$(\prod_{i=1}^{q} m_i^{\lambda_i}) \cdot r(T) \cdot (\prod_{i=1}^{r} P_i^{e_i}) \subset (f_1, \ldots, f_s)$$

$$\text{with:} \quad N = \sum_1^r e_i \le 3 \cdot d_1 \ldots d_\mu \quad \text{where} \quad \mu = \min(s, n)$$

$$\text{and:} \quad N + L = \sum_1^r e_i + \sum_1^q \lambda_i \le (d_1 + H) \ldots (d_v + H) \quad \text{where} \quad v = \min(s, n+1)$$

Proof of corollary 1)

In this case $m = 3$ so that two kinds of prime ideals m_i with height greater than 2 appear:

  - with height 3, thus (T_1, T_2, T_3)-primary

  - with height 3, thus $(T_a - x \cdot T_b, T_a - y \cdot T_c)$-primary for some couple (x, y) of scalars in k and some permutation $\{a, b, c\} = \{1, 2, 3\}$.

Therefore we can choose in each m_i an equation $s_i(T)$ of degree 1 in T, and define:

$$s(T) := r(T) \cdot \Pi s_i(T)$$

so that $s(T)$ has degree bounded by:

$$\deg r(T) + (N + L) \le 3^{v-1} \cdot d_1 \ldots d_v \cdot \left(\sum_1^v \frac{1}{d_i}\right) \cdot H + (d_1 + H) \ldots (d_v + H)$$

moreover: $\quad (a(T) \cdot f^M \in I) \Rightarrow (f^M \in I \cdot K[X]) \Rightarrow (f \in rad(I \cdot K[X]))$

But each P_i contains $rad(I)$, thus f^N belongs to $(\prod_{i=1}^r P_i^{e_i})$ and as a consequence $s(T)$ belongs to $(\prod_{i=1}^q m_i^{\lambda_i}) \cdot r(T) \cdot (\prod_{i=1}^r P_i^{e_i}) \subset (f_1, \ldots, f_s)$. As for the other bounds, they are a consequence of the graded structure in T and X.

Proof of corollary 2)

a) Homogeneizing $k[T_1, T_2]$ by T_3, and $R[X_1, \ldots, X_n]$ by X_0,

$$(a(T) \cdot f^M \in I) \text{ becomes: } ((T_3^t \cdot A(T)) \cdot (X_0 \cdot F)^{M'} \in (F_1, \ldots, F_s))$$

for some integers t and M and the change to capitals indicates the homogenized polynomial. Then we use corollary 1) applied to $(X_0 \cdot F)$, and afterwards we specialize $T_3 := 1$ and $X_0 := 1$ in the result to obtain a).

b) We take $f = 1$ and we can apply a) above.

Proof of corollary 3)

$m = 1$ so that there are no ideals m_i. We put $s(T) = r(T)$ and then we proceed as in corollary 1). Only the X gradation gives a bound for the degrees of the $A_i \cdot f_i$. As for the growing of the degree of $r(T)$, it stays linear in H in this case.

References

[1] C.A. Berenstein and D.C. Struppa "Recent improvements in the complexity of the effective nullstellensatz" , Linear Algebra and its Applications (1991) p203-215.

[2] C.A. Berenstein and D.C. Struppa "Small degree solutions for the polynomial Bezout equation" , Linear Algebra Appl. **98** (1988) p41-55.

[3] C.A. Berenstein and A.Yger : "Effective Bezout identities in $Q[z_1, \ldots, z_n]$", Acta Mathematica, **166** (1991), p 69-120. II

[4] W.Dale Brownawell : "Bounds for the degre in the nullstellensatz" , Ann. of Math.(2), **126** (1987), p577-591.

[5] W.Dale Brownawell : "A prime power product version of the nullstellensatz", Mich. J. Math.

[6] L.Caniglia, A.Galligo, J.Heintz : "Borne simple exponentielle pour les degrés dans le théorème des zéros sur un corps de caractéristique quelconque", C.R.A.S , Paris **307** (1988) , p255-258.

[7] Greta Hermann : "Die frage der endliche vielen schritte in der theorie der polynomideale." , Math. Ann., **95** (1926), p736-788.

[8] J. Kollár : "Sharp effective nullstellensatz" , Journal of the American Mathematical Society **1** (1988) p 963-975.

[9] Noaï Fitchas and André Galligo : "Nullstellensatz effectif and conjecture de Serre (Théorème de Quillen-Suslin) pour le calcul formel" , Math. Nachr. **149** (1990), p 231-253.

[10] Ju.V. Nesterenko : "Transcendental Numbers" , Izv. Akad. Nauk. SSSR, Ser. Mat. Tom **41** n$^{\underline{o}}$2 (1977), p243.

[11] Patrice Philippon : "Dénominateurs dans le théorème des zéros de Hilbert", Acta Arithmetica **58** (1) (1991), p1-25.

[12] Patrice Philippon : "Critères pour l'indépendance algébrique" , Publications Mathmatiques de l'I.H.E.S n$^{\underline{o}}$**64** (1986), p5-52.

[13] Bernard Teissier : "Résultats récents d'algèbre commutative effective", Séminaire Bourbaki 42ième année **718** (1989-90). Astérisque **189-190** (1991), p107-131.

Frédéric Smietanski
URA 168 du CNRS
Université de Nice-Sophia Antipolis, Faculté des sciences,
B.P 71 06108 Nice Cedex 2 France.

An Elimination Method Based on Seidenberg's Theory and Its Applications

D. Wang

Abstract. In this paper we present an elimination method for algebraically closed fields based on Seidenberg's theory. The method produces, for any pair $[PS, QS]$ of sets of multivariate polynomials, a sequence of triangular forms $TF_1, \ldots, TF_e$ and polynomial sets $US_1, \ldots, US_e$ such that the difference set of common zeros of PS and QS is the same as the union of the difference sets of common zeros of TF_i and US_i. Moreover, the triangular forms TF_i and polynomial sets US_i can be so computed as to give a necessary and sufficient condition for the given system to have algebraic zeros for some prescribed variables. This method has a number of applications such as to solving systems of polynomial equations and inequalities, mechanical theorem proving in geometry, irreducible decomposition of algebraic varieties and constructive proof of Hilbert's Nullstellensatz which are partially discussed in the paper. Preliminary experiments show that the efficiency of this method is at least comparable with that of the well-known methods of characteristic sets and Gröbner bases for some applications. A few illustrative yet encouraging examples performed by a draft implementation of the method are given.

1. Introduction

It has been widely known that systems of multivariate polynomials can be triangularized by such methods as those based on resultants [7, 14], characteristic sets [15, 16, 27, 28] and Gröbner bases [1, 2]. These methods proceed in successively eliminating the variables in some manner and are called elimination methods in general. A fundamental application of elimination methods is to solving systems of polynomial equations, which is of practical importance and has been extensively investigated by many prominent researchers, see [1-3, 5, 6, 10-13, 20, 28, 29] for instance. Nevertheless, it is little known that A. Seidenberg [17, 18] also published an elimination method for algebraically closed fields in 1956 (2 years later than the publication of his well known decision method for real closed fields). The goal of Seidenberg was to decide whether a system of (differential) polynomial equations and inequalities ($\neq$) has a solution in some extension of the ground field. This was motivated from an observation on the relationship between elimination methods and Hilbert's Nullstellensatz. His method was then developed to give a constructive proof of the Nullstellensatz in various cases and to "throw light on the Nullstellensatz". This motivation and goal led Seidenberg to put his emphasis mainly on the theory (together with its completeness) without taking real computation into account. Though his method is constructive, apparently it is neither efficient nor quite appropriate for nowadays diverse applications. Fortunately, there are several remarkable ideas underlying Seidenberg's theory which can be adopted for developing an efficient procedure. Such ideas include performing the elimination top-down (analogous to the procedure of Gauss elimination) rather than bottom-up, splitting the polynomial sets at each time of the pseudo-division according to the vanishing and non-vanishing of the leading coefficient of the dividing polynomial, and preserving the equivalence of the solvability between the given system and the disjunction of the eliminated systems by performing the so-called projection which has been re-introduced recently by Wu [30]. The purpose of this paper is to present an elimination method by adopting these basic ideas of Seidenberg and some ideas from the theory of characteristic sets, which, we believe, is more efficient and appropriate for applications. Our method produces, for any pair $[PS, QS]$ of sets of multivariate polynomials, a sequence of triangular forms $TF_1, \ldots, TF_e$ and polynomial sets $US_1, \ldots, US_e$ such that the difference set of common zeros of PS and QS is the same as the union of the difference sets of common zeros of TF_i and US_i. Moreover, the triangular forms TF_i and polynomial sets US_i can be so computed as to give a necessary and sufficient condition for the given system to have algebraic zeros for some prescribed variables. Our work has been inspired by the theory of Seidenberg [17, 18], the characteristic set method of Ritt-Wu [15, 16, 27-29] and Wu's recent projection theorem [30]. However, our elimination method is presented in a form different from each of them. Also, the purpose and applicability of our method are all different from those of Seidenberg's. We are interested not only in deciding the existence of solutions of polynomial equations and inequations but also and even more in exactly finding all possible solutions, whereas the latter was not considered by Seidenberg at all. As the same as Seidenberg's original method, our method, applied to the solvability problem of systems of polynomial equations and inequations, does not require polynomial factorization[1], and it works for ordinary polynomials, differential polynomials and partial differential polynomials over the basic fields of characteristic 0 and of characteristic $p \neq 0$. In the present paper we restrict ourselves to ordinary polynomials over the ground field of

---

[1] This claim is of value only in theory. In practice, polynomial factorization may be heuristically employed to reduce the size of occurring polynomials and to simplify the output. Moreover, polynomial factorization over algebraic extension fields is required if we apply the method to the irreducible decomposition of algebraic varieties (cf. Subsection 4.3).

characteristic 0. The method in other cases will be presented in the sequel.

We shall discuss some applications of our method including solving systems of polynomial equations and inequations, mechanical theorem proving in geometry, irreducible decomposition of algebraic varieties and constructive proof of Hilbert's Nullstellensatz. The author has made a draft implementation of the presented method in Maple. Preliminary experiments show that the method is very efficient and the extent of its efficiency is at least comparable with that of the well-known methods of characteristic sets and Gröbner bases for some applications. A few illustrative yet encouraging examples performed by the draft implementation will be given.

The proofs are given in Appendix A.

2. Preliminaries

Let K be a basic field of characteristic 0 and $x_1 \prec \cdots \prec x_n$, abbreviated sometimes to x, be a set of n variables with a fixed ordering. By a polynomial we shall mean an element in the ring $K[x_1, \ldots, x_n]$. For any polynomial P and a variable x_i the degree of P in x_i will be denoted by $deg(P, x_i)$. If $P \notin K$ (i.e., it is not a constant), we shall call the variable x_p which occurs in P with biggest subscript p the *leading variable* of P, denoted by $lvar(P)$, and the leading coefficient of P with respect to $lvar(P)$ the *initial* of P, denoted by $ini(P)$. We define the leading variable of any constant to be x_0 which is $\prec x_1$.

Let P be a non-constant polynomial with $x_p = lvar(P)$, $I = ini(P)$ and $m = deg(P, x_p)$. $P - I x_p^m$ will be called the *reductum* of P, denoted by $red(P)$. If Q is any other polynomial, the pseudo-remainder R of Q with respect to P in x_p will be denoted by $prem(Q, P)$.

By a polynomial set we shall mean a finite, non-empty set of non-zero polynomials in $K[x_1, \ldots, x_n]$. For any polynomial set PS, the set of those polynomials in PS which involve the variables $x_1, \ldots, x_k$ only is denoted by

$$PS^{(k)} = PS \cap K[x_1, \ldots, x_k]$$

as usual. If $a_1, \ldots, a_k$ are k elements in some extension field of K, then the set of polynomials obtained form PS by substituting $a_1, \ldots, a_k$ respectively for $x_1, \ldots, x_k$ is denoted as

$$PS^{(a,k)} = PS|_{x_1 = a_1, \ldots, x_k = a_k}.$$

Let QS be any other polynomial set. We denote, by $\tilde{K}$-$Zero(PS/QS)$, the set of those common zeros of polynomials in PS which are not zeros of any polynomial in QS in some extension field $\tilde{K}$ of K, called the *difference set* of common zeros of PS and QS. If the field $\tilde{K}$ is not specified, then $\tilde{K}$-$Zero(PS/QS)$ will be simply denoted by $Zero(PS/QS)$. From now on, while writing $Zero(PS/QS) \neq \emptyset$, we shall always mean $\tilde{K}$-$Zero(PS/QS) \neq \emptyset$ unless explained otherwise, that is, the corresponding system of polynomial equations and inequations has solutions in some extension field $\tilde{K}$

of K. We shall write $Zero(P/QS)$ for $Zero(\{P\}/QS)$ and $Zero(PS/Q)$ for $Zero(PS/\{Q\})$, and write $Zero(PS)$ for $Zero(PS/QS)$ if $QS = \emptyset$ or all polynomials in QS are constant.

In what follows, we shall call an ordered set a *list*, denoted by putting its elements into a pair of square brackets instead of a pair of braces. In particular, ϕ and $\emptyset$ stand respectively for *empty list* and *empty set*. All operations of sets are applied to lists in the usual way except for preserving the order. For example, $[b, a] \cup [c] = [b, a, c]$ but neither $[a, b, c]$ nor $[a, c, b]$. For a non-empty set or list S, its first element will be denoted by $first(S)$.

Definition 1. *A finite, non-empty list* TF: $[T_1, T_2, \ldots, T_r]$ *of non-constant polynomials is called a* triangular form *if* $lvar(T_1) \prec lvar(T_2) \prec \cdots \prec lvar(T_r)$. *A triangular form* TF *with* $I_i = ini(T_i)$ *is said to be* fine *if either* $r = 1$, *or* $r > 1$ *while* $prem(I_i, [T_1, \ldots, T_{i-1}]) \neq 0$ *for* $i = 2, \ldots, r$. *It is said to be* perfect *if either* $r = 1$, *or* $r > 1$ *while* $Zero(TF/\{I_1, \ldots, I_r\}) \neq \emptyset$.

For example, the triangular form $TF = [x_1^2 - 2, x_2^2 - 2x_1x_2 + 2, (x_2 - x_1)x_3+1]$ is fine but not perfect because any zero of the first two polynomials in TF makes the vanishing of the initial $x_2 - x_1$ of the third polynomial (actually, $Zero(TF) = \emptyset$), whereas the triangular $[x_1^2 - 1, (x_1^3 - 1)x_2 + 1]$ is perfect. It is easy to show that any perfect triangular form is fine.

In the remaining part of this section we state three simple lemmas which are fundamental for the proof of our main theorem. The formal proofs of these lemmas are given in [26].

Lemma 1. *Let* $T \in K[x_1, \ldots, x_i]$ *be a polynomial with* $lvar(T) = x_i, ini(T) = I$, *$PS, QS \subset K[x_1, \ldots, x_n]$ be two polynomial sets and RS be the set of all non-zero pseudo-remainders of polynomials in PS with respect to T in x_i. Then*

$$Zero(PS \cup \{T\}/QS) = Zero(RS \cup \{T\}/QS \cup \{I\})$$

$$\cup Zero(PS \cup \{I, red(T)\}/QS). \qquad (2.1)$$

Lemma 2. *Let* $PS \subset K[x_1, \ldots, x_i]$ *and* $QS \subset K[x_1, \ldots, x_n]$ *be two polynomial sets, where* PS *is not necessarily non-empty. Suppose there are polynomials, say* $H_1, \ldots, H_h$ *in all, in* QS *such that their leading variables are* $\succ x_i$. *Let* $H_{l1}, \ldots, H_{lm_l}$ *be all non-zero coefficients of the power products in* H_l *with respect to those variables which are* $\succ x_i$ *and set* $H_{l0} = 0$ *for* $l = 1, \ldots, h$. *Then*

$$Zero(PS/QS) \neq \emptyset \iff \bigcup_{1 \leq j_1 \leq m_1, \ldots, 1 \leq j_h \leq m_h} Zero(PS \cup HS_{j_1 \cdots j_h}/QS_{j_1 \cdots j_h}) \neq \emptyset$$

$$(2.2)$$

in some extension field of K, *and*

$$Zero(PS/QS) = \bigcup_{1 \leq j_1 \leq m_1, \ldots, 1 \leq j_h \leq m_h} Zero(PS \cup HS_{j_1 \cdots j_h}/QS'_{j_1 \cdots j_h}), \qquad (2.3)$$

where $HS_{j_1\cdots j_h} = \{H_{lj}|\ 0 \le j \le j_l - 1, 1 \le l \le h\} \setminus \{0\}$, $QS_{j_1\cdots j_h} = QS \setminus$ $\{H_1, \ldots, H_h\} \cup \{H_{1j_1}, \ldots, H_{hj_h}\}$ and $QS'_{j_1\cdots j_h} = QS \cup \{H_{1j_1}, \ldots, H_{hj_h}\}$.

Lemma 3. *Let $T \in K[x_1, \ldots, x_i]$ be a polynomial with $lvar(T) = x_i$, $ini(T)$ $= I$ and $deg(T, x_i) = d$, and $PS \subset K[x_1, \ldots, x_{i-1}], QS \subset K[x_1, \ldots, x_i]$ be two polynomial sets.*

a) If all polynomials in QS have their leading variables $\prec x_i$, then

$$Zero(PS \cup \{T\}/QS \cup \{I\}) \ne \emptyset \Longleftrightarrow Zero(PS/QS \cup \{I\}) \ne \emptyset \qquad (2.4)$$

in some extension field of K.

b) Suppose there are polynomials, say $H_1, \ldots, H_h$ in all, in QS such that their leading variables are x_i. Let R be the pseudo-remainder of $(H_1 \cdots H_h)^d$ with respect to T (i.e., $R = prem((H_1 \cdots H_h)^d, T))$ and $QS' = QS \setminus \{H_1, \ldots, H_h\} \cup \{I, R\}$. Then

$$Zero(PS \cup \{T\}/QS \cup \{I\}) \ne \emptyset \Longleftrightarrow Zero(PS/QS') \ne \emptyset \qquad (2.5)$$

in some extension field of K, and

$$Zero(PS \cup \{T\}/QS \cup \{I\}) = Zero(PS \cup \{T\}/QS'). \qquad (2.6)$$

We end this section by defining the concept of quintuplets which will be through used in our presentation of the algorithms.

Definition 2. *Let the integers k and n $(0 \le k \le n)$ be given. A quintuplet of level i $(1 \le i \le n)$ is a list $[PS, QS, TF, US, i]$ of five elements, where $PS \subset K[x_1, \ldots, x_i]$, TF is a triangular form with $lvar(first(TF)) = x_p$ for some $p > i$, $QS \subset K[x_1, \ldots, x_{max(p,k)}]$ and $US \subset K[x_1, \ldots, x_n]$ with $lvar(U) \succ x_{max(p,k)}$ for all $U \in US$.*

3. The Main Algorithm

Our main result of this paper is the following theorem which provides an elimination method for algebraically closed fields. The idea underlying this method is rather simple, whereas the presentation in an algorithmic form is quite involved. Some details are taken into account mainly for the sake of efficiency.

Theorem 1. *For any finite, non-empty set PS, any finite set QS of non-zero polynomials in $K[x_1, \ldots, x_n]$ and any integer k $(0 \le k \le n)$, the algorithm TRIANGULARIZE described below either determines $Zero(PS/QS) = \emptyset$, or computes a sequence of triangular forms $TF_1, \ldots, TF_e$ and finite sets $US_1, \ldots, US_e$ of non-zero polynomials such that*

a) $Zero(PS/QS) = \bigcup_{i=1}^{e} Zero(TF_i/US_i),$ \qquad (3.1) where $Zero(ini(T)/US_i) = \emptyset, T \in TF_i$ *(in any extension field of K) for all i, and moreover $prem(U, TF_i) \neq 0, U \in US_i$ for all i if $k = n$ or 0;*

b) if $k < n$, then for k elements $a_1, \ldots, a_k$ in any extension field $\tilde{K}$ of K

$$(a_1, \ldots, a_k) \in \bigcup_{i=1}^{e} Zero(TF_i^{(k)}/US_i^{(k)}) \Longleftrightarrow Zero(PS^{(a,k)}/QS^{(a,k)}) \neq \emptyset \quad (3.2)$$

in some algebraic extension field of $\tilde{K}$;

c) if $k < n$, then for each i and any $(a_1, \ldots, a_k) \in Zero(TF_i^{(k)}/US_i^{(k)})$

$$Zero(TF_i^{*(a,k)}/US_i^{*(a,k)}) \neq \emptyset \quad (3.3)$$

in some algebraic extension field of $K(a_1, \ldots, a_k)$, and in particular the triangular form $TF_i \setminus TF_i^{(k)}$ is perfect, where $PS^{(a,k)} = PS|_{x_1=a_1,\ldots,x_k=a_k}$, $TF_i^{(k)} = TF_i \cap K[x_1, \ldots, x_k]$, $TF_i^{*(a,k)} = (TF_i \setminus TF_i^{(k)})|_{x_1=a_1,\ldots,x_k=a_k}$ and similarly for $QS^{(a,k)}$, $US_i^{(k)}$ and $US_i^{*(a,k)}$.

The sequence of triangular forms $TF_1, \ldots, TF_e$ and polynomial sets $US_1, \ldots, US_e$ computed from $[PS, QS]$ in Theorem 1 will be denoted by $\Psi = \{[TF_1, US_1], \ldots, [TF_e, US_e]\}$. When $Zero(PS/QS) = \emptyset$ is determined, we say $e = 0$, or $\Psi = \emptyset$. The algorithms below will be described in such a form that can be easily transformed into the programming code. The used notations and phrases should be understandable without detailed explanation. They follow essentially the paradigm of the Maple language. We omit the **endif** and **enddo** declarations and use the block-structure instead to keep the presentation brief.

Algorithm TRIANGULARIZE (Input: PS, QS, k; Output: Ψ)

> *where $PS, QS \subset K[x_1, \ldots, x_n], 0 \leq k \leq n$, $\Psi = \emptyset$ or*
> *$\{[TF_1, QS_1], \ldots, [TF_e, QS_e]\}$ satisfying a), b) and c) in Theorem 1.*

1	$\Psi := \emptyset$; $\Phi := \{[PS, QS, \phi, \emptyset, n]\}$
2	while $\Phi \neq \emptyset$ do
3	$\quad [PS', QS', TF, US, m] := first(\Phi)$; $\Phi := \Phi \setminus \{[PS', QS', TF, US, m]\}$
4	$\quad$ for i from m to 1 step -1 do
5	$\quad\quad$ if $PS' \cap K \neq \emptyset$ then go to 3
6	$\quad\quad$ if $\exists P \in PS'$ such that $lvar(P) = x_i$ then
7	$\quad\quad\quad [T, PS', QS', \Delta] :=$ELIMINATE$(PS', QS', TF, US, i)$
8	$\quad\quad\quad \Phi := \Phi \cup \Delta$
9	$\quad\quad\quad$ if $PS' \cap K \neq \emptyset$ then go to 3
10	$\quad\quad$ else
11	$\quad\quad\quad$ if $i \neq k$ then go to 4 *(for next i)*
12	$\quad\quad\quad$ else $T := 0$
13	$\quad\quad$ if $i \geq k$ then
14	$\quad\quad\quad [QS', US, \Delta] :=$PROJECT_A$(PS' \cup \{T\}, QS', TF, US, i)$
15	$\quad\quad\quad \Phi := \Phi \cup \Delta$
16	$\quad\quad\quad$ if $T = 0$ then go to 4 *(for next i)*
17	$\quad\quad QS' := REDUCE\_PROJECT(T, QS', i, k)$
18	$\quad\quad$ if $0 \in QS'$ then go to 3
19	$\quad\quad TF := [T] \cup TF$
20	$\quad \Psi := \Psi \cup \{[TF, QS' \cup US]\}$

In the while-loop (Lines 2-20) — the main body of this algorithm, the set Φ of quintuplets increases and decreases, while the pairs $[TF, QS' \cup US]$ of triangular forms and polynomial sets are produced. This procedure terminates as soon as the set Φ becomes empty. Within the while-loop, for each quintuplet $[PS', QS', TF, US, m]$ taken from Φ the variables are eliminated, inductively from x_m to x_1, by the subalgorithm ELIMINATE. The function of the subalgorithms PROJECT_A and REDUCE_PROJECT is to make the produced pairs $[TF, QS' \cup US]$ to satisfy the properties a), b) and c) in Theorem 1.

The main purpose of the subalgorithm ELIMINATE is to eliminate the variable x_i from those polynomials in the non-empty subset MS of PS' whose leading variables are x_i, so that among the eliminated polynomials, only one, say T, has x_i as its leading variable, while all the others, say put into the set PS', have their leading variables lower than x_i. This is done recursively by pseudo-dividing those polynomials of higher degree in MS by a polynomial T with the lowest degree in x_i (Line 9), resetting MS to be the set of T and those remainders whose leading variables are x_i (Line 11), and adjoining those non-zero remainders whose leading variables are lower than x_i to PS' (Line 10). At each time of the pseudo-division, the initial of the dividing polynomial is assumed to be non-zero, being adjoined to the polynomial set QS' (Line 8), while the case that the initial happens to be 0 is considered disjunctively by replacing the corresponding polynomial in

PS' with its initial and its reductum, forming a quintuplet and adjoining it to Δ (Line 7).

Subalgorithm ELIMINATE (Input: PS, QS, TF, US, i; Output: $[T, PS', QS', \Delta]$)

> where $[PS, QS, TF, US, i]$ is a quintuplets of level i,
> $T \in K[x_1, \ldots, x_i]$ with $lvar(T) = x_i$, $PS' \subset K[x_1, \ldots, x_{i-1}]$,
> $QS' \subset K[x_1, \ldots, x_{\max(p,k)}]$, p is the subscript of $lvar(first(TF))$ if
> $TF \neq \phi$, and n otherwise, Δ is a set of quintuplets of level i.

1	$QS' := QS$; $\Delta := \emptyset$; $T := 0$
2	$MS := \{P \in PS \mid lvar(P) = x_i\}$; $PS' := PS \setminus MS$
3	while $MS \neq \{T\}$ do
4	$\quad T :=$ an element of MS with lowest degree in x_i
5	$\quad I :=$SIMPLIFY$(ini(T), QS')$
6	$\quad$ if $I \notin K$ then
7	$\quad\quad \Delta := \Delta \cup \{[(MS \cup PS' \setminus \{T\}) \cup \{red(T), I\}, QS', TF, US, i]\}$
8	$\quad\quad QS' := QS' \cup \{I\}$
9	$\quad RS := \{R \mid R =$SIMPLIFY$(prem(M, T), QS'), R \neq 0, M \in MS \setminus \{T\}\}$
10	$\quad PS' := PS' \cup \{R \in RS \mid lvar(R) \prec x_i\}$
11	$\quad MS := \{T\} \cup \{R \in RS \mid lvar(R) = x_i\}$

The subalgorithm PROJECT_A is an implementation of Lemma 2 which splits the system $[PS, QS]$ into a set of subsystems, of which all but one are stored in Δ. Those polynomials corresponding to the $H_1, \ldots, H_h$ in Lemma 2 are removed from QS but adjoined to US, forming the new sets QS' and US' (Line 2). One of the produced subsystems is taken as output of the subalgorithm (Lines 12-13).

The first part (Lines 1-4) of the subalgorithm REDUCE_PROJECT implements Lemma 3. The case that all polynomials in QS have their leading variables $\prec x_i$ is trivial. If QS contains polynomials whose leading variables are the same as the leading variable x_i of T, then the remainder of the dth power of the product of all those polynomials with respect to T is formed (Line 4), where $d = deg(T, x_i)$. This process together with that in PROJECT_A, called *projection*, is to guarantee the equivalence of the emptiness of zero sets between the given system and the disjunction of the produced systems during the successive elimination of variables. The second part (Lines 1-3 and 5) of REDUCE_PROJECT is simply to compute the pseudo-remainders of polynomials in QS with respect to T in order to remove some empty components and to satisfy the property a) in Theorem 1.

Subalgorithm PROJECT_A (Input: PS, QS, TF, US, i; Output: $[QS', US', \Delta]$)

> where $[PS, QS, TF, US, i]$ is quintuplet of level i, $QS' \subset K[x_1, \ldots, x_i]$,
> $US' \subset K[x_1, \ldots, x_n]$ with $lvar(U) \succ x_i$ for all $U \in US'$,
> Δ is a set of quintuplets of level i.

1	$HS := \{Q \in QS \mid lvar(Q) \succ x_i\}$
2	$QS' := QS \setminus HS$; $US' := US \cup HS$; $\Delta := \emptyset$
3	if $HS \neq \emptyset$ then
4	$\quad h := 0$
5	$\quad$ for each $H \in HS$ do
6	$\qquad V := \{x_j \mid i < j \leq n, deg(H, x_j) > 0\}$; $h := h + 1$
7	$\qquad \{H_{h1}, \ldots, H_{hm_h}\} :=$ set of coefficients of H w.r.t. V
8	$\qquad$ if $\exists j$ such that $H_{hj} \in K$ then $h := h - 1$
9	$\quad$ if $h > 0$ then
10	$\qquad$ for l from 1 to h do $H_{l0} := 0$
11	$\qquad \Delta := \{[PS \cup \{H_{lj} \mid 0 \leq j \leq j_l - 1, 1 \leq l \leq h\} \setminus \{0\},$
	$\qquad\quad QS' \cup \{H_{1j_1}, \ldots, H_{hj_h}\}, TF, US', i] \mid$
	$\qquad\quad 1 \leq j_1 \leq m_1, \ldots, 1 \leq j_h \leq m_h\}$
12	$\qquad QS' := QS' \cup \{H_{11}, \ldots, H_{h1}\}$
13	$\qquad \Delta := \Delta \setminus \{[PS \setminus \{0\}, QS', TF, US', i]\}$

Subalgorithm REDUCE_PROJECT (Input: T, QS, i, k; Output: QS')

> where $T \in K[x_1, \ldots, x_i]$ with $lvar(T) = x_i$, $0 \leq k \leq n$,
> $QS, QS' \subset K[x_1, \ldots, x_i]$ with $Zero(ini(T)/QS) = \emptyset$.

1	$KS := \{Q \in QS \mid lvar(Q) = x_i\}$
2	if $KS = \emptyset$ then $QS' := QS$
3	else
4	$\quad$ if $i > k$ then $QS' := (QS \setminus KS) \cup \{prem(\prod_{K \in KS} K^{deg(T, x_i)}, T)\}$
5	$\quad$ else $QS' := (QS \setminus KS) \cup \{prem(K, T) \mid K \in KS\}$

Note that k and n in the specification of the subalgorithms refer to the global variables mentioned in Theorem 1.

The subalgorithm SIMPLIFY is rather simple. It is used mainly for removing some redundant factors in order to control the expansion of polynomial size and for avoiding some unnecessary computations.

Subalgorithm SIMPLIFY (Input: M, QS; Output: M')

> where $M, M' \in K[x_1, \ldots, x_n], QS \subset K[x_1, \ldots, x_n]$

1	$M' := M$
2	if $M' \neq 0$ then
3	$\quad$ for each $Q \in QS \setminus \{1\}$ do
4	$\qquad$ while $Q \mid M'$ do $M' := M'/Q$

The termination of the main algorithm TRIANGULARIZE is guaranteed if the subalgorithms and the while-loop terminate. The termination of

ELIMINATE, PROJECT_A, REDUCE_PROJECT and SIMPLIFY is all quite obvious. To see the termination of the while-loop, we define, for any pair $[PS, QS]$ of polynomial sets, a triple $Index(PS/QS) = [d, m, p]$, where $m =$ the minimal j such that $PS \subset K[x_1, \ldots, x_j]$, $d =$ the lowest degree (in x_m) of those polynomials in PS whose leading variable is x_m, $p = \max(m, p')$ and $p' =$ the minimal j such that $QS \subset K[x_1, \ldots, x_j]$. We order two triples as $[d_1, m_1, p_1] \prec [d_2, m_2, p_2]$ if $p_1 < p_2$, or $p_1 = p_2$ while $m_1 < m_2$, or $p_1 = p_2, m_1 = m_2$ while $d_1 < d_2$. Now, for the pair $[PS', QS']$ of the first two polynomial sets in any quintuplet taken from Ψ (Line 3) in the algorithm TRIANGULARIZE, let PS^* and QS^* be the first two polynomial sets in some quintuplet of the Δ produced by ELIMINATE (Line 7) or by PROJECT_A (Line 11). Then we always have $Index(PS^*/QS^*) \prec Index(PS'/QS')$. From this one may immediately conclude that the while-loop can have only a finite number of iterations.

To see the correctness of TRIANGULARIZE (i.e., the computed Ψ satisfies the properties a), b) and c) in Theorem 1), we may note the zero relation

$$Zero(PS \cup TF/QS \cup US) = Zero(PS' \cup TF'/QS' \cup US) \cup$$

$$\bigcup\nolimits_{[PS^*, QS^*, TF, US, i] \in \Delta} Zero(PS^* \cup TF/QS^* \cup US)$$

for ELIMINATE according to (2.1), the zero relation

$$Zero(PS \cup TF/QS \cup US) = Zero(PS \cup TF/QS' \cup US') \cup$$

$$\bigcup\nolimits_{[PS^*, QS^*, TF, US', i] \in \Delta} Zero(PS^* \cup TF/QS^* \cup US')$$

for PROJECT_A according to (2.2), and the zero relation (2.6). From them one can easily obtain the property a). The properties b) and c) can be proved by using Lemmas 2 and 3 stated in Section 2. All the formal proofs which are given in [26] are quite trivial but involve some details.

We note that the usage of the quintuplets which are produced during the splitting of the systems has increased somewhat the complication of our algorithms in the presentation but it helps to keep the amount of information needed for the subsequent computation. In particular, for each quintuplet $[PS', QS', TF, US, m]$ the already produced triangular from TF is retained, so that the subsequent elimination for it needs to start only from the variable x_m (but not from x_n).

With respect to the integer k in Theorem 1 (and in TRIANGULARIZE), we shall call $n - k$ the *dimension* of projection and say that the elimination is performed with *full* projection if the dimension is n, *without* projection if the dimension is 0.

4. Applications

The elimination method described in the preceding section has several immediate applications. We shall mention some of them below without full details. More complete application results will be reported in several forecoming papers of the author under preparation.

4.1. Solving Systems of Polynomial Equations and Inequations

Since for a triangular form TF and a polynomial set US the zero set $Zero(TF/US)$ can be determined by solving unary polynomial equations and by some simple substitutions and verifications, the presented elimination method leads to a natural application to solving systems of polynomial equations and inequalities. This may be done by first computing the sequence of triangular forms $TF_1, \ldots, TF_e$ and polynomial sets $US_1, \ldots, US_e$ with or without projection and then solving $TF_i = 0, US_i \neq 0$ for each i. If we want only to determine the solvability of the system without computing exactly the solutions, then we may perform the elimination with full projection and apply the following theorem.

Theorem 2. Let $\Psi = \{[TF_1, US_1], \ldots, [TF_e, US_e]\}$ be computed by the algorithm TRIANGULARIZE from $[PS, QS]$ with full projection. Then

   a) $Zero(PS/QS) = \emptyset$ if and only if $e = 0$, i.e., $\Psi = \emptyset$;

   b) $Zero(PS/QS)$ is finite but not empty if and only if $e \geq 1$ and the number of polynomials in each TF_i is exactly n ($n =$ the number of variables), and if so, $Zero(TF_i/US_i) \neq \emptyset$ for all i;

   c) $Zero(PS/QS)$ is infinite if and only if $e \geq 1$ and there is an i such that the number of polynomials in TF_i is less than n.

4.2. Mechanical Theorem Proving in Geometry

Following Wu [27], let a geometric theorem $T(HYP, CON)$ be algebraized with its hypothesis and conclusion being expressed as polynomial relations $HYP: H_1(u_1, \ldots, u_d, x_1, \ldots, x_n) = 0, \ldots, H_s(u_1, \ldots, u_d, x_1, \ldots, x_n) = 0$ and $CON: C_1(u_1, \ldots, u_d, x_1, \ldots, x_n) = 0, \ldots, C_t(u_1, \ldots, u_d, x_1, \ldots, x_n) = 0$, respectively, with coefficients in the geometry-associated field K, where $u_1, \ldots, u_d$ are parameters and $x_1, \ldots, x_n$ are some geometric dependents. We assume that the choice of parameters and geometric dependents are proper, so that the geometric configuration (algebraic variety) defined by the hypothesis relations $H_1 = 0, \ldots, H_s = 0$ over K is of dimension d. A condition $D(u_1, \ldots, u_d, x_1, \ldots, x_n) \neq 0$ under which the geometric theorem is true will be called a *non-degenerate* condition, if the configuration defined by $H_1 = 0, \ldots, H_s = 0, D = 0$ over K is of dimension less than d. A geometric theorem is said to be *generally* true if it is true under some non-degenerate conditions. The proving of geometric theorems by using the described elimination method is based on the following theorem.

Theorem 3. Let $HYP = \{H_1, \ldots, H_s\}, CON = \{C_1, \ldots, C_t\}$ and $\Psi_i = \{[TF_1, US_1], \ldots, [TF_e, US_e]\}$ be computed by the algorithm TRIANGULARIZE from $[HYP, \{C_i\}]$ with full projection in $K(u_1, \ldots, u_d)[x_1, \ldots, x_n]$. Then the geometric theorem $T(HYP, CON)$ is generally true if and only if $e = 0$, i.e., $\Psi_i = \emptyset$, for all i.

It should be noted that, as we want to prove true theorems, the elimination may be first performed without projection. If $\Psi_i = \emptyset$, then the theorem is proved already. Only in the case $\Psi_i \neq \emptyset$ we need to further consider the full projection, in order to reach the conclusion that the theorem is not generally true. It is also easy to get from the elimination procedure the exact non-degenerate conditions, which are actually among the negations of initials of the dividing polynomials occurring during the pseudo-division. To deal with non-degenerate conditions there may be two ways of using our method. One way is to modify the algorithm TRIANGULARIZE so that the initials occurring during the pseudo-division are always assumed to be nonzero. Then we analyze whether the vanishing of these initials corresponds to some degenerate cases of the theorem from geometric consideration. The other is just to use the algorithm TRIANGULARIZE which considers the vanishing of each initial and detects whether it corresponds to some degenerate case by producing a polynomial in the variables u only from the algebraic definition.

We note finally that the restriction on the proper choice of parameters and geometric dependents can be removed as it is shown in [4, 9].

4.3. Irreducible Decomposition of Algebraic Varieties

A fine triangular form TF: $[T_1(u, y_1), T_2(u, y_1, y_2), \ldots, T_r(u, y_1, \ldots, y_r)]$, where u stands for $u_1, \ldots, u_d$, is said to be *irreducible* if T_1, as a polynomial in y_1, is irreducible over the field $K(u_1, \ldots, u_d)$, and for any i ($2 \leq i \leq r$), T_i, considered as a polynomial in y_i, is irreducible over the field $K(u_1, \ldots, u_d, y_1, \ldots, y_{i-1})$ in which the algebraic elements $y_1, \ldots, y_{i-1}$ have $T_1, \ldots, T_{i-1}$ as their minimal polynomials. If a fine triangular TF is reducible, then there should be some integer k and polynomials[2] $D, T_k^{(1)}, \ldots, T_k^{(h)}$ with $lvar(D) \prec lvar(T_k), lvar(T_k^{(1)}) = \cdots = lvar(T_k^{(h)}) = lvar(T_k)$ and each $T_k^{(j)}$ is irreducible over the field $K(u_1, \ldots, u_d, y_1, \ldots, y_{k-1})$ such that $prem(D \cdot T_k - T_k^{(1)} \cdots T_k^{(h)}, [T_1, \ldots, T_{k-1}]) = 0$. That is, there are some polynomials $Q_i \in K[u_1, \ldots, u_d, y_1, \ldots, y_k]$ and integers q_i such that

$$I_1^{q_1} \cdots I_{k-1}^{q_{k-1}} (D \cdot T_k - T_k^{(1)} \cdots T_k^{(h)}) = \sum_{i=1}^{k-1} Q_i T_i.$$

Now, suppose $[TF, US]$ is some $[TF_i, US_i]$ in Theorem 1 with the elimination being performed without projection and let $TF^{(j)}$ be $[T_1, \ldots, T_{k-1}, T_k^{(j)}, T_k, \ldots, T_r]$ for $j = 1, \ldots, h$. We note that $US \neq 0$ implies $I_i \neq 0$ for all i. Let $US^{(j)} = \{prem(U, [T_1, \ldots, T_{k-1}, H_j]) \mid U \in US \cup \{D\}\}$ for each i. If $US^{(j)}$ does not contain 0, then $TF^{(j)}$ is still a fine triangular form.

Let $PS^* = \{D, T_1, \ldots, T_r\}$. Then by applying the algorithm ELIMINATE (with a slight modification by taking the triangularized form $[T_1, \ldots,$

---

[2]The elimination method presented in Section 3 does not require polynomial factorization. However, to get the D and $T_k^{(j)}$ here polynomial factorization over algebraic extension fields is necessary.

T_r] into account for the sake of efficiency) to PS^* and $QS^* = US$ we shall get a sequence of triangular forms $TF_1^*, \ldots, TF_t^*$ and polynomial sets $US_1^*, \ldots, US_t^*$ such that

$$Zero(PS^*/QS^*) = \bigcup_{i=1}^{t} Zero(TF_i^*/US_i^*).$$

Therefore, we can replace $Zero(TF/US)$ in (3.1) by

$$\bigcup_{1 \leq j \leq h, 0 \notin US^{(j)}} Zero(TF^{(j)}/US^{(j)}) \cup \bigcup_{i=1}^{t} Zero(TF_i^*/US_i^*).$$

Proceeding in this way we should obtain a zero decomposition of the form (3.1) with each TF_i being irreducible. Take in particular $QS = \emptyset$. Then from each pair $[TF_i, US_i]$ we can compute, by methods like the one in [21], a finite set PS_i of polynomials which defines an irreducible variety. Therefore, we shall get an irreducible decomposition of the algebraic variety defined by PS.

4.4. Constructive Proof of Hilbert's Nullstellensatz

Hilbert's Nullstellensatz asserts that, for any polynomial set $PS = \{P_1, \ldots, P_s\}$ and a polynomial Q in $K[x_1, \ldots, x_n]$, if $Zero(PS/Q) = \emptyset$ in some extension field of K then there exists an integer q and some polynomials $A_1, \ldots, A_s \in K[x_1, \ldots, x_n]$ such that $Q^q = \sum_{i=1}^{s} A_i P_i$. From our elimination method with full projection it is easy to devise a constructive proof of the Nullstellensatz. Consider the procedure of determining the emptiness of $Zero(PS/Q)$ which is processed roughly by eliminating all the variables from x_n to x_1 for all produced quintuplets. For a quintuplet $[PS', QS', TF, US, m]$, the emptiness of $Zero(PS'/QS')$ is determined when a non-zero constant in some $\widetilde{PS'}$ is produced (see Lines 5 and 9 of **TRIANGU-LARIZE**) or 0 in some $\widetilde{QS'}$ is produced (Line 18 of **TRIANGULARIZE**), where $Zero(PS'/QS') = \emptyset$ if and only if $Zero(\widetilde{PS'}/\widetilde{QS'}) = \emptyset$, and the zero sets corresponding to the subsequently generated quintuplets are empty. For $\widetilde{PS'}$ and $\widetilde{QS'}$, it is trivial to represent some power of the product of polynomials in $\widetilde{QS'}$ as a linear combination of polynomials in $\widetilde{PS'}$ with polynomial coefficients. Now, by back-tracking the various reductions in the elimination process for getting $\widetilde{PS'}, \widetilde{QS'}$ from PS', QS' and thus from PS, Q, and seeing how to write down the desired form for the original system if that for the reduced systems is known, we would be able to write down the representation of some power of the product of polynomials in QS' as a linear combination of polynomials in PS' and thus the representation of some power of Q as a linear combination of polynomials in PS with polynomial coefficients. The proof is very similar to that of Seidenberg [17]. We need only to note a few variants such as how to represent some power of $G_1 \cdots G_t$ as a linear combination of $P_1, \ldots, P_s$ if we know such a representation for each G_i.

4.5. Other Applications in Geometric Reasoning

In [24] we have shown how Wu's projection method can be used to determine the necessary and sufficient condition for an algebraic hypersurface to have singularities of arbitrary multiplicity, to derive the locus equations of a motion in an n-dimensional space and to compute the image (implicitization) of the parametric equations. As our elimination method with full projection has the same function as Wu's, it can be naturally applied to all these problems. We shall explain such applications in detail later. The second example in the next section may serve as an illustration of our method to the implicitization problem.

5. Experiments and Examples

For implementing the described elimination method several details should be carefully taken into account for the sake of efficiency and tidy output. For example, the execution should exit the while-loop as soon as PS' occurs to contain a constant, the subalgorithm SIMPLIFY for removing redundant factors should be heuristically used, some power of polynomials such as $x_1^3 x_3^2$ should be freed as to $x_1 x_3$ (but squarefree computation is too expensive), the polynomial set QS' should be compressed, some unary polynomials in QS may be delete optionally or heuristically by verifying the GCD of them and the unary polynomial in TF, and 0 should removed from PS' and constants from QS'. The author has made a draft implementation of the method in Maple 4.3. Preliminary experiments show that it is very efficient. The extent of its efficiency is at least comparable with the well-known methods of characteristic sets and Gröbner bases. Actually, we have tried our algorithm for the same set of 50 test examples as experimented in [23]. It is faster than all six variants of Wu's algorithm (with the author's implementation [22]) for computing zero decompositions (characteristic series) for 30 of the 39 successful examples. For the limitation of pages we are not going to explain the relations and differences between our method and other well known ones in this section. We shall do so and make a systematic comparison on the efficiency and applicability of these methods together with experimental results elsewhere. Below we shall only give a few illustrative examples which appear to be rather encouraging. The experiments were made on an Apollo DN10000 under a UNIX operating system. The timings are given in CPU seconds and include the time for garbage collection.

Example 1. The following set of 3 polynomials

$$PS = \{x^{31} - x^6 - x - y, x^8 - z, x^{10} - t\}$$

is taken from a paper by Traverso and Donati (Proc. ISSAC'89, 192-198) on the experimentation of Gröbner bases. Fix the order of variables as $t \prec z \prec y \prec x$. Let us see how our algorithm runs for this example.

Consider first the maximal variable x. Let MS be the set of those polynomials in PS which have x as their common leading variable; then actually $MS = PS$. Take now $R_0 = x^8 - z$ from MS which has the lowest degree 8 in x and the constant 1 as its initial. Pseudo-dividing the other two polynomials $x^{31} - x^6 - x - y, x^{10} - t$ in MS by R_0, we get two non-zero remainders

$$R_1 = z^3 x^7 - x^6 - x - y, \quad R_2 = z x^2 - t,$$

of which both have x as their leading variable. Reset $MS = \{R_0, R_1, R_2\}$. Then R_2 has the lowest degree 2 in x with its initial z. Let $\Delta = \{[\{R_0, R_1, z, -t\}, \emptyset, \phi, \emptyset, 4]\}$ and $QS = \{z\}$. Again, pseudo-dividing R_0 and R_1 by R_2, we get their remainders

$$R_3 = -z^5 + t^4, \quad R_4 = -z^3 x - y z^3 + x t^3 z^3 - t^3,$$

with $lvar(R_3) = z$ and $lvar(R_4) = x$. Let $PS' = \{R_3\}$ and $MS = \{R_2, R_4\}$. Now R_4, with its initial $t^3 z^3 - z^3$ being simplified to $t^3 - 1$ by z in QS, is of degree 1 lower than that of R_2 in x. Adjoin the quintuplet $[\{R_3, R_2, -z^3 y - t^3, t^3 - 1\}, \{z\}, \phi, \emptyset, 4]$ to Δ and $t^3 - 1$ to QS. Pseudo-dividing R_2 by R_4 we obtain its remainder

$$R_5 = -z^5 t + 2 z^5 t^4 - t^7 z^5 + y^2 z^6 + 2 y z^3 t^3 + t^6$$

with $lvar(R_5) = y$. Hence, we have $TF = [R_4]$ and $PS' = \{R_3, R_5\}$. Let us first perform the elimination without projection. Clearly, the remainders of the two polynomials in QS with respect to TF are themselves.

Consider next the variables y and z. Since the two polynomials in PS' have respectively y and z as their leading variables, no elimination is necessary. Thereby, we obtain $TF_1 = [R_3, R_5, R_4]$ and $US_1 = \{z, t^3 - 1\}$.

Now there are two quintuplets in Δ which need to be further considered. For the first $[\{R_0, R_1, z, -t\}, \emptyset, \phi, \emptyset, 4]$, the two polynomials R_0 and R_1 have their leading variable x, in which R_1 has a lower degree 7 and its initial $z^3 \sim z$. We may split R_1 into two cases $z = 0$ and $z \neq 0$ by strictly following the described algorithm, which is somewhat complicated. Actually, we may simplify R_0 and R_1 by z and $-t$, as we did in the implementation, and make the obtained polynomials power-freed. Then we get immediately a triangular from $TF_2 = [-t, z, y, x]$ with $US_2 = \emptyset$. For the second quintuplet $[\{R_3, R_2, -z^3 y - t^3, t^3 - 1\}, \{z\}, \phi, \emptyset, 4]$, the polynomials $R_2, -z^3 y - t^3, R_3, t^3 - 1$ have respectively their leading variables x, y, z, t and thus constitute a triangular form already. Hence we obtain $TF_3 = [t^3 - 1, R_3, -z^3 y - t^3, R_2]$ and $US_3 = \{z\}$. The above elimination was carried out by our implementation in 0.65 seconds.

To perform the elimination with full projection, for $z \in US_1$ we need to compute the pseudo-remainder of z^5, instead of the pseudo-remainder of z, with respect to R_3. It is $-t^4 \sim t$, so US_1 is replaced by $\{t, t^3 - 1\}$. As the same, for $z \in US_3$ we need to compute the pseudo-remainder of z^5 with respect to R_3, which is $-t^4 \sim t$, and then the pseudo-remainder of t^3 with respect to $t^3 - 1$, which is the constant 1. Hence, $\mathcal{US}_3$ is replaced by $\emptyset$. Now the elimination takes 0.7 seconds.

However, with respect to the purely lexical ordering of the computation of the Gröbner basis of PS by using the built in **grobner** package of Maple 4.3 requires 898.266 seconds. With respect to the same variable ordering, the computation of the zero decomposition takes 158.150 seconds by using Ritt's original characteristic set algorithm and 67.467, 67.650, 81.467 seconds respectively by using three different variants of Wu's improved algorithm. By a strategy proposed in [25] the times for Wu's algorithm may be reduced to 3.35, 3.06, 20.334 seconds, respectively.

Example 2. Determine the implicit form (in variables x and y) of the curve given by the following system of equations

$$(x - u)^2 + (y - v)^2 - 1 = 0,$$
$$v^2 - u^3 = 0,$$
$$2v(x - u) + 3u^2(y - v) = 0,$$
$$(3wu^2 - 1)(2wv - 1) = 0.$$

This is a formulation of an offset to the curve $y^2 - x^3 = 0$. The example was communicated by P. Vermeer from the Department of Computer Science, Purdue University. The author was told that it ran out of swap space (280 MB) before completing the computation by using the Gröbner basis implementation available in Macsyma on a Symbolic machine for getting the solution. We have tried to compute the Gröbner basis of the corresponding polynomial set in Maple 4.3 on our Apollo DN10000 without success within 2000 seconds. We have also tried to determine the implicit equations by using the characteristic set method with Wu's projection theorem. The computation of the characteristic set (with respect to the ordering $x \prec y \prec u \prec v \prec w$) is rather easy (in 10.1 seconds), the computation of the zero decomposition is however not easy. We tried to compute it with six variants, of which four did not succeed within 2000 CPU seconds. For the two successful variants, 11 ascending sets (of which several contain polynomials with very big integers as coefficients) were produced in 937.3 and 1003.883 seconds, respectively [23]. The projection of these ascending sets using the author's implementation of Wu's algorithm took more than 2000 seconds. The author was able to exclude some redundant ascending sets by technical proof so as to get the implicit equations [24].

By using our algorithm, we perform the elimination with projection of dimension 3. Then 5 triangular forms can be computed in 18.534 seconds. From them with some simplification we are able to determine the implicit equations as

$$729x^8 + 216x^7 + 729x^6y^2 - 2900x^6 - 1458x^5y^2 - 2376x^5$$
$$-2619x^4y^2 + 3870x^4 + 4072x^3 - 1458y^4x^3 - 4892x^3y^2$$
$$-1188x^2 - 297x^2y^2 + 729y^4x^2 + 5814xy^2 - 1656x - 4158xy^4 \qquad (5.1)$$

$$+427y^2 - 1685y^4 + 729y^6 + 529 = 0, \quad x \neq 0,$$

or

$$729y^4 - 956y^2 - 529 = 0, \quad x = 0. \tag{5.2}$$

Example 3 [24]. Let $ABCD$ be a square on the plane, CG be parallel to the diagonal BD, E be on CG such that $BE = BD$ and F be the intersection point of BE and DC. Prove $DF = DE$.

In an algebraic formulation of this geometric theorem, the set of hypothesis-polynomials is $HYP = \{x_1^2 + x_2^2 - 2u_1x_1 - u_1^2, x_2 + x_1 - 2u_1, x_2x_3 - u_1x_2 - u_1x_1 + u_1^2\}$ and the conclusion-polynomial is $C = x_3^2 - x_2^2 + 2u_1x_2 - x_1^2 - u_1^2$. By using our method, the theorem can be proved in 0.217 seconds by providing the subsidiary condition $x_2 \neq 0$, which may be examined by some analysis to be equivalent to the condition $u_1 \neq 0$, and in 0.55 seconds by providing directly the non-degenerate condition $u_1 \neq 0$. Clearly, the condition $u_1 \neq 0$ corresponds to the geometric assumption that the square does not degenerate to a point. In that degenerate case, the theorem is no longer true and, actually, it has no meaning.

The theorems which have been proved by using our method include Simson's theorem, Pappus' theorem, Feuerbach's theorem, the butterfly theorem, Morley's theorem and Brianchon's theorem. The range of computing time by using our method is similar to that by using Wu's method.

Appendix A. Proofs of Lemmas and Theorems

Proof of Lemma 1. For any polynomial $P \in PS$, we can pseudo-divide P by T with respect to x_i so as to get a remainder formula of the form

$$I^s P = AT + R,$$

where $A, R \in K[x_1, \ldots, x_n]$ and the integer $s > 0$. For any $\bar{x} \in Zero(PS \cup \{T\}/QS)$, we have $T(\bar{x}) = 0$ and $P(\bar{x}) = 0$ for all $P \in PS$, so $R(\bar{x}) = 0$ for all $R \in RS$. Clearly, $Q(\bar{x}) \neq 0$ for all $Q \in QS$. If $I(\bar{x}) \neq 0$, then $\bar{x} \in Zero(RS \cup \{T\}/QS \cup \{I\})$. Otherwise, we have $I(\bar{x}) = 0$ and thus $red(I)(\bar{x}) = 0$; therefore $\bar{x} \in Zero(PS \cup \{I, red(T)\}/QS)$. This shows that the left-hand side of (2.1) is contained in the right-hand side. To show the opposite, we see that if $\bar{x} \in Zero(PS \cup \{I, red(T)\}/QS)$, then $T(\bar{x}) = 0$ and thus $\bar{x} \in Zero(PS \cup \{T\}/QS)$. If $\bar{x} \in Zero(RS \cup \{T\}/QS \cup \{I\})$, then by the remainder formula we have $P(\bar{x}) = 0$ for all $P \in PS$, so $\bar{x} \in Zero(PS \cup \{T\}/QS)$ as well. ∎

Proof of Lemma 2. We first prove (2.2). If $Zero(PS/QS) \neq \emptyset$, i.e., there is an $\bar{x} = (\bar{x}_1, \ldots, \bar{x}_n) \in Zero(PS/QS)$ in some extension field of K, then $H_{l1}(\bar{x}_1, \ldots, \bar{x}_i), \ldots, H_{lm_l}(\bar{x}_1, \ldots, \bar{x}_i)$ cannot be all 0 for every l. Let j_l' be the smallest such that $H_{lj_l'}(\bar{x}_1, \ldots, \bar{x}_i) \neq 0$ for each l. Then

$$(\bar{x}_1, \ldots, \bar{x}_i) \in Zero(PS \cup HS_{j_1' \cdots j_h'}/QS_{j_1' \cdots j_h'}). \tag{2.7}$$

On the other hand, if the right-hand side of (2.2) holds, then there must be some $\bar{x}_1, \ldots, \bar{x}_i$ in some extension field of K and some indices $j'_1, \ldots, j'_h$ such that (2.7) holds. Therefore, $H_l|_{x_1=\bar{x}_1, \ldots, x_i=\bar{x}_i} \not\equiv 0$ for all l, so there are $\bar{x}_{i+1}, \ldots, \bar{x}_n$ such that $H_l(\bar{x}_1, \ldots, \bar{x}_n) \neq 0$ for all l. Hence $(\bar{x}_1, \ldots, \bar{x}_n)$ is in $Zero(PS/QS)$, that is, $Zero(PS/QS) \neq \emptyset$.

To prove (2.3), we first see that its right-hand side is obviously contained in its left-hand side. This is simply because $Zero(PS \cup HS_{j_1 \cdots j_h}/QS'_{j_1 \cdots j_h})$ is contained in $Zero(PS/QS)$ for each set of $j_1, \cdots, j_h$. On the other hand, for any $\bar{x} = (\bar{x}_1, \ldots, \bar{x}_n) \in Zero(PS/QS)$ let j'_l be the smallest such that $H_{lj'_l}(\bar{x}_1, \ldots, \bar{x}_i) \neq 0$ for each l as before. Then $\bar{x} \in Zero(PS \cup HS_{j'_1 \cdots j'_h}/QS'_{j'_1 \cdots j'_h})$ and thus $\bar{x}$ belongs to the right-hand side of (2.3). ∎

Proof of Lemma 3. *a)* Since all polynomials in QS have their leading variables $\prec x_i$ (i.e., $QS \subset K[x_1, \ldots, x_{i-1}]$), $Zero(PS \cup \{T\}/QS \cup \{I\}) \neq \emptyset$ if and only if $Zero(PS/QS \cup \{I\}) \neq \emptyset$ and $T|_{x_1=\bar{x}_1, \ldots, x_{i-1}=\bar{x}_{i-1}}$ has a zero $\bar{x}_i$ for x_i in some extension field of K for any $(\bar{x}_1, \ldots, \bar{x}_{i-1}) \in Zero(PS/QS \cup \{I\})$. The last condition is naturally satisfied according to the fundamental theorem of algebra, so (2.4) is proved.

b) To prove (2.5), we first suppose $Zero(PS \cup \{T\}/QS \cup \{I\}) \neq \emptyset$, i.e., there is an $\bar{x} = (\bar{x}_1, \ldots, \bar{x}_i) \in Zero(PS \cup \{T\}/QS \cup \{I\})$. Then $T(\bar{x}) = 0$, $I(\bar{x}) \neq 0$ and $H_1 \cdots H_h(\bar{x}) \neq 0$, so by the remainder formula (for some integer $s \geq 0$)

$$I^s(H_1 \cdots H_h)^d = AT + R, \tag{2.8}$$

we have $R(\bar{x}) \neq 0$. Therefore, $\bar{x} \in Zero(PS/QS')$, i.e., $Zero(PS/QS') \neq \emptyset$.

Now suppose $Zero(PS/QS') \neq \emptyset$, i.e., there is an $\bar{x} = (\bar{x}_1, \ldots, \bar{x}_i) \in Zero(PS/QS')$. We note that, while regarding $T, H_1, \ldots, H_h$ as polynomials in x_i with coefficients in $K(x_1, \ldots, x_{i-1})$, T contains a factor not occurring in any of $H_1, \ldots, H_h$ if and only if $R \neq 0$. Since $R(\bar{x}) \neq 0$, $T|_{x_1=\bar{x}_1, \ldots, x_{i-1}=\bar{x}_{i-1}}$ must contain a factor, say T', which is not a factor of any $H_i|_{x_1=\bar{x}_1, \ldots, x_{i-1}=\bar{x}_{i-1}}$ $(1 \leq i \leq h)$. Hence, there must be an $\tilde{x}_i$ in some extension field of $K(\bar{x}_1, \ldots, \bar{x}_{i-1})$ and thus of K such that $T(\bar{x}_1, \ldots, \bar{x}_{i-1}, \tilde{x}_i) = 0$ while $H_1 \cdots H_h(\bar{x}_1, \ldots, \bar{x}_{i-1}, \tilde{x}_i) \neq 0$ (actually, any zero of T' does). Therefore, $(\bar{x}_1, \ldots, \bar{x}_{i-1}, \tilde{x}_i) \in Zero(PS \cup \{T\}/QS \cup \{I\})$, so $Zero(PS \cup \{T\}/QS \cup \{I\}) \neq \emptyset$. This completes the proof of (2.5).

Finally, from the remainder formula (2.8) it is easy to see that under the condition $I \neq 0$, $H_1 \cdots H_h \neq 0$ if and only if $R \neq 0$. Hence (2.6) holds true. ∎

Subalgorithm SIMPLIFY. *The input to this subalgorithm consists of a polynomial $M \in K[x_1, \ldots, x_n]$ and a polynomial set $QS \subset K[x_1, \ldots, x_n]$. The algorithm computes another polynomial $M' \in K[x_1, \ldots, x_n]$ such that*

$$Zero(PS \cup \{M\}/QS) = Zero(PS \cup \{M'\}/QS), \tag{3.4}$$

$$Zero(PS/QS \cup \{M\}) = Zero(PS/QS \cup \{M'\}), \qquad (3.5)$$

for any polynomial set PS.

Proof. The termination of this subalgorithm is obvious, because QS is a finite set, and for any $Q \neq 1$ and $M \neq 0$, M contains only a finite power of Q as its factor. Since M' is different from M only by a factor that is a product of polynomials in QS, the zero relations hold trivially. ∎

Subalgorithm ELIMINATE. *The input to this subalgorithm is a quintuplet* $[PS, QS, TF, US, i]$ *of level i. The algorithm computes a polynomial $T \in K[x_1, \ldots, x_i]$ with $lvar(T) = x_i$, a polynomial set $PS' \subset K[x_1, \ldots, x_{i-1}]$, a polynomial set $QS' \subset K[x_1, \ldots, x_{\max(p,k)}]$ and a set Δ of quintuplets of level i, where p is the subscript of $lvar(first(TF))$ if $TF \neq \phi$, and n otherwise, such that*

$$Zero(PS/QS) = Zero(PS' \cup \{T\}/QS') \cup$$

$$\bigcup_{[PS^*, QS^*, TF, US, i] \in \Delta} Zero(PS^*/QS^*). \qquad (3.6)$$

Proof. First we see that every substep within the while-loop of the algorithm terminates. Since in each iteration of the loop the degree of the polynomial T in x_i decreases at least by 1, after a finite number of steps all non-zero remainders of polynomials in $MS \setminus \{T\}$ with respect to T should have their leading variables $\prec x_i$. Then the set MS becomes $\{T\}$ and the while-loop terminates.

The zero relation (3.6) follows directly from a repeated application of the zero relation (2.1) in Lemma 1 to the pseudo-division and the zero relations (3.4) and (3.5) to SIMPLIFY called in the algorithm. ∎

Subalgorithm PROJECT_A. *The input to this subalgorithm is a quintuplet $[PS, QS, TF, US, i]$ of level i. It computes a polynomial set $QS' \subset K[x_1, \ldots, x_i]$, a polynomial set $US' \subset K[x_1, \ldots, x_n]$ with $lvar(U) \succ x_i$ for all $U \in US'$ and a set Δ of quintuplets of level i such that*

$$Zero(PS/QS) \neq \emptyset \Longleftrightarrow Zero(PS/QS') \cup$$

$$\bigcup_{[PS^*, QS^*, TF, US', i] \in \Delta} Zero(PS^*/QS^*) \neq \emptyset, \qquad (3.7)$$

$$Zero(PS/QS) = Zero(PS/QS' \cup HS) \cup$$

$$\bigcup_{[PS^*, QS^*, TF, US', i] \in \Delta} Zero(PS^*/QS^* \cup HS), \qquad (3.8)$$

where $HS = \{Q \in QS | \ lvar(Q) \succ x_i\}$ and $US \cup HS = US'$.

Proof. Since there is no loop in the algorithm, its termination is obvious.

As we have mentioned, this subalgorithm is just an implementation of Lemma 2. To prove (3.7) and (3.8), we note that $[PS, QS']$ here corresponds to the subsystem in Lemma 2 for the indices $j_1 = 1, \ldots, j_h = 1$, while the $[PS^*, QS^*]$'s corresponding to those subsystems in Lemma 2 for all other indices are stored in Δ. Therefore, (3.7) and (3.8) are actually an alternative form of (2.2) and (2.3) in Lemma 2, so they hold.   ∎

Subalgorithm REDUCE_PROJECT. *The input to this subalgorithm consists of two integers i and k, a polynomial $T \in K[x_1, \ldots, x_i]$ with $lvar(T) = x_i$ and a polynomial set $QS \subset K[x_1, \ldots, x_i]$ with $Zero(ini(T)/QS) = \emptyset$. The algorithm computes a polynomial set $QS' \subset K[x_1, \ldots, x_i]$ such that*

$$Zero(PS \cup \{T\}/QS) = Zero(PS \cup \{T\}/QS'), \qquad (3.9)$$

and if $i > k$, then

$$Zero(PS \cup \{T\}/QS) \neq \emptyset \Longleftrightarrow Zero(PS/QS') \neq \emptyset, \qquad (3.10)$$

for any polynomial set $PS \subset K[x_1, \ldots, x_{i-1}]$.

Proof. As no loop is involved in this subalgorithm, its termination is obvious.

Since $Zero(ini(T)/QS) = \emptyset$ and $lvar(ini(T)) \prec x_i$, $Zero(ini(T)/QS') = \emptyset$. Hence, $Zero(PS \cup \{T\}/QS \cup \{ini(T)\}) = Zero(PS \cup \{T\}/QS)$ and $Zero(PS/QS' \cup \{ini(T)\}) = Zero(PS/QS')$. Thus, (3.9) and (3.10) follow simply from (2.4)-(2.6) in Lemma 3 for $i > k$.

Since $Zero(ini(T)/QS) = \emptyset$, by (2.1) we have (3.9) for $i \leq k$.   ∎

Algorithm TRIANGULARIZE. *The input to this algorithm consists of two polynomial sets $PS, QS \subset K[x_1, \ldots, x_n]$ and an integer k ($0 \leq k \leq n$). The algorithm computes a set $\Psi = \{[TF_1, QS_1], \ldots, [TF_e, QS_e]\}$ of triangular forms and polynomial sets such that either $\Psi \neq \emptyset$ when $Zero(PS/QS) = \emptyset$ in any extension field of K, or the properties a), b) and c) in Theorem 1 are satisfied.*

Proof. To prove the termination of the algorithm TRIANGULARIZE, we need only to prove that the while-loop terminates. For this purpose, we define, for any pair $[PS, QS]$ of two polynomial sets, a triple $Index(PS/QS) = [d, m, p]$, where $m = $ the minimal j such that $PS \subset K[x_1, \ldots, x_j]$, $d = $ the lowest degree (in x_m) of those polynomials in PS whose leading variables are x_m, $p = \max(m, p')$ and $p' = $ the minimal j such that $QS \subset K[x_1, \ldots, x_j]$. We order two triples as $[d_1, m_1, p_1] \prec [d_2, m_2, p_2]$ if $p_1 < p_2$, or $p_1 = p_2$ while $m_1 < m_2$, or $p_1 = p_2, m_1 = m_2$ while $d_1 < d_2$. Now, for the pair $[PS', QS']$ of the first two polynomial sets in any quintuplet taken from Ψ (Line 3) in the algorithm TRIANGULARIZE, let PS^* and QS^* be the first two polynomial sets in some quintuplet of the Δ produced by ELIMINATE (Line 7) or by PROJECT_A (Line 11). Then we always have

$Index(PS^*/QS^*) \prec Index(PS'/QS')$. Since each component of the triple $Index(PS/QS)$ is a positive integer, any steadily decreasing sequence of triples cannot be infinite. From this we can immediately conclude that the while-loop may have only a finite number of iterations.

Now we want to prove the correctness of **TRIANGULARIZE**, that is, to prove that the computed Ψ satisfies the properties a), b) and c) in Theorem 1.

Proof of Theorem 1. The algorithm **TRIANGULARIZE** can be viewed as for computing a multiple branch tree with the pair $[PS, QS]$ of polynomial sets associated at the root. With each node j of the tree there is a pair $[PS_j, QS_j]$ of polynomial sets associated, and with each leaf i of the tree there is a pair $[\alpha_i, \beta_i]$ associated such that either α_i is a triangular form and β_i is a polynomial set, or α_i is a non-zero constant, or β_i is 0. It is clear that new branches are generated by the subalgorithms **ELIMINATE** and **PROJECT_A** with the zero relations (3.6) and (3.8) being preserved.

We note that, for any polynomial sets HS, PS_i, QS_i, if $Zero(PS/QS) = \bigcup_i Zero(PS_i/QS_i)$, then $Zero(PS \cup HS/QS) = \bigcup_i Zero(PS_i \cup HS/QS_i)$ and $Zero(PS/QS \cup HS) = \bigcup_i Zero(PS_i/QS_i \cup HS)$. According to this, it follows from the zero relations (3.6), (3.8) and (3.9) respectively that

$$Zero(PS \cup TF/QS \cup US) = Zero(PS' \cup TF'/QS' \cup US)\cup$$

$$\bigcup_{[PS^*,QS^*,TF,US,i]\in\Delta} Zero(PS^* \cup TF/QS^* \cup US), \quad (3.6')$$

$$Zero(PS \cup TF/QS \cup US) = Zero(PS \cup TF/QS' \cup US')\cup$$

$$\bigcup_{[PS^*,QS^*,TF,US',i]\in\Delta} Zero(PS^* \cup TF/QS^* \cup US'), \quad (3.8')$$

$$Zero(PS \cup TF'/QS \cup US) = Zero(PS \cup TF'/QS' \cup US), \quad (3.9')$$

where $TF' = [T] \cup TF$ and $US' = US \cup HS$. By a recursive application of these relations we have

$$Zero(PS/QS) = \bigcup_i Zero(\alpha_i/\beta_i).$$

Trivially, we can remove all such components that α_i is a non-zero constant, or β_i is 0 from this zero decomposition. If all the components are removed in this way, that is, $\Psi = \emptyset$, then $Zero(PS/QS) = \emptyset$ over any extension field of K. Otherwise, the computed set Ψ satisfies the properties a), b) and c) in Theorem 1 to be proved as follows.

a) From the above discussion the zero relation (3.1) is obtained immediately. That $Zero(ini(T)/US_i) = \emptyset$ for all $T \in TF_i$ is simply because each $ini(T)$ has been adjoined by the subalgorithm **ELIMINATE** (Line 8) with some simplification to the set corresponding to US_i. If $k = n$, then all polynomials in US_i are actually the pseudo-remainders of some polynomials with respect to TF_i for every i. Hence we have $prem(U, TF_i) \neq 0$ for all $U \in US_i$. If $k = 0$, then $prem(U, TF_i) \neq 0$ for all $U \in US_i$ is ensured simply by the property $c)$ to be proved below.

$b)$ We first suppose the left-hand side of (3.2) holds, so there is an i such that $(a_1, \ldots, a_k) \in Zero(TF_i^{(k)}/US_i^{(k)})$. By (3.3) to be proved, there should be some $(a_{k+1}, \ldots, a_n) \in Zero(TF_i^{*(a,k)}/US_i^{*(a,k)})$. Hence $(a_1, \ldots, a_n) \in Zero(TF_i/US_i)$. By (3.1), $(a_1, \ldots, a_n) \in Zero(PS/QS)$. It follows that $(a_{k+1}, \ldots, a_n) \in Zero(PS^{(a,k)}/QS^{(a,k)})$, i.e., $Zero(PS^{(a,k)}/QS^{(a,k)}) \neq \emptyset$.

Now suppose $Zero(PS^{(a,k)}/QS^{(a,k)}) \neq \emptyset$, i.e., there is some $(a_{k+1}, \ldots, a_n) \in Zero(PS^{(a,k)}/QS^{(a,k)})$. Then $(a_1, \ldots, a_n) \in Zero(PS/QS)$. By (3.1), there must be an i such that $(a_1, \ldots, a_n) \in Zero(TF_i/US_i)$. In particular, we should have $(a_1, \ldots, a_k) \in Zero(TF_i^{(k)}/US_i^{(k)})$. Therefore, $(a_1, \ldots, a_k) \in \bigcup_{i=1}^{e} Zero(TF_i^{(k)}/US_i^{(k)})$ and (3.2) is proved.

$c)$ We first see how and when the projections are performed. Consider one branch of the tree with whose leaf the pair $[TF, US]$ is associated. Let the number of polynomials in TF be r and the leading variable of the jth polynomial in TF be x_{p_j}, and let j' be the smallest j such that $k < p_j$ if it exists. Then the subalgorithm PROJECT_A is called at Line 14 of TRIANGULARIZE for $i = p_r, \ldots, p_{j'}, k$, and Lines 1-4 in the subalgorithm REDUCE_PROJECT are executed for $i = p_r, \ldots, p_{j'}$.

To prove (3.3), we consider any $(a_1, \ldots, a_k) \in Zero(TF^{(k)}/US^{(k)})$ (with the subscript i of TF and US being omitted for brevity). The case both $TF \setminus TF^{(k)} = \phi$ and $US \setminus US^{(k)} = \emptyset$ is trivial. Now suppose $TF \setminus TF^{(k)} = \phi$, while $US \setminus US^{(k)} \neq \emptyset$. Then according to the algorithm TRIANGULARIZE the branch $[TF, US]$ must be obtained from some quintuplet of the form $[\widetilde{PS} \cup \{T\}, \widetilde{QS}, \phi, \emptyset, k]$ corresponding to the state value of $[PS' \cup \{T\}, QS', TF, US, i]$ in the algorithm TRIANGULARIZE after executing Lines 6-12, where $\widetilde{PS} \subset K[x_1, \ldots, x_{k-1}]$, $\widetilde{QS} \setminus \widetilde{QS}^{(k)} = US \setminus US^{(k)}$ and T is either 0 (in this case it is simply removed from the polynomial set) or a polynomial with $lvar(T) = x_k$, in the following way: First call the subalgorithm PROJECT_A at Line 14 of TRIANGULARIZE for $i = k$, by which the polynomial sets $\widetilde{QS}' \subset K[x_1, \ldots, x_k]$, $\widetilde{US}' = US \setminus US^{(k)}$ and the set Δ of quintuplets are produced, satisfying the relation corresponding to (3.7). For some subsystem, say $[\widetilde{PS} \cup \{T\}, \widetilde{QS}']$ without loss of generality, produced at the preceding step, let us consider the subtree with whose root this subsystem is associated. Then, by a repeated application of the zero relations (3.6) and (3.9) along the elimination for this subtree we should get a zero decomposition of the form

$$Zero(\widetilde{PS} \cup \{T\}/\widetilde{QS}') = \bigcup_{j=1}^{t} Zero(TF_j'/US_j'). \qquad (3.11)$$

Then $[TF^{(k)}, US^{(k)}]$ is one of the pairs $[TF_1', US_1'], \ldots, [TF_t', US_t']$ in this decomposition. Therefore $(a_1, \ldots, a_k) \in Zero(\widetilde{PS} \cup \{T\}/\widetilde{QS}')$. By (3.7), $Zero(\widetilde{PS} \cup \{T\}/\widetilde{QS}) \neq \emptyset$. From the proof of (2.2) we see easily that there are moreover some $a_{k+1}, \ldots, a_n$ such that $(a_1, \ldots, a_n) \in Zero(\widetilde{PS} \cup \{T\}/\widetilde{QS})$. Hence $(a_1, \ldots, a_n)$ is not a zero of any polynomial in $\widetilde{QS}$, of course, nor a zero of any polynomial in $US \setminus US^{(k)}$. That is, $Zero(TF^{*(a,k)}/US^{*(a,k)}) = Zero(\phi/US^{*(a,k)}) \neq \emptyset$.

Next, we suppose $TF \setminus TF^{(k)} \neq \phi$. Then the first polynomial in $TF \setminus TF^{(k)}$ can be written as $T(x_1, \ldots, x_p)$, where $lvar(T) = x_p$ and $k < p = p_{j'} \leq n$. In this case, according to the algorithm TRIANGULARIZE the branch $[TF, US]$ must be obtained from some quintuplet of the form $[\widetilde{PS} \cup \{T\}, \widetilde{QS}, \widetilde{TF}, \widetilde{US}, p]$ corresponding to the state value of $[PS' \cup \{T\}, QS', TF, US, i]$ in the algorithm TRIANGULARIZE after executing Lines 13-16, where $\widetilde{PS} \subset K[x_1, \ldots, x_{p-1}]$, $\widetilde{QS} \subset K[x_1, \ldots, x_p]$, T is a polynomial with $lvar(T) = x_p$, $\widetilde{TF} = TF \setminus TF^{(p)}$, $\widetilde{US} = US \setminus US^{(p)}$ and $\widetilde{QS} \setminus \widetilde{QS}^{(k)} = US^{(p)} \setminus US^{(k)}$, in the following way: First call the subalgorithm REDUCE_PROJECT at Line 17 of TRIANGULARIZE for $i = p$, in which Lines 1-4 are executed and by which the polynomial set $\widetilde{QS}'$ is produced, satisfying the relation corresponding to (3.10). Then the subalgorithm PROJECT_A is called at Line 14 of TRIANGULARIZE for $T = 0$ and $i = k$, by which the polynomial sets $\widetilde{QS}'' \subset K[x_1, \ldots, x_k]$, $\widetilde{US}'' = US \setminus US^{(k)}$ and the set Δ of quintuplets are produced, satisfying the relation corresponding to (3.7). Take a subsystem, say $[\widetilde{PS}, \widetilde{QS}'']$ without loss of generality, produced at the preceding step, and consider the subtree with whose root this subsystem is associated. Again, by a repeated application of the zero relations (3.6) and (3.9) along the elimination for this subtree we should get a decomposition of a form similar to (3.11) for $Zero(\widetilde{PS}/\widetilde{QS}'')$. Then $[TF^{(k)}, US^{(k)}]$ is one of the pairs $[TF_1', US_1'], \ldots, [TF_t', US_t']$ in the corresponding decomposition. Therefore $(a_1, \ldots, a_k) \in Zero(\widetilde{PS}/\widetilde{QS}'')$. By (3.7) and (3.9), $Zero(\widetilde{PS} \cup \{T\}/\widetilde{QS}) \neq \emptyset$. From the proof of (2.2), we see that there are some $a_{k+1}, \ldots, a_p$ such that $(a_1, \ldots, a_p) \in Zero(\widetilde{PS}/\widetilde{QS}')$. From the proof of (2.4) and (2.5) we see that there is moreover an $\tilde{a}_p$ such that $(a_1, \ldots, a_{p-1}, \tilde{a}_p) \in Zero(\widetilde{PS} \cup \{T\}/\widetilde{QS})$. Hence $(a_1, \ldots, a_{p-1}, \tilde{a}_p)$ is a zero of T, but not a zero of any polynomial in $\widetilde{QS}$, of course, nor a zero of any polynomial in $US^{(p)} \setminus US^{(k)}$. That is,

$$(a_1, \ldots, a_{p-1}, \tilde{a}_p) \in Zero(T/(US^{(p)} \setminus US^{(k)}) \cap K[x_1, \ldots, x_p]).$$

Let us redenote $\tilde{a}_p$ by a_p; then $(a_1, \ldots, a_p) \in Zero(TF^{(p)}/US^{(p)})$.

Finally, we consider p as k and make additional use of the relations (3.8) and (3.9) in the course of computing the decomposition corresponding to (3.11). Then, (3.3) is proved by induction.

Since $Zero(ini(T)/US_i) = \emptyset$ for any $T \in TF_i$ and $Zero(TF_i^{*(a,k)}/US_i^{*(a,k)})$ $\neq \emptyset$, by definition the triangular form $TF_i \setminus TF_i^{(k)}$ is perfect. ∎

Proof of Theorem 2. a) If $e = 0$ (i.e., $\Psi = \emptyset$), from the discussions before a) in the proof of Theorem 1 we know that $Zero(PS/QS) = \emptyset$ in any extension field of K. Now suppose $Zero(PS/QS) = \emptyset$. If $e \neq 0$, by c) of Theorem 1 (for $k = 0$) each $Zero(TF_i/US_i)$ is non-empty in some extension field of K. This is a contradiction according to (3.1).

b) Suppose $Zero(PS/QS)$ is finite but not empty. By a) just proved, $e \geq 1$. By c) of Theorem 1, $Zero(TF_i/US_i) \neq \emptyset$ for each i. We want to prove that the number of polynomials in TF_i is exactly n. If this is not the case, then there must be an l such that $p_l = l$ while $p = p_{l+1} > l + 1$, where p_j is the subscript of the leading variable of the jth polynomial in TF_i. By c) of Theorem 1 (for $k = 0$), $Zero(TF_i^{(l)}/US_i^{(l)}) \neq \emptyset$. Let $(a_1, \ldots, a_l) \in Zero(TF_i^{(l)}/US_i^{(l)})$. Now, look at c) in the proof of Theorem 1 above and the proof of (2.2) and (2.5). We can see that there are infinitely many choices of $a_{l+1}, \ldots, a_{p-1}$ for $T = T_{l+1}$ to have a zero in some extension field of $K(a_1, \ldots, a_l)$ which does not vanish any polynomial in $US_i^{*(a,l)} \cap K(a_1, \ldots, a_l)[x_{l+1}, \ldots, x_p]$. For each $(a_1, \ldots, a_l, \ldots, a_p)$ of these infinitely many zeros, $Zero(TF_i^{*(a,p)}/US_i^{*(a,p)}) \neq \emptyset$ by looking at c) in the proof of Theorem 1 again. Therefore, $Zero(TF_i/US_i)$ is infinite and thus by (3.1), $Zero(PS/QS)$ is infinite. The leads to a contradiction. Hence, the number of polynomials in TF_i is exactly n. On the contrary, if the number of polynomials in TF_i is exactly n, then the leading variable of the jth polynomial T_j in TF_i is x_j for all j. As the polynomial obtained from T_j by substituting any $a_1, \ldots, a_{j-1}$ for $x_1, \ldots, x_{j-1}$ can have only a finite number of (at most $deg(T_j, x_j)$) zeros for x_j, $Zero(TF_i/US_i)$ is finite for all i. Therefore, $Zero(PS/QS)$ is finite.

c) Since the number of polynomials in a triangular form cannot be more than n, by a) and b) just proved $Zero(PS/QS)$ is infinite if and only if $e \geq 1$ and the number of polynomials in TF_i is less than n for some i. ∎

Proof of Theorem 3. Without loss of generality, we assume $t = 1$ and write C for C_1. If the theorem T is generally true, then there is a polynomial $D(u_1, \ldots, u_d, x_1, \ldots, x_n)$ such that $Zero(HYP \cup \{D\}) = \emptyset$ and $Zero(HYP/\{D, C\}) = \emptyset$ in any extension field of $K(u_1, \ldots, u_d)$ according to the definition. We want to prove $Zero(HYP/C) = \emptyset$ in any extension field of $K(u_1, \ldots, u_d)$. Otherwise, there is an $\bar{x}$ such that $H(\bar{x}) = 0$ for all $H \in HYP$ and $C(\bar{x}) \neq 0$. If $D(\bar{x}) = 0$, then $\bar{x} \in Zero(HYP \cup \{D\})$. This is impossible, so $D(\bar{x}) \neq 0$. This implies that $\bar{x} \in Zero(HYP/\{D, C\})$, which is a contradiction again. Hence $Zero(HYP/\{C\}) = \emptyset$. By a) of Theorem 2, $e = 0$, i.e., $\Psi = \emptyset$. On the other hand, if $e = 0$, then $Zero(HYP/C) = \emptyset$ in any extension field of $K(u_1, \ldots, u_d)$. Therefore, by Hilbert's Nullstellensatz, there are polynomials $A_1, \ldots, A_s \in K(u_1, \ldots, u_d)[x_1, \ldots, x_n]$ and some integer q such that

$$C^q = \sum_{i=1}^{s} A_i H_i.$$

Multiplying the above formula by the common denominator $D(u_1, \ldots, u_d)$ of $A_1, \ldots, A_s$, we obtain an expression of the form

$$D \cdot C^q = \sum_{i=1}^{s} \bar{A}_i H_i,$$

where $\bar{A}_i \in K[u_1, \ldots, u_d, x_1, \ldots, x_n]$ for each i. Therefore, under the condition $D \neq 0$, $H_1 = 0, \ldots, H_s = 0$ imply $C = 0$. Clearly, the geometric configuration defined by $H_1 = 0, \ldots, H_s = 0, D = 0$ over K is of dimension less than d. Hence, the theorem is generally true. $\blacksquare$

Appendix B. Comparison with Ritt-Wu's Zero Decomposition

The table below shows the timings of our algorithm TRIANGULARIZE (without projection) in comparison with those of Ritt-Wu's zero decomposition in six variants on a set of 50 test examples (cf. [23]). The experiments were made by using a draft implementation of the algorithms described in this paper and the **charsets** package of this author in Maple 4.3 running on an Apollo DN10000 under a Unix operating system OS 10.3. All the timings are given in CPU seconds and include the time for garbage collection, where the garbage collection occurs approximately after every 100,000 words ($= 400,000$ bytes) used. For each computation, if the CPU time is more than 2000 seconds, indicated by >2000, it is interrupted manually. *** means that the computation was rejected by the Maple system for the reason "object too large". For the examples indicated by ♣, our algorithm is faster than all six variants of Ritt-Wu's decomposition algorithm. This is 30 among the 39 complete test cases.

	ics	qics		mcs			TRIANG-
Ex	charsetn	charsetn	wcharsetn	charsetn	wcharsetn	triser	ULARIZE
♣1	141.167	107.800	89.167	54.350	19.450	23.400	1.817
♣2	341.150	341.700	165.600	198.300	99.534	483.067	77.066
3	***	>2000	>2000	>2000	>2000	***	> 2000
5	40.633	7.500	5.450	8.667	8.800	8.416	7.434
♣6	6.267	6.300	5.933	3.833	3.316	3.134	1.050
♣8	1.950	2.400	1.550	1.217	1.000	3.233	0.400
9	11.100	10.917	10.000	7.017	6.050	6.033	12.567
♣10	0.333	0.333	0.333	0.333	0.767	0.334	0.150
♣11	2.716	3.333	1.983	1.067	0.650	1.033	0.316
12	26.767	20.217	27.233	20.500	17.650	27.367	39.634
♣13	0.250	0.250	0.233	0.133	0.134	0.134	0.083
14	>2000	>2000	>2000	>2000	>2000	>2000	> 2000
♣15	17.350	20.317	24.983	30.000	231.200	13.816	3.384
16	>2000	>2000	>2000	>2000	>2000	>2000	> 2000
17	>2000	>2000	>2000	>2000	>2000	>2000	> 2000
♣18	3.650	4.566	4.383	2.050	1.783	1.817	0.617
19	18.550	18.750	325.133	***	***	***	***
♣20	14.333	8.667	5.000	3.950	3.116	2.416	0.900
21	>2000	>2000	>2000	>2000	>2000	>2000	> 2000
♣22	173.800	176.166	189.583	1733.050	>2000	126.216	120.050
♣23	69.617	72.183	9.467	13.300	5.400	7.883	1.250
♣24	57.733	44.384	49.833	1811.700	1451.533	54.584	14.734
♣25	11.366	11.367	17.183	12.566	14.434	13.500	1.500
♣26	8.533	8.650	7.417	21.967	15.617	19.467	1.784
♣27	14.467	15.383	11.633	10.450	5.750	7.733	1.383
♣28	11.683	11.783	9.850	15.583	23.316	8.133	3.217
♣29	2170.567	165.267	164.250	67.650	67.467	81.467	0.650
30	4.216	4.433	3.334	4.434	3.567	6.200	5.650
♣32	9.300	11.600	10.816	12.383	7.800	11.967	2.700
♣33	288.917	322.450	190.900	428.317	176.733	949.150	70.317
34	>2000	>2000	>2000	>2000	>2000	>2000	> 2000
35	88.450	65.816	38.700	73.083	44.084	54.466	84.400
♣36	0.934	1.066	0.299	0.917	0.233	0.233	0.134
♣37	1.267	1.350	1.384	1.050	0.600	0.617	0.367
♣38	12.233	12.150	10.950	18.100	19.617	7.833	3.100
♣39	744.283	747.450	760.367	>2000	>2000	>2000	144.966
♣41	187.733	187.400	237.816	140.267	95.267	149.616	29.084
♣42	2.333	2.583	2.334	2.033	1.783	1.984	0.467
43	>2000	113.350	110.933	>2000	>2000	>2000	151.266
♣44	1628.384	1183.533	510.017	>2000	>2000	>2000	27.367
♣45	15.616	17.600	20.400	10.383	386.283	14.350	7.600
46	3.584	3.884	3.833	3.666	2.933	4.333	4.516
♣47	>2000	937.300	1003.883	>2000	>2000	>2000	6.400
♣48	5.366	5.683	3.583	2.083	1.750	2.100	0.567
49	21.517	20.917	17.900	18.017	13.867	15.433	50.384

Acknowledgments. This work is supported in part by the Austrian

Ministry of Science and Research under ESPRIT Basic Research Action 3125 (MEDLAR). The author wishes to thank the referees for their valuable comments on improving this paper.

References

[1] B. Buchberger, *An Algorithmical Criterion for the Solvability of Algebraic Systems of Equations (in German)*, Aequantiones Math. **4**, 374-383 (1970).

[2] B. Buchberger, *Gröbner Bases: An Algorithmic Method in Polynomial Ideal Theory*, Multidimensional Systems Theory (N. K. Bose, ed.), D. Reidel Publishing Company, Dordrecht-Boston, 184-232 (1985).

[3] J. F. Canny, E. Kaltofen and L. Yagati, *Solving Systems of Non-Linear Polynomial Equations Faster*, Proc. ISSAC'89 (G. H. Gonnet, ed.), 121-128 (1989).

[4] G. Carrà Ferro and G. Gallo, *A Procedure to Prove Geometrical Statements*, Proc. AAECC-5, Lecture Notes in Comput. Sci. **356**, 141-150 (1987).

[5] D. Yu. Grigoriev and A. L. Chistiv, *Subexponential-Time Solving Systems of Algebraic Equations*, Preprints LOMI E-9-83 and E-10-83, Leningrad (1983).

[6] D. Yu. Grigoriev and A. L. Chistiv, *Fast Decomposition of Polynomials into Irreducible Ones and the Solution of Systems of Algebraic Equations*, Soviet Math. Dokl. (AMS Translation) **29**, 380-383 (1984).

[7] W. V. D. Hodge and D. Pedoe, *Methods of Algebraic Geometry*, Vol. I, II, Cambridge University Press, Cambridge (1947/1952).

[8] M. Kalkbrener, *Three Contributions to Elimination Theory*, Ph.D thesis, RISC-LINZ, Johannes Kepler University, Austria (1991).

[9] D. Kapur, *Using Gröbner Bases to Reason About Geometry Problems*, J. Symb. Comput. **2**, 399-408 (1996).

[10] D. Lazard, *Systems of Algebraic Equations*, Proc. EUROSAM'79 (E. W. Ng, ed.), 88-94 (1979).

[11] D. Lazard, *Resolution des systemes d'equation algebriques*, Theoretical Comput. Sci. **15**, 77-110 (1981).

[12] D. Lazard, *A New Method for Solving Algebraic Systems of Positive Dimension*, Discrete Applied Math. **32**, to appear (1991).

[13] D. Lazard, *Solving Zero-dimensional Algebraic Systems*, J. Symb. Comput. **13**, 117-131 (1992).

[14] F. S. MaCaulay, *The Algebraic Theory of Modular Systems*, Stechert-Hafner Service Agency, New York-London (1964).

[15] J. F. Ritt, *Differential Equations from the Algebraic Standpoint*, Amer. Math. Soc., New York (1932).

[16] J. F. Ritt, *Differential Algebra*, Amer. Math. Soc., New York (1950).

[17] A. Seidenberg, *Some Remarks on Hilbert's Nullstellensatz*, Arch. Math. **7**, 235-240 (1956).

[18] A. Seidenberg, *An Elimination Theory for Differential Algebra*, Univ. California Publ. Math. (N.S.) **3**(2), 31-66 (1956).

[19] A. Seidenberg, *On k-Constructable Sets, k-Elementary Formulae, and Elimination Theory*, J. reine angew. Math. **239/240**, 256-267 (1969).

[20] B. Sturmfels, *Sparse Elimination Theory*, Presented at the Meeting on Comput. Algebraic Geometry and Commutative Algebra (Cortona, June 1991).

[21] D. M. Wang, *Characteristic Sets and Zero Structure of Polynomial Sets*, Lecture Notes, RISC-LINZ, Johannes Kepler University, Austria (1989).

[22] D. M. Wang, *An Implementation of the Characteristic Set Method in Maple*, Proc. DISCO'92 (Bath, England, April 13-15, 1992), to appear (1992).

[23] D. M. Wang, *On Wu's Method for Solving Systems of Algebraic Equations*, RISC-Linz Series no. 91-52.0, Johannes Kepler University, Austria (1991).

[24] D. M. Wang, *Reasoning about Geometric Problems Using Algebraic Methods*, MEDLAR 24-Month Deliverables, to appear (1991).

[25] D. M. Wang, *A Strategy for Speeding up the Computation of Characteristic Sets*, Proc. 17th Int. Symp. Math. Foundations of Comput. Sci. (Prague, August 24-28, 1992), to appear (1992).

[26] D. M. Wang, *An Elimination Method for Polynomial Systems*, Preprint, RISC-LINZ, Johannes Kepler University, Austria (1992).

[27] W. T. Wu, *Basic Principles of Mechanical Theorem Proving in Elementary Geometries*, J. Sys. Sci. & Math. Scis. 4, 207-235 (1984); J. Automated Reasoning 2, 221-252 (1986).

[28] W. T. Wu, *On Zeros of Algebraic Equations — An application of Ritt principle*, Kexue Tongbao 31, 1-5 (1986).

[29] W. T. Wu, *A Zero Structure Theorem for Polynomial Equations-Solving*, MM Research Preprints, No. 1, 2-12 (1987).

[30] W. T. Wu, *On a Projection Theorem of Quasi-Varieties in Elimination Theory*, Chinese Ann. Math. Ser. B 11(2), 220-226 (1990).

Dongming Wang,
Research Institute for Symbolic Computation,
Johannes Kepler University,
A-4040 Linz, Austria

Progress in Mathematics

Edited by:

J. Oesterlé
Département de Mathématiques
Université de Paris VI
4, Place Jussieu
75230 Paris Cedex 05, France

A. Weinstein
Department of Mathematics
University of California
Berkeley, CA 94720
U.S.A.

Progress in Mathematics is a series of books intended for professional mathematicians and scientists, encompassing all areas of pure mathematics. This distinguished series, which began in 1979, includes authored monographs and edited collections of papers on important research developments as well as expositions of particular subject areas.

We encourage preparation of manuscripts in some form of TeX for delivery in camera-ready copy which leads to rapid publication, or in electronic form for interfacing with laser printers or typesetters.

Proposals should be sent directly to the editors or to: Birkhäuser Boston, 675 Massachusetts Avenue, Cambridge, MA 02139, U. S. A.